Histamine and H₁-Receptor Antagonists in Allergic Disease

Compliments of

CLINICAL ALLERGY AND IMMUNOLOGY

Series Editor

MICHAEL A. KALINER, M.D.

Medical Director
Institute for Asthma and Allergy
Washington, D.C.

Histamine and H_1-Receptor Antagonists in Allergic Disease

edited by

F. Estelle R. Simons
University of Manitoba
Winnipeg, Manitoba, Canada

Marcel Dekker, Inc.   New York • Basel • Hong Kong

Library of Congress Cataloging-in-Publication Data

Histamine and H_1-receptor antagonists in allergic disease / edited by F. Estelle R. Simons.
    p. cm. — (Clinical allergy and immunology ; 7)
    Includes index.
    ISBN 0-8247-9509-1 (alk. paper)
    1. Antihistamines. 2. Allergy — Chemotherapy. 3. Histamine — Receptors.
4. Histamine. I. Simons, F. E. R. II. Series. [DNLM: 1. Hypersensitivity —
immunology. 2. Histamine H_1 Antagonists — immunology. 3. Receptors, Histamine
— immunology. 4. Histamine — immunology. W1 CL652 v.7 1996 / QW 900
S588 1996]
RM666.A5H55 1996
616.97'061—dc20
DNLM/DLC
for Library of Congress   96-4746
                                                     CIP

The publisher offers discounts on this book when ordered in bulk quantities. For more information, write to Special Sales/Professional Marketing at the address below.

This book is printed on acid-free paper.

Marcel Dekker, Inc.
270 Madison Avenue, New York, New York 10016

Current printing (last digit):
10 9 8 7 6 5 4 3 2

PRINTED IN THE UNITED STATES OF AMERICA

Dedicated with deepest gratitude to my father,
Francis Raymond Edward Davies
(1910–1995).

Series Introduction

The decision to initiate a series of books on clinical allergy and immunology was based upon the need to create a library of texts useful for clinicians and scientists in these rapidly enlarging fields. There already are excellent textbooks providing overviews of the fields of allergy and immunology, and the scientific journals attempt to provide concise reviews of selected topics of interest. However, there is no library of books that takes relevant topics and expands them into texts for the express purpose of making them of interest to practicing physicians and researchers. Thus, this new series.

Clinical Allergy and Immunology will develop into the premier series of texts for our field. The initial books have addressed sinusitis, eosinophils, molecular biology, neuropeptides, provocation testing, and now histamine and its receptors. The benefactors of this series should be patients with allergic disorders because their physicians will now have a source of concise but authoritative summaries of a field. The other major benefactors will be clinicians. Despite the fact that allergy is the single most prevalent chronic disease suffered by humans, only a small number of medical schools incorporate allergy into their curricula, and many medical students are notably deficient in their knowledge of these diseases. As allergy is largely an outpatient specialty, few house officers see allergic patients other than those experiencing acute asthma or anaphylaxis. To clinicians, this series offers the chance to develop an extensive knowledge about selected topics, each chosen for its clinical relevance.

Histamine holds a special place in allergy. It was the first mediator to be associated with allergic reactions, and was the first whose inhibition proved useful in the management of disease. Much success has accompanied the use of antihistamines, and scientists have continued to create new and safer antihistamines for clinical use. This book provides the most current and useful information on histamine and its antagonists. As with all others in the series, it satisfies needs in both clinical and research areas and should prove very helpful for providing new insights and new directions. We welcome this work to the series.

Michael A. Kaliner

Preface

Yet all experience is an arch wherethro'
Gleams that untravell'd world . . .
Tennyson

During the past decade, there have been major advances in our understanding of histamine receptors, histamine, and histamine H_1-receptor antagonists. This book is designed to provide comprehensive, up-to-date information about progress in these areas for specialists in allergy, dermatology, otolaryngology, and pulmonology. It has been divided into four parts: "Histamine Receptors and Histamine"; "H_1-Receptor Antagonists: Basic Science"; "H_1-Receptor Antagonists: Clinical Science"; and "H_1-Receptor Antagonists: Adverse Effects."

In Part I (Chapters 1–3), fundamental information about histamine receptors and histamine is reviewed. For many years, histamine receptors and their agonists and antagonists have been characterized using a pharmacological approach. Recently, the gene encoding the H_1-receptor has been cloned in human leukocytes and the structure of the H_1-receptor protein has been deduced. The precise requirements for H_1-receptor activation by H_1-receptor agonists and H_1-receptor antagonists are now an active area of study. This new information is presented in Chapters 1 and 2.

Histamine itself, reviewed in Chapter 3, has fascinated and challenged researchers for most of the twentieth century. Its importance as a major chemical mediator of inflammation in allergic disorders such as rhinoconjunctivitis, asthma, urticaria, and anaphylaxis has long been recognized. More recently, its role in neurotransmission has been defined, and its role as a ubiquitous, extracellular messenger has been promulgated.

In Part II (Chapters 4–6), basic scientific information about H_1-receptor antagonists is reviewed. Their chemical structure and classification is discussed in Chapter 4. All first-generation H_1-receptor antagonists contain a substituted ethylamine group and have some structural similarity to histamine. Many of the newer H_1-receptor antagonists are larger molecules in which the structural similarity to histamine is less readily apparent, and many of these medications do not fit the traditional classification system perfectly.

Chapters 5 and 6 are devoted to the antiallergic effects of H_1-receptor antagonists, which have generated considerable interest during the past few years. These effects are not related to H_1-blockade, but rather to ionic changes in the cell membrane. In vitro they are concentration-dependent and in vivo they are dose-dependent, generally requiring higher doses than those recommended for symptom relief in allergic rhinoconjunctivitis or urticaria.

In Part III (Chapters 7–12), the clinical scientific basis of H_1-receptor antagonist use is explored in detail. Chapter 7 focuses on pharmacokinetics and pharmacodynamics. Most of the older antihistamines used worldwide were introduced before regulatory agencies required information about absorption, distribution, elimination, and metabolism, or about onset, duration, and offset of action. The newer, second-generation H_1-receptor antagonists developed during the past decade are more thoroughly studied with regard to pharmacokinetics and pharmacodynamics than their predecessors. Investigation of the interactions of H_1-receptor antagonists with macrolide antibiotics, antifungals, or other medications that may be administered concurrently has assumed considerable importance recently.

The rationale for the use of H_1-receptor antagonists in treatment of allergic rhinoconjunctivitis, asthma, urticaria, or atopic dermatitis includes the following: in these disorders local challenge with histamine produces some of the symptoms; challenge with antigen or another relevant stimulus may result in local or systemic increase in histamine concentrations; histamine concentrations may increase spontaneously during active disease; pretreatment with an H_1-receptor antagonist prevents symptoms after challenge with histamine, antigen, or another relevant stimulus; and, most importantly, H_1-receptor antagonists relieve naturally occurring symptoms. This is the rationale for using H_1-receptor antagonists in the treatment of such disorders.

In Chapter 8, the role of H_1-receptor antagonists in the treatment of allergic rhinoconjunctivitis is reviewed. From the consumer's point of view, these are first-line medications in the treatment of this disorder, and they are available in many countries without prescription. Topical application of H_1-receptor antagonists to the nasal mucosa or conjunctiva in allergic rhinoconjunctivitis is currently of considerable interest. The advantages of this route of administration include prompt onset of action, minimal systemic absorption, and little systemic toxicity.

The fascinating history of the role of H_1-receptor antagonists in the treatment of chronic asthma is presented in Chapter 9. Use of these medications in asthma is no longer viewed with concern; indeed, some of them are recognized as having a modest bronchodilator effect, although they are not generally recommended as primary medications for the treatment of asthma.

Providing optimal relief of symptoms in urticaria remains a challenge. H_1-receptor antagonists continue to be the most important medications available for the treatment of this disorder, as reviewed in Chapter 10. Until recently, there were few studies of histamine-receptor antagonists in human anaphylaxis. Progress in this area is highlighted in Chapter 11.

Children with allergic disorders are often the last group of sufferers to receive the benefits of new medications. The newer, nonsedating H_1-receptor antagonists are no exception, as in many countries pediatric formulations of these medications are still not available for use. The older, sedating H_1-receptor antagonists, widely marketed in these formulations, are perceived as being both safe and effective in children and even in infants, when in fact many of them are neither safe nor effective in this population. This topic is thoroughly examined in Chapter 12.

In Part IV (Chapters 13 and 14), the adverse effects of H_1-receptor antagonists are reviewed. After ingestion of older H_1-receptor antagonists such as diphenhydramine, somnolence and impairment of performance are surprisingly common. Unfortunately, as pointed out in Chapter 13, although these problems are readily detectable using electroencephalogram techniques or standardized performance tests, they may not be perceptible to the person who has ingested the medication. The major advantage of the second-generation H_1-receptor antagonists over their predecessors, when given in manufacturers' recommended doses, is absence of somnolence and other central nervous system adverse effects.

After astemizole and terfenadine had been in use worldwide for many years, rare association with adverse cardiovascular effects was reported, including prolongation of the QTc interval, torsade de pointes, and other ventricular dysrhythmias. These adverse effects, discussed in depth in Chapter 14, are, for the most part, predictable and preventable. The underlying mechanism involves excessive delay of repolarization and induction of early after-depolarizations. Astemizole or terfenadine should not be used by persons with preexisting cardiovascular or hepatic disorders, or by those concomitantly receiving cytochrome P_{450} inhibitors such as erythromycin or ketoconazole.

In summary, this book brings together for the first time a perspective on histamine receptors, histamine, and H_1-receptor antagonists. We sincerely thank the scientists and clinician-scientists who helped to make it a reality. All the senior authors are internationally recognized as experts in their respective fields who have made substantial contributions to advances in histamine receptor research, histamine research, or H_1-receptor antagonist research during the past decade. With their co-authors, they have provided outstanding state-of-the-art reviews of these important areas.

F. Estelle R. Simons

Contents

Contributors

James N. Baraniuk, M.D.   Georgetown University, Washington, D.C.

Fuad M. Baroody, M.D.   The University of Chicago, Chicago, Illinois

Jean Bousquet, M.D.   Montpellier University, Montpellier, France

A. M. Campbell, Ph.D.   Montpellier University, Montpellier, France

C. W. Canonica, M.D.   Genoa University, Genoa, Italy

Badrul Alam Chowdhury, M.D., Ph.D.   University of Tennessee College of Medicine, Memphis, Tennessee

Martin K. Church, Ph.D., D.Sc.   University of Southampton, Southampton, England

Andrea D. Collinson, B.Sc.   University of Southampton, Southampton, England

Jonathan M. Corne, M.A.(Camb), M.R.C.P.   University of Southampton, Southampton, England

J. Andrew Grant, M.D.   University of Texas Medical Branch, Galveston, Texas

Stephen T. Holgate, M.D., D.Sc., F.R.C.P.   University of Southampton, Southampton, England

Peter Honig, M.D., M.P.H.   Georgetown University, Washington, D.C.

Peter Howarth, D.M., F.R.C.P.   University of Southampton, Southampton, England

Michael A. Kaliner, M.D.   Institute for Asthma and Allergy, Washington Hospital Center, Washington, D.C.

Rob Leurs, Ph.D.   Leiden/Amsterdam Center for Drug Research, Vrije Universiteit, Amsterdam, The Netherlands

Philip L. Lieberman, M.D.   University of Tennessee, Cordova, Tennessee

Eli O. Meltzer, M.D.   University of California, San Diego, California

Robert M. Naclerio, M.D.   The University of Chicago, Chicago, Illinois

Yoshimichi Okayama, M.D., Ph.D.   University of Southampton, Southampton, England

Tommy C. Sim, M.D.   University of Texas Medical Branch, Galveston, Texas

F. Estelle R. Simons, M.D., F.R.C.P.C.   The University of Manitoba, Winnipeg, Manitoba, Canada

Keith J. Simons, M.Sc., Ph.D.   The University of Manitoba, Winnipeg, Manitoba, Canada

Martine J. Smit, Ph.D.   Leiden/Amsterdam Center for Drug Research, Vrije Universiteit, Amsterdam, The Netherlands

Hendrik Timmerman, Ph.D.   Leiden/Amsterdam Center for Drug Research, Vrije Universiteit, Amsterdam, The Netherlands

Martha V. White, M.D.   Institute for Asthma and Allergy, Washington Hospital Center, Washington, D.C.

Michael J. Welch, M.D.   University of California, San Diego, California

Stephen L. Winbery, M.D., Ph.D.   Methodist Central Hospital Teaching Practice, Memphis, Tennessee

Histamine and H_1-Receptor Antagonists in Allergic Disease

1

Histamine Receptors: Specific Ligands, Receptor Biochemistry, and Signal Transduction

Rob Leurs, Martine J. Smit, and Hendrik Timmerman
Leiden/Amsterdam Center for Drug Research,
Vrije Universiteit, Amsterdam, The Netherlands

I. INTRODUCTION

Multicellular organisms respond to several endogenous and exogenous stimuli with the development of highly coordinated responses. An essential contribution to this communication system is made by the action of various hormones and neurotransmitters. These endogenously released molecules transfer the signals from one cell to another by interaction with highly specific receptor proteins, which are usually embedded in the plasma membrane. One main group of receptor proteins are the so-called G protein-coupled receptors. Upon receptor activation by the appropriate ligands, these proteins activate a G protein, which in turn will modulate the activity of regulatory enzymes or ion channels. Since several different G proteins and target proteins exist, the class of G protein-coupled receptors influences the cellular activity in various distinct ways. The mono-amine histamine is one of the many neurotransmitters that act via G protein-coupled receptors. In the central nervous system (CNS), this amine is synthesized in a restricted population of neurons located in the tubero-mammillary nucleus of the posterior hypothalamus. These neurons project diffusely to most cerebral areas and have been implicated in various functions of the brain of mammalian species (e.g., sleep/wakefulness, hormonal secretion, cardiovascular control) (1). In various peripheral tissues histamine is stored in mast cells, basophils, enterochromaffin cells and,

1

probably, in specific neurons as well. In the gastric mucosa, for example, depending on the species, histamine is stored either in enterochromaffin-like cells or specific mast cells that release it in response to gastrin and acetylcholine (2). Histamine in turn stimulates parietal cell gastric acid secretion by interaction with specific receptors.

Histamine stored in mast cells plays a very important role in the pathogenesis of various allergic conditions. After degranulation of the mast cells by a variety of stimuli, release of histamine leads to well-known symptoms of allergic disorders in the skin and airways (3, 4). Based on these observations, histamine is considered to be one of the important mediators of allergy and inflammation.

Research in the histamine field initially focused completely on the role of histamine in allergic diseases and resulted in the development of several potent "antihistamines" (e.g., mepyramine or pyrilamine), which were useful in inhibiting certain symptoms of allergic conditions. The observation that these antihistamines could not antagonize histamine-induced effects on the stomach and the heart led Ash and Schild in 1966 to the hypothesis that histamine should act via at least two distinct receptor subtypes (5). This hypothesis became generally accepted in 1972, when Black and his co-workers succeeded in synthesizing a series of new compounds (e.g., burimamide) that were able to block the effects of histamine on the stomach and the heart (6). These H_2-receptor antagonists proved to be very useful in the therapy of gastric ulcers. In recent years it has become apparent that histamine is not only a mediator of several pathophysiological conditions, but also functions as a neurotransmitter (1). As for many other neurotransmitter systems, a presynaptic histamine receptor (H_3) exists for histamine. The identification of this new histaminergic receptor subtype in 1983 by Arrang and colleagues (7) gave rise to a new field of interest both for pharmacologists and medicinal chemists. During the last 7 years extensive research in several laboratories has shown that the H_3-receptor can be regarded as a general regulatory system and a potential target for new therapies (8–11).

In this chapter we will describe molecular pharmacological properties of the various histamine receptor subtypes. Attention will be given to the development of specific pharmacological tools, the biochemical aspects of the receptor proteins, and the different biochemical processes that are triggered in the various target cells/tissues after receptor activation.

II. SELECTIVE LIGANDS FOR THE HISTAMINE RECEPTORS

In this section we will describe the various histaminergic agents that can be used as selective pharmacological tools for the study of the three receptor

subtypes. For a more detailed description of the pharmacochemical aspects of the ligands for the receptor subtypes, the interested reader is
referred to recent detailed reviews by Ganellin (12), Leurs et al. (13), and
Van der Goot et al. (14).

A. Histamine H_1-Receptor Ligands

Many pharmacological tools are available for studying the H_1-receptor,
although highly potent receptor agonists are not yet known. Nevertheless,
2-thiazolylethylamine and substituted 2-phenylhistamines (12, 13, 15, 16)
are rather selective agents for the H_1-receptor (Fig. 1) and could be used
for this purpose. The 2-phenylhistamines show relatively high affinity for
the H_1-receptor, but on various test systems appear to be partial agonists
(17, 18).

Many potent and selective receptor antagonists are available to study
H_1-receptors (12, 13). However, one should be aware of possible muscarinic and serotoninergic receptor antagonist properties and local anesthetic
properties of many older H_1-receptor antagonists at concentrations that
are usually much higher than those needed for blocking the H_1-receptor.
Currently mepyramine is the most commonly used H_1-receptor antagonist

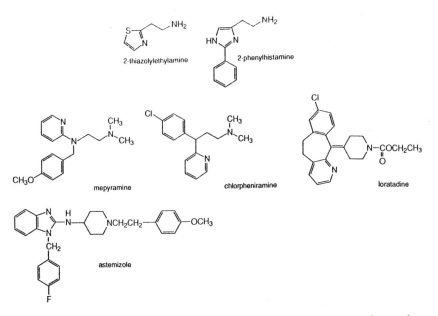

Figure 1 Structures of various selective histamine H_1-receptor agonists and antagonists. Mepyramine = pyrilamine.

$(pA_2 = 9)$ in pharmacological studies, whereas the d- and l-enantiomers of chlorpheniramine are also very effective for receptor classification (Fig. 1). These compounds pass readily through the blood–brain barrier and can also be used for in vivo CNS studies. Recently many new H_1-receptor antagonists have been developed (e.g., astemizole, loratadine) (Fig. 1), which do not cross the blood–brain barrier as readily. Finally, some irreversible antagonists have also been described. Two strongly related photoaffinity probes ($[^{125}I]$-iodoazidophenpyramine and $[^3H]$-azidobenzamide) have been used for irreversible labeling of the H_1-receptor binding proteins in guinea pig brain and heart (19) and bovine adrenal medulla (20), respectively.

For competitive labeling of H_1-receptors, tritium-labeled mepyramine is often the first choice. However, some precautions should be taken in using $[^3H]$-mepyramine for labeling the H_1-receptor. This compound might be taken up in cells (21) and has been shown to label various non-H_1-receptor binding sites, for example, in guinea pig lung (22) and rat liver (23). If a very high sensitivity is required, the iodonated ligand $[^{125}I]$-iodobolpyramine (24) might be of interest. Due to its high sensitivity this ligand might also be used for autoradiographic localization of H_1-receptors (25). However, it shows remarkable species differences (24), labels sites on human lymphocytes with a rather unexpected and as yet not "understood" pharmacology (26), and has not been extensively used thus far. Finally, a quaternary ligand ($[^3H]$-(+)-N-methyl-4-methyldiphenhydramine), has been introduced for the specific labeling of cell-surface H_1-receptors (27). Such a tool could be very useful indeed, but although this ligand labels H_1-receptors in guinea pig brain (27), unfortunately no information has been presented so far for the actual labeling of H_1-receptors on whole cells.

B. Histamine H_2-Receptor Ligands

For the histamine H_2-receptor both agonists and antagonists are available for proper pharmacological characterization. Dimaprit, impromidine, and sopromidine are potent H_2-receptor agonists (12–14), but it should be acknowledged that these drugs are moderate or even potent H_3-receptor antagonists as well (11, 28). The use of sopromidine and its S-enantiomer might be very valuable, since the S-enantiomer shows no H_2-receptor agonistic activity, but is a moderately active H_2-receptor antagonist ($pA_2 = 5.6$) (29). Recently, we developed a completely new class of H_2-receptor agonists. Based on a powerful combination of quantum chemical calculations and pharmacological data on dimaprit derivatives, we suggested a new model for the activation of the H_2-receptor (30, 31). Predictions with this model led to the synthesis of a series of potent and selective agents, from which amthamine (2-amino-5-(2-aminoethyl)-4-methylthiazole) and amselamine (2-amino-5-(2-aminoethyl)-4-methyl-1,3-selenazole) (Fig. 2) combine a high H_2-receptor selectivity with a potency that is slightly higher than histamine's [(32), unpublished data from our laboratory].

After the introduction of the first H_2-receptor antagonist by Black et al. in 1972 (6), many compounds with potent H_2-receptor antagonistic properties have been described (see Refs. 12–14 for extensive reviews). Nowadays compounds such as cimetidine, ranitidine, and tiotidine are usually applied as selective tools for functional studies of the H_2-receptor (Fig. 2). Due to its lack of specificity, the use of cimetidine for binding studies proved to be unsuccessful (33). Tritiated tiotidine has been used for labeling the H_2-receptor in various tissues. Although in several instances H_2-receptor binding can be measured, the very high level of nonspecific binding hinders widespread use of this radioligand as well [see Garbarg et al. (33) and references therein]. The recently introduced [^{125}I]-iodoaminopotentidine is more interesting (34). This ligand labels H_2-receptors in guinea pig brain with a high affinity and sensitivity (34) and has been used for the characterization of cloned rat and human H_2-receptors, which were expressed in Chinese hamster ovary (CHO) cells (35, 36). Moreover, it has also been successfully used in autoradiographic studies (31). An azido-derivative ([^{125}I]-iodoazidopotentidine) of this potent H_2-

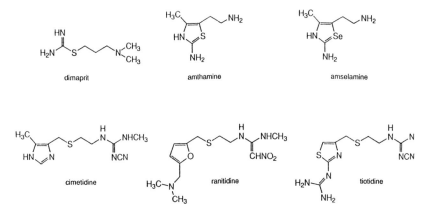

Figure 2 Structures of various selective histamine H_2-receptor agonists and antagonists.

receptor antagonist has been introduced to label the H_2-receptor protein in guinea pig brain irreversibly (34).

C. Histamine H_3-Receptor Ligands

Soon after the initial description of the H_3-receptor (7), Arrang et al. (37) described highly potent and selective ligands for this receptor subtype: R-(α)-methylhistamine and thioperamide as, respectively, H_3-receptor agonist and antagonist, are valuable tools for receptor identification (Fig. 3). The use of R-(α)-methylhistamine in combination with its less active enantiomer S-(α)-methylhistamine is very effective for the characterization of H_3-receptor-mediated effects. The dimethylated histamine analogue R-(α,β)-dimethylhistamine has also recently been shown to be a potent H_3-receptor agonist (38). Moreover, isothioureum analogues of histamine resulted in the development of both very potent H_3-receptor agonists and antagonists (11). The unsubstituted isothioureum analogue S-[2-(4(5)-imidazolyl)ethyl]isothiourea (VUF 8325 or imetit), (Fig. 3) was found to be a very potent agonist (39–41). Moreover, recently Vollinga et al. (42) incorporated the amine function in a piperidine ring. The resulting 4(5)-1H-imidazolylmethylpiperidine (immepip) (Fig. 3) also appears to combine high H_3-receptor selectivity with high agonistic potency (42). Due to its unique structure, this compound might become a valuable tool for future quantitative structure-activity relationship (QSAR) studies.

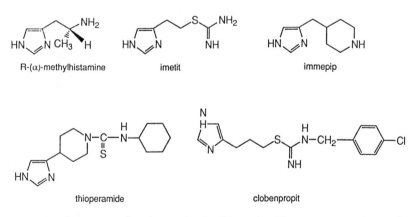

Figure 3 Structures of various selective histamine H_3-receptor agonists and antagonists.

Within a series of N-substituted S-[3-(4(5)-imidazolyl)propyl]isothiourea derivatives it was observed that N-substitution of the isothioureum group leads to compounds with antagonist properties (40). The introduction of a 4-chlorobenzyl group, as a substituent of a nitrogen of the isothiourea moiety, results in the most potent H_3-receptor antagonist known so far (VUF 9153, clobenpropit, $pA_2 = 9.9$) (Fig. 3) (11, 40). Despite its high in vitro activity, in rats clobenpropit shows a similar in vivo activity when compared to thioperamide (43). Since it was shown by ex vivo binding that clobenpropit does not easily penetrate the blood–brain barrier (43) pharmacokinetic differences between thioperamide and clobenpropit should explain the observed differences between the in vitro and in vivo activities of these two compounds. In contrast, in mice clobenpropit indeed appears to be 10 times more potent than thioperamide in vivo, which indicates important species differences (44).

Due to the availability of various compounds with a reasonably high affinity, it is not very surprising that some radioligands for the H_3-receptor are also available (33). With the initial description of R-(α)-methylhistamine as a selective ligand for the H_3-receptor (37), Arrang et al. (37) also presented the tritiated compound as a suitable radioligand for the H_3-receptor. As a consequence of the relatively low abundance of the H_3-receptor in the various tissues, its use is rather limited. Recently, to overcome this problem, tritiated N^α-methylhistamine has been introduced (45, 46). This compound is labeled to a threefold higher specific activity and labels H_3-receptors, but shares another intrinsic problem with R-(α)-methylhistamine. Since both compounds are H_3-receptor agonists, complex binding characteristics can be expected and have indeed been found (33, 46–48). The introduction of a highly potent, iodonated radioligand by our laboratory can therefore be seen as a major breakthrough in this field (49–51). In a series of N-substituted isothioureum analogues, N-substitution of the isothioureum group with a 4-iodophenylethyl group leads to a potent H_3-receptor antagonist ($pA_2 = 9.2$, guinea pig ileum). This compound, iodophenpropit (S-[3-(4(5)-imidazolyl)propyl]-N-2-(4-iodophenyl)-ethyl] isothiouronium hydrogen sulphate), shows a high selectivity toward the H_3-receptor (affinities for H_1- and H_2-receptors are more than 1000-fold lower) and has therefore been radiolabeled with [^{125}I]-iodine (49). The resulting [^{125}I]-iodophenpropit shows saturable, readily reversible, high affinity binding to rat cortex membranes ($K_D = 0.32$ nM, $B_{max} = 209$ fmol/mg protein) (Fig. 1), which can be displaced by thioperamide in a monophasic way ($K_i = 5$ nM) (50, 51). In contrast, displacement of the specific [^{125}I]-iodophenpropit binding by various H_3-receptor agonists results in biphasic displacement curves; in the presence of nonhydrolyzable GTP-analogues the agonist displacement curves are shifted rightward, indicating the coupling to G proteins (50, 51). The radiolabeled antagonist has also been used for autoradiographic studies in rat brain (51). A heterogeneous distribution of [^{125}I]-iodophenpropit binding sites was found with highest densities in the cerebral cortical layers, caudate-putamen complex, olfactory tubercules, hippocampal formation, amygdala complex, and hypothalamus (51).

III. MOLECULAR ASPECTS OF THE HISTAMINE RECEPTORS

A. The H$_1$-Receptor Protein

Using [^{125}I]-iodoazidophenpyramine, Ruat et al. irreversibly labeled H$_1$-receptor proteins in rat, guinea pig, and mouse brain (19, 52). Following SDS-PAGE analysis of the labeled proteins, two main polypeptides (56 and 41–47 kDa) were found to be specifically labeled (52). Based on experiments with protease inhibitors, it was suggested that the H$_1$-receptor binding protein was represented by the 56-kDa peptide, whereas the other labeled peptide was probably a result of protease action (52). Using the already mentioned [^{3}H]-azidobenzamide, Yamashita et al. (20) recently found receptor peptides of similar size (53–58 kDa) to be labeled in bovine adrenal medulla membranes. Whereas the 56-kDa peptide was also found in guinea pig lung and ileum, a peptide with a substantially higher molecular weight (68 kDa) was labeled in guinea pig heart tissue (19). Although at present no pharmacological differences have been observed between the H$_1$-receptors from guinea pig heart and brain tissue, these results suggest the occurrence of several isoforms for the H$_1$-receptor. Moreover, the molecular weight found for the H$_1$-receptor after photoaffinity labeling is in sharp contrast with the reported weight (38–40 kDa) of a purified [^{3}H]-mepyramine binding protein from DDT1-MF2 smooth muscle cells (53). Since several highly potent H$_1$-receptor antagonists possess only a moderate affinity for this binding protein, it is not yet clear whether the binding of [^{3}H]-mepyramine to these cells really represents H$_1$-receptor binding (53).

Recently, Yamashita et al. (54) succeeded in the cloning of the gene encoding for the bovine H$_1$-receptor. Using the approach of expression cloning (injection of mRNA in Xenopus oocytes), a cDNA clone from a cDNA library of bovine adrenal medulla was isolated and expressed in both Xenopus oocytes and COS-7 cells (54). The expressed receptor protein specifically binds [^{3}H]-mepyramine and shows in displacement studies typical H_1-receptor characteristics (54). The cloned gene encodes for a 491 amino acid receptor protein (apparent molecular weight 56 kDa) with all of the structural features of a G protein-coupled receptor (Fig. 4; 7 transmembrane domains, N-terminal glycosylation sites, phosphorylation sites for protein kinase A and C) (54). Since based on biochemical studies it is known that the H_1-receptor protein from guinea pig brain is glycosylated (55), the predicted molecular weight of 56 kDa is certainly underestimated. Using the sequence information of the bovine H_1-receptor gene, the genes for the rat (56), guinea pig (57), and human (58–59) H_1-receptor have now also been cloned. These genes are all intronless and encode for proteins of 486 (rat), 488 (guinea pig), and 487 (human) amino acids, respectively. The homology between the several receptor proteins is quite high in the intracellular domains (ca. 90%), but is significantly lower in the intracellular and extracellular parts. The various cloned genes should, in our view, be regarded as real species homologues, although some small changes in pharmacology are noted (54, 56–59).

With the availability of the cloned genes new possibilities have emerged for neurobiological research. Complementary to measurement of expression of the H_1-receptor protein by autoradiography, it is now also possible to localize the H_1-receptor mRNA by in situ hybridization. Accordingly, such studies have been reported for guinea pig brain tissue (57). The localization of receptor mRNA correlates very well with the presence of receptor proteins. Yet, in specific regions of both the hippocampus and the cerebellum, different cellular structures appear to contain the mRNA and receptor protein, respectively (57). These important findings will be very valuable for a proper understanding of the role of histamine in brain function.

Using the information from molecular biological studies, it is now also possible to investigate the structure–function relationships of the receptor protein. This type of information is of great importance, since, for example, it could lead to a rational design of new and better ligands. On the basis of results of site-directed mutagenesis studies of various other aminergic receptors it is generally accepted that the binding of these small neurotransmitters mainly occurs in the transmembrane domains (60). With respect to the H_1-receptor recognition of histamine it is apparent that, as for the receptors of other biogenic amines, an aspartate is highly conserved in the third transmembrane domain (Fig. 5). As shown for other aminergic receptors, this aspartate residue will probably function as a binding site for the protonated amino function of histamine. In the fifth transmembrane domain a threonine and asparagine (Fig. 5) have been suggested to be

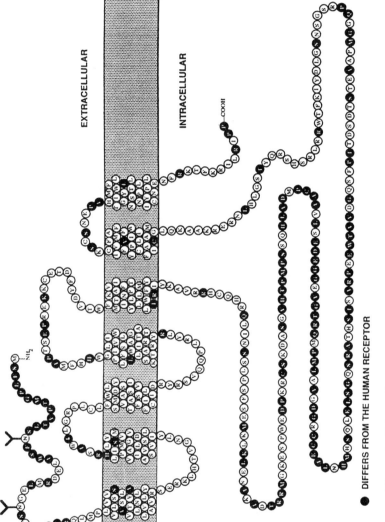

Figure 4 Schematic representation of the guinea pig histamine H_1-receptor protein as deduced from the guinea pig gene (57). The amino acids that are different from the human H_1-receptor protein (59) are indicated by the black amino acids. The receptor protein is probably N-glycosylated at the amino terminus.

● DIFFERS FROM THE HUMAN RECEPTOR

responsible for binding the imidazole moiety (61). However, one should be aware that this hypothesis was based solely on the homologous positions of these residues compared to the amino acid residues that were shown to be involved in the binding of catecholamines to the β-receptor (60). We have now investigated the role of these amino acids in detail by site-directed mutagenesis to nonfunctional alanines (Ref. 18; Table 1). From this study the threonine residue does not seem to be very important for the interaction with either agonist or antagonist, although for some of the tested antagonists a slight increase in affinity was noticed for the mutated receptor protein (18). In contrast, the asparagine residue appears to be very important for the interaction with histamine and 2-methylhistamine, a rather selective H_1-receptor agonist (18). The affinity of the selective agonist 2-(3-bromophenyl)histamine and the nonimidazole agonists 2-

TM 3

H_1 human	F	W	L	S	M	D	Y	V	A	S	T	A	S	I	F	S	V	F	I	L	C	I
H_1 guinea-pig	F	W	L	S	M	D	Y	V	A	S	T	A	S	I	F	S	V	F	I	L	C	I
H_1 bovine	F	W	L	S	M	D	Y	V	A	S	T	A	S	I	F	S	V	F	I	L	C	I
H_1 rat	F	W	L	S	M	D	Y	V	A	S	T	A	S	I	F	S	V	F	I	L	C	I
H_2 human	I	Y	T	S	L	D	V	M	L	C	T	A	S	I	L	N	L	F	M	I	S	L
H_2 canine	I	Y	T	S	L	D	V	M	L	C	T	A	S	I	L	N	L	F	M	I	S	L
H_2 rat	I	Y	T	S	L	D	V	M	L	C	T	A	S	I	L	N	L	F	M	I	S	L
Beta 1 human	L	W	T	S	V	D	V	L	C	V	T	A	S	I	E	T	L	C	V	I	A	L

(asterisk below the D column)

TM 5

H_1 human	W	F	K	V	M	T	A	I	I	N	F	Y	L	P	T	L	L	M	L	W	F	Y
H_1 guinea-pig	W	F	K	V	M	T	A	I	I	N	F	Y	L	P	T	L	L	M	L	W	F	Y
H_1 bovine	W	F	K	V	M	T	A	I	I	N	F	Y	L	P	T	L	L	M	L	W	F	Y
H_1 rat	W	F	K	I	M	T	A	I	I	N	F	Y	L	P	T	L	L	M	L	W	F	Y
H_2 human	G	L	-	-	V	D	G	L	V	T	F	Y	L	P	L	L	I	M	C	I	T	Y
H_2 canine	G	L	-	-	V	D	G	L	V	T	F	Y	L	P	L	L	V	M	C	I	T	Y
H_2 rat	G	L	-	-	V	D	G	L	V	T	F	Y	L	P	L	L	I	M	C	I	T	Y
Beta 1 human	A	Y	A	I	A	S	S	V	V	S	F	Y	V	P	L	C	I	M	A	F	V	Y

(asterisks below the M and F columns)

Figure 5 Sequence alignment of the putative third and fifth transmembrane (TM) domain of the histamine H_1- and H_2-receptor proteins. The asterisks indicate the amino acids that are thought to be involved in the interaction with histamine. For comparison the transmembrane domains of the human β_1-receptor are shown.

12

pyridylethylamine and 2-thiazolylamine for the Asn^{207} Ala mutant was only slightly affected (18). These data indicate that asparagine^{207} probably interacts with the N^τ-nitrogen of the imidazole ring of histamine. This interaction is very important for histamine and 2-methylhistamine receptor binding, but seems to be only of minor importance for 2-(3-bromophenyl)-histamine. As was already suggested on the basis of pharmacochemical studies (11), the aromatic 2-substituent apparently interacts in a specific way with the H_1-receptor and this interaction results in high affinity binding to the receptor protein. Another interesting conclusion can be drawn for the nonimidazole agonists. Whereas on the basis of the agonist actions of these ligands, the N^τ-nitrogen of the imidazole ring of histamine has always been regarded as nonfunctional, these ideas should now be reconsidered. Apparently, histamine and the nonimidazole agonists use different functionalities to bind to the receptor. One could hypothesize that the N^π-nitrogen of histamine, 2-pyridylethylamine and 2-thiazolylamine is involved in receptor activation by these ligands, but not in receptor binding. More detailed studies should identify the amino acid residues of the H_1-receptor protein that are involved in the interaction with the aromatic ring of the 2-phenylhistamines and the N^π-nitrogen of the H_1-receptor agonists.

B. The H$_2$-Receptor Protein

Detailed biochemical information for the H_2-receptor has also been obtained by photoaffinity labeling studies. Using [^{125}I]-iodoazidopotentidine Ruat et al. (34) showed that the H_2-receptor binding peptide is probably a 59-kDa protein, although the purification of a [^{3}H]-tiotidine binding protein from human HGT-1 cells resulted in the identification of a 70-kDa protein (62). Yet, the H_2-receptor nature of the [^{3}H]-tiotidine binding protein from these cells has not clearly been defined.

The gene encoding for the H_2-receptor has been cloned in several species, including humans (63–65). Using the polymerase chain reaction with degenerate oligonucleotides, based on the known sequence homology of various G protein-coupled receptors and canine gastric parietal cDNA, Gantz et al. obtained the gene encoding for the canine H_2-receptor (63). Using the information of the canine gene, the genes encoding for the rat and human H_2-receptor were rapidly cloned thereafter (64, 65). The various genes show considerable homology (80–90%) and are probably real species homologues (63–65). The genes encode for a 359 (dog, man) or a 358 (rat) amino acid receptor protein, with many of the structural features of G protein-coupled receptors and apparent molecular weights of approximately 40 kDa. Since in the N-terminal extracellular tail a consensus sequence for N-linked glycosylation is present, the actual molecular

Table 1 K_i-values of various H_1-receptor agonists for the wild type guinea pig histamine H_1 receptor and the $Thr^{203}Ala$ and the $Asn^{207}Ala$ mutant stably expressed in CHO cells. K_i-values were determined by displacement of $^3[H]$mepyramine binding. Data shown are mean ± SEM of three to four independent experiments.

H_1-agonist	Structure	Wild type	K_i-value Thr^{203}Ala	Asn^{207}Ala
histamine		12 ± 2 μM (3)	28 ± 4 μM* (3)	> 1000 μM* (4)
2-methylhistamine		42 ± 6 μM (3)	25 ± 5 μM (3)	> 1000 μM* (3)
2-(3-bromophenyl)histamine		0.67 ± 0.15 μM (3)	0.44 ± 0.09 μM (3)	4.3 ± 1.1 μM* (4)
2-thiazolylethylamine		24 ± 3 μM (3)	28 ± 5 μM (3)	147 ± 25 μM* (3)
2-pyridylethylamine		38 ± 3 μM (3)	44 ± 3 μM (3)	129 ± 13 μM* (3)

* Indicates a significant difference compared to WT receptors ($p < 0.05$).

weight of the receptor will be significantly higher. Expression of the rat and human receptor gene in Chinese hamster ovary cells was followed by an extensive pharmacological characterization (35, 36). Using the highly selective and potent radiolabel [^{125}I]-iodoaminopotentidine, Traiffort et al. (35) and Leurs et al. (36) revealed that both the rat and human H_2-receptors show a similar pharmacological profile to the H_2-receptor in guinea pig heart, the commonly accepted H_2-receptor reference system, or human brain. Although for the dog receptor protein extensive pharmacological data are not yet available, it is very likely that this gene also encodes for a classical H_2-receptor.

With respect to receptor recognition, the H_2-receptor contains an aspartate residue in the third transmembrane domain (Fig. 5). Since this residue is found in both the H_2-receptor and the H_1-receptor, this again suggests the involvement of this aspartate in the binding of the amino function of histamine. Site-directed mutagenesis of the canine H_2-receptor by Gantz et al. (66) showed indeed that the aspartate residue in the third transmembrane domain is involved in the binding of histamine and H_2-receptor antagonists. The threonine and asparagine residues in the fifth transmembrane domain of the H_1-receptor protein have been replaced by an aspartate and a threonine residue, respectively (Fig. 5). Mutations of these two residues to alanines affected the pharmacological properties of the receptor protein; a loss of the [^{3}H]-tiotidine binding was noticed when the aspartate residue was eliminated. Moreover, the histamine-induced cAMP accumulation was severely reduced, although the EC_{50}-value for histamine was not altered (66). Changing the threonine residue to alanine resulted in a loss of affinity for [^{3}H]-tiotidine and a reduction of both the EC_{50} and the maximal cAMP response of histamine (66). These data indicate that the threonine and aspartate residues in the fifth transmembrane domain of the H_2-receptor likely interact with histamine and the antagonist [^{3}H]-tiotidine. Based on studies with other biogenic amine receptors, the interaction of the two residues with the antagonist is, however, rather unexpected and certainly deserves future attention. Moreover, the interaction of the two amino acids with histamine is not well defined; the mutated receptors require more detailed investigations with, for example, the recently developed nonimidazole agonists amthamine and amselamine.

C. The H_3-Receptor Protein

Detailed biochemical studies of the histamine H_3-receptor have been hampered by the low abundance of this receptor protein and the lack of highly sensitive radioligands. Using [^{3}H]-histamine, Zweig et al. recently reported the solubilization of the H_3-receptor from bovine brain tissue (67). Size-exclusion chromatography revealed an apparent molecular weight of 220,000 kDa for the digitonin-solubilized receptor protein. Due to the protein-associated digitonin this value is clearly overestimated. Recently, a value of 70 kDa has been reported as the molecular weight for the human H_3-receptor (68). Human gastric tumor cell line HGT-1 contains [^{3}H]-N^α-methylhistamine high affinity binding sites, which were effectively solubilized with Triton-X-100 (68). Using thioperamide-sepharose affinity chromatography the protein was effectively purified and still showed high affinity for [^{3}H]-N^α-methylhistamine (68).

No information is yet available on the molecular biological aspects of the H_3-receptor, but some recent developments in this field might soon give rise to interesting findings. First, the described purification of the receptor protein (68) and/or the development of the highly sensitive radiolabel [^{125}I]-iodophenpropit (50, 51) are both important steps for an expression-cloning strategy to obtain the gene encoding the H_3-receptor. Second, recent advances in the field of the histamine H_1- and H_2-receptor might also contribute in this respect. Yet, a homology approach based on these developments could be unsuccessful as there is no actual need for an extensive homology between the H_3-receptor and the other members of the histamine receptor family. Also, there is only a limited homology between the H_1- and H_2-receptor genes.

IV. TRANSMEMBRANE SIGNALING

In this section we describe the signal transduction pathways that can be activated by the various histaminergic receptor subtypes. With the identification of an increasing number of G proteins and effector systems, histamine receptors are shown to be coupled to various different signal transduction systems. For a good understanding of the action of histamine in (patho)physiological processes, detailed knowledge of the intracellular pathways activated by histamine is indispensable.

A. Signal Transduction of the H_1-Receptor

The histamine H_1-receptor is, among other G protein-coupled receptors, a Ca^{2+}-mobilizing receptor. It is widely accepted that activation of Ca^{2+}-mobilizing receptors is associated with the phospholipase C catalyzed hydrolysis of membrane inositide phospholipids (69). Receptor stimulation leads to the hydrolysis of phosphatidyl 4,5-biphosphate resulting in the formation of inositol 1,4,5-triphosphate (IP_3) and 1,2-diacylglycerol (DAG). IP_3 was shown to mobilize Ca^{2+} from intracellular stores, whereas DAG was shown to activate protein kinase C (PKC). For a detailed description of the various functions of the second messengers, the interested reader is referred to a recent review by Berridge (70).

In accordance with the above, histamine has been shown to induce production of inositol phosphates in several tissues, including brain, airway smooth muscle, intestinal smooth muscle, vascular smooth muscle, and heart tissue (1, 4, 71–74). Moreover, in guinea pig brain, regions with the highest density of H_1-receptors were found to display the largest phosphoinositide response (75, 76). However, in some tissues (guinea pig ileum, and neonatal brain), the H_1 response to histamine itself appeared to be masked by an H_1-antagonist-insensitive component (77, 78).

In membrane preparations of both rat cerebral cortex and 1321N1 astrocytoma cells, H_1-receptor-stimulated phospholipid hydrolysis was found to be dependent on the presence of guanine nucleotides (73, 79). Studies in 1321N1 astrocytoma (73), HeLa (80), and CHO cells stably expressing H_1-receptors (81) showed the inositol phosphate response to be insensitive to pertussis toxin treatment, indicating the involvement of a pertussis toxin-insensitive G protein. The actual nature of the pertussin toxin-insensitive G protein remains unclear. However, the recent cloning of the H_1-receptor as well as identification of various G proteins will allow a detailed investigation of this issue.

One of the physiological consequences of the production of inositol phosphates is the elevation of intracellular Ca^{2+}. The use of fluorescent Ca^{2+} indicators in various isolated cell systems has clearly demonstrated an increase of the intracellular Ca^{2+} concentration upon H_1-receptor activation (81–90). The histamine-induced Ca^{2+} response is characterized by a rapid transient rise of the intracellular Ca^{2+} concentration, which is followed by a sustained elevation of the Ca^{2+} concentration. Experiments in Ca^{2+}-free medium and with inorganic Ca^{2+} antagonists suggest that the sustained response is highly dependent on the influx of extracellular calcium, whereas the transient increase is caused by the release of Ca^{2+} from intracellular Ca^{2+} stores (81–90).

Since Ca^{2+} is involved in the regulation of many cellular functions, the increase of intracellular Ca^{2+} concentration can explain a wide variety of pharmacological responses induced after stimulation of the H_1-receptor. First of all, elevation of intracellular Ca^{2+} levels leads to further stimulation of phospholipase C, most likely the phospholipase Cβ subtype, as it is known to be sensitive to Ca^{2+} (91). Moreover, the histamine-induced production of inositol phosphates in both brain and tracheal slices (92, 93) and CHO cells stably expressing the guinea pig H_1-receptor (81) was shown to be highly dependent on the influx of extracellular Ca^{2+}.

Besides the activation of phospholipase C, histamine-induced increase of Ca^{2+} seems to induce the production of nitric oxide. In various vascular preparations, endothelium-dependent relaxation is observed upon H_1-receptor activation, which was found to be related to the production of nitric oxide (94–97). In cultured bovine aortic cells the actual generation of nitric oxide could be measured after stimulation of histamine (98). In a variety of airway and heart preparations, H_1-receptor activation was shown to induce the production of cGMP, which may be ascribed to generation of nitric oxide (99–104). Thus, vascular responses to histamine may be explained by the generation of nitric oxide, which appears to be highly dependent on the mobilization of intracellular Ca^{2+}.

Another consequence of the histamine-induced mobilization of Ca^{2+} is the generation of arachidonic acid metabolites, prostacyclin and thromboxane A_2, due to liberation of arachidonic acid from the phospholipids (81, 105, 106). In CHO transfected with the guinea pig H_1-receptor and HeLa cells, the release of arachidonic acid appeared to be phospholipase A_2-dependent (81). In these CHO cells the histamine-induced activation of phospholipase A_2 was found to be partially sensitive to pertussis toxin. In contrast, in these cells the regulation of phospholipase C was completely insensitive to pertussis toxin (81). These findings are in agreement with the observations described by Murayama et al. (107), who reported that in human and rabbit platelets histamine stimulates the release of arachidonic acid via a pertussis toxin-sensitive pathway without the generation of inositol phosphates. In HeLa cells, on the other hand, the phospholipase A_2-dependent release of arachidonic acid appeared to be insensitive to pertussis toxin (81). Thus, the H_1-receptor-mediated activation of phospholipase A_2 in transfected CHO cells may be ascribed to two different pathways; arachidonic acid can be stimulated via an elevation of the intracellular Ca^{2+} concentration (as in HeLa cells) and via interaction of phospholipase A_2 with a pertussis toxin-sensitive G protein (as in platelets).

Last, the increase of intracellular Ca^{2+} is most likely involved in the regulation of cAMP levels. H_1-receptor stimulation seems to modulate cAMP responses via histamine H_2-, adenosine A_2- and VIP-receptors (108–112). The exact mechanism involved in the elevation of cAMP remains to be determined; however, a role for both Ca^{2+} and protein kinase C has been suggested (113). Yet recent findings in CHO cells expressing the guinea pig H_1-receptor showed that neither protein kinase C nor Ca^{2+} accounts for the observed elevation of cAMP levels (81). The latter observations could possibly be explained by stimulation of adenylyl cyclase by G protein $\beta\gamma$-subunits.

In summary, the histamine H_1-receptor is coupled to the phospholipase C-dependent inositol phosphate pathway, thereby inducing a rise in the intracellular Ca^{2+} concentration. The latter can explain the histamine-induced production of other second messengers such as cGMP, cAMP, and arachidonic acid metabolites. Yet it appears that responses such as modulation of cAMP and phospholipase A_2 activity can also be regulated by phospholipase C-independent pathways. How activation of the H_1-receptor leads to these responses requires further detailed mechanistic investigation.

B. Signal Transduction of the H₂-Receptor

It is generally accepted that the histamine H_2-receptor is coupled to the adenylate cyclase system. A large number of reports have shown that histamine increases the levels of cAMP in brain, stomach, and heart tissue of several species, including humans (1, 114–122). Since the H_2-receptor mediates its response through activation of adenylate cyclase in membrane fractions in a guanyl nucleotide-sensitive manner (122), an activation model similar to the well-studied β-adrenergic receptor system is often presented (122). However, it should be emphasized that, even though such a model is highly probable, no direct evidence is at present available for the interaction with a G_s protein. In this respect it is very interesting that recently Ozawa and co-workers (115, 121) suggested that an H_2-receptor-stimulated phospholipid methylation is responsible for a subsequent activation of adenylate cyclase. In both rat brain (115, 123) and guinea pig heart (121) histamine was shown to rapidly stimulate the phospholipid methylation. The cAMP response in both tissues was significantly slower, completely dependent on the presence of the methyl donor S-adenosyl-L-methionine and reduced by an inhibitor of the phospholipid methyltransferases (115, 121). Moreover, in rat brain the regional distribution of the histamine-induced phospholipid methylation closely paralleled the H_2-receptor-mediated cAMP response (115). These data are intriguing and suggest that the phospholipid response might be necessary for an effective coupling of the receptor to the G protein responsible for the activation of the cyclase. Also interesting in this respect is the fact that similar suggestions have already been reported by Hirata and Axelrod in 1980 for the β-adrenergic receptor (124). Yet it should be noted that, currently, these observations are not drawing much attention at all, even within the field of the β-adrenergic receptor.

Although the linkage of the H_2-receptor to the adenylate cyclase is rather well accepted, some findings argue against a universal role for cAMP. In guinea pig brain, the regional distribution of H_2-receptor binding sites does not parallel the observed H_2-receptor-mediated cyclase activation (34). Moreover, Haas et al. (125) observed a denervation hypersensitivity to histamine at the electrophysiological level, whereas under the same conditions the cAMP response was unaltered. In view of the above findings it is very interesting that, recently, new signaling pathways have been described for the H_2-receptor. In differentiated HL-60 and HEPA cells transfected with the canine H_2 receptor, an H_2-receptor-mediated increase of the intracellular Ca^{2+} concentration was observed (126–128). The increase of intracellular Ca^{2+} was found to result from the release of Ca^{2+} from intracellular stores (126, 128). The H_2-receptor-dependent Ca^{2+} mobilization is probably due to the activation of phospholipase C, since histamine was also found to increase the levels of inositol 1,4,5-trisphosphate. In both cell lines these effects were found to be inhibited by cholera toxin but not by pertussis toxin, suggesting the involvement of a G protein (126, 128). Recent reports in hamster DDT1 MF-2 cells and bovine trachea smooth muscle have demonstrated a cAMP-mediated inhibition of the production of inositol phosphates (129–131). The cross-talk between the cAMP cascade and phosphoinositide system could well explain the observed inhibitory effects induced by cholera toxin. Yet, in the HEPA cells, forskolin did not inhibit the histamine-induced effects, suggesting the involvement of another mechanism (128).

Recently, another cAMP-independent response was described for the cloned rat H_2-receptor expressed in CHO cells (35). Besides a massive production of cAMP upon H_2-receptor activation, an inhibition of the release of arachidonic acid induced by either constitutive purinergic receptors or a Ca^{2+}-ionophore was observed. The histamine-induced inhibition was potently inhibited by ranitidine ($K_i = 0.16 \ \mu M$) and also induced by dimaprit. Forskolin, prostaglandin E_1, or 8-bromo-cAMP did not mimic the effects induced by histamine, indicating the involvement of a cAMP-independent process (35). Histamine did not significantly modify the features of the Ca^{2+}-responses to ATP. At present the actual mechanism for this response is still unknown. Yet one should be aware that this coupling to a new signaling system could be due to the high expression of receptors in these cells. In contrast, expression of the gene encoding the human H_2-receptor in CHO cells was shown to be functionally coupled to adenylate cyclase, but did not influence the inositol phosphate turnover or arachidonic acid release (36). The expression level of the human H_2-receptor resembles a more physiological level of expression (36). However, the observed sequence differences, particularly in the intracellular parts, between the rat and human receptor (64, 65) might also be a possible explanation for the observed differences in signal transduction. In summary, the H_2-receptor is primarily coupled to the adenylate cyclase-dependent production of cAMP. Yet, the observed breakdown of phosphoinositides, elevation of intracellular Ca^{2+} levels and phospholipase A_2 activity upon H_2-receptor activation seem to be regulated via other cAMP-independent pathways.

C. Signal Transduction of the H_3-Receptor

The H_3-receptor is also thought to belong to the family of G protein-coupled receptors. Binding studies (47, 48, 51, 67, 68, 132, 133), as well as inhibitory effects induced by pertussis toxin and cholera toxin (68, 134, 135) have supported the involvement of a G protein in the H_3-receptor-mediated responses. Little is known about the signal transfer mechanism of the H_3-receptor. Recently, a highly interesting negative coupling to phospholipase C was shown in HGT-1 gastric tumor cell (68). Upon stimulation of the H_3-receptor, basal and muscarinic receptor-mediated IP_3 production were inhibited. Cholera and pertussis toxin were able to inhibit this response (68). Recently, a PTX-sensitive G protein inhibiting phospholipase C activity was identified in bovine brain (136). It is not clear whether the above findings may be explained by a direct effect on phospholipase C or rather by inhibition of the influx of extracellular Ca^{2+}, which in some cases can activate phospholipase C (91). As frequently observed with other neurotransmitters, the release of [^{3}H]-histamine from the neurons is highly dependent upon the presence of extracellular Ca^{2+} (7, 137). It has been shown that the H_3-receptor effects diminish as the frequency of the electrical stimulation is increased (138) or as the extracellular Ca^{2+}-concentration is changed (139). Although the Ca^{2+}-regulation of H_3-agonist binding (33) is intriguing, the above-mentioned observations are more likely to be explained by an H_3-receptor-mediated inhibitory effect on the entry of extracellular Ca^{2+}.

Reduction of Ca^{2+}-entry may be accomplished by several mechanisms. Modulation of the N-type Ca^{2+}-channels leading to an inhibition of the Ca^{2+} influx has been described for some presynaptic receptors (140). Recently, Takemura et al. (141) reported that histamine release from rat hypothalamic slices (K^+ depolarization) could be effectively inhibited by the N-type Ca^{2+}-channel-blocker ω-conotoxine, but not by the L-type Ca^{2+}-antagonist nilvaldipine (141). The development of appropriate models to study the effects of H_3-receptor activation on neuronal Ca^{2+}-

currents will have to be awaited in order to elucidate a possible modulation of N-type voltage-dependent Ca^{2+} by the H_3-receptor. However, very recently Yang and Hatton provided the first evidence for an H_3-receptor-mediated modulation of ion permeability of neurons (142). Using isolated magnocellular histaminergic neurons from rat posterior hypothalamus, histamine was shown to depress both spontaneous and current-evoked spikes and to hyperpolarize these cells (142). These effects were mimicked by (R)-α-methylhistamine and inhibited by impromidine, strongly suggesting the involvement of H_3-receptors (142). The hyperpolarization was blocked by the K^+-channel blockers 4-aminopyridine and TEA, indicating that the H_3-receptor might be linked to hyperpolarizing K^+-channels (142). These exciting data suggest that histamine might reduce neuronal Ca^{2+}-entry and subsequent histamine release by inhibiting N-type Ca^{2+}-channels via a hyperpolarization of the neurons. At present it is not clear how the H_3-receptor would be linked to this K^+-channel. Whether or not the K^+-channels are modulated directly via a G protein or indirectly via the production of diffusible second messengers will be one of the questions to be answered soon.

In contrast to the Ca^{2+}-related activities, a negative linkage to the cAMP-producing adenylate cyclase system has been suggested for receptors inhibiting neurotransmitter release. However, the H_3-receptor does not seem to control intracellular cAMP levels (143, 144).

Besides a role in regulating neurotransmitter release, the H_3-receptor is probably also involved in some postsynaptic action of histamine. In these cases, other signal transduction mechanisms have been suggested. In 1992, Ea Kim et al. (145) reported the endothelium-dependent relaxation of rabbit middle cerebral artery after activation of the H_3-receptor. Although no direct evidence for the production of nitric oxide was presented, the removal of the endothelium and the use of the NO-synthase inhibitors N^G-monomethyl-L-arginine and N^G-nitro-L-arginine could effectively prevent the H_3-receptor-mediated relaxation (145). However, inhibitors of various enzymes involved in the production of prostacyclin could also partially prevent the relaxation (145). Complete inhibition was only obtained by the simultaneous inhibition of NO-synthase and the prostacyclin metabolism (145). Although the actions of the inhibitors of the prostacyclin pathway could be explained by antioxidant properties with respect to nitric oxide, these data seem to suggest that H_3-receptor-mediated relaxation arises from the production of nitric oxide and/or the arachidonic acid metabolite prostacyclin (145). These mediators may be produced separately and cause vasodilation independently or may be involved in a more complex scheme where, for example, H_3-receptor-mediated prostacyclin formation causes a subsequent production of nitric oxide. Previously, such a complex interaction has been shown to occur in the coronary artery of the pig (146). These data suggest that the H_3-receptor might be linked to the activation of NO-synthase and/or the arachidonic acid metabolism. However, the direct link between these two systems and H_3-receptor activation is not clear at this moment (Ca^{2+} mobilization ?). Further studies are awaited in this intriguing H_3-receptor model system.

The presence of H_3-receptors on blood vessels was substantiated by a study of Oike et al. (135). Using whole patch-clamp measurements on dispersed smooth muscle cells from the rabbit saphenous artery it was shown that H_3-receptor activation augmented the voltage-dependent Ca^{2+} current by an activation of L-type Ca^{2+}-channels (135). Intracellular application of GTP clearly enhanced the H_3-receptor effects, whereas GDPβS completely abolished the histamine-induced augmentation (135). However, in the presence of nonhydrolyzable GTP analogues, the H_3-receptor-mediated augmentation was reversed to an H_3-receptor-mediated inhibition of the current. Similar observations were made for noradrenaline and angiotensin II, indicating a rather general mechanism in this preparation. Both the inhibitory and stimulatory responses were not affected by treatment with pertussis toxin (135). These data show for the first time rather directly the involvement of pertussis toxin-insensitive G proteins in H_3-receptor effects. At present it is not known how the involved G protein would affect the Ca^{2+} current, although a role of protein kinase C or A is highly unlikely (135). Moreover, it is also not known how activation of the H_3-receptor could affect the Ca^{2+} current in two distinct ways.

The signal transfer mechanism(s) of the histamine H_3-receptor is far from clear as yet. Maybe the application of the newly-synthesized potent and selective H_3-agonists and antagonists will allow us to complete the picture in the near future.

V. SUMMARY

In recent years, several new developments in the area of the molecular pharmacology of histamine receptors have occurred. Only with a good understanding of the receptor targets will therapeutic intervention via these receptors improve. In this chapter we have described the various selective agonists, antagonists, and radioligands that are currently available for studying the three histamine receptor subtypes. Moreover, with the introduction of molecular biology techniques in the field of receptor research, important new biochemical insights into the receptor proteins have become available. We have documented the current biochemical information about the histamine receptor proteins and discussed the various possible transmembrane signaling pathways for the three histamine receptor subtypes.

REFERENCES

1. Schwartz JC, Arrang JM, Garbarg M, Pollard H, Ruat M. Histaminergic transmission in mammalian brain. Physiol Rev 1991;71:1–51.
2. Rangachari PK. Histamine: mercurial messenger in the gut. Am J Physiol 1992;262:G1–G13.
3. Ring J, Sedlmeier F, Van der Helm H, Mayr T, Walz U, Ibel H, Riepel H, Przybilla B, Reimann HJ, Dorsch W. Histamine and Allergic Diseases. In: Ring J, ed. New Trends in Allergy. Berlin: Springer Verlag, 1985;2:44–77.
4. Barnes PJ. Histamine receptors in the lung. In: Timmerman H, Van der Goot H, eds. New Perspectives in Histamine Research. Basel:Birkhauser Verlag, 1991:103–122.
5. Ash ASF, Schild HO. Receptors mediating some actions of histamine. Br J Pharmacol 1966;27:427–439.
6. Black JW, Duncan WAM, Durant GJ, Ganellin CR, Parsons ME. Definition and antagonism of histamine H_2-receptor. Nature 1972;236:385–390.
7. Arrang JM, Garbarg M, Schwartz JC. Autoinhibition of brain histamine release by a novel class (H_3) of histamine receptor. Nature 1983;302: 832–837.
8. Van der Werf JF, Timmerman H. The histamine H_3-receptor: a general presynaptic histaminergic regulatory mechanism? TIPS 1989;10:159–162.
9. Timmerman H. Histamine H_3 ligands: just pharmacological tools or potential therapeutic agents? J Med Chem 1990;33:4–11.
10. Schwartz JC, Arrang JM, Garbarg M, Pollard H. A third histamine receptor subtype: characterization, localization and functions of the H_3-receptor. Agents Actions 1990;30:13–23.
11. Leurs R, Timmerman H. The histamine H_3-receptor: A target for developing new drugs. In: Jucker E, ed. Progress in Drug Research. Basel: Birkhauser Verlag, 1992;39:128–165.
12. Ganellin CR. Pharmacochemistry of H_1- and H_2-receptors. In: Schwartz JC, Haas H, eds. The Histamine Receptor. New York: Wiley-Liss, 1992: 1–56.
13. Leurs R, Van der Goot H, Timmerman H. Histaminergic agonists and antagonists: recent developments. In: Testa B, ed. Advances in Drug Research. London:Academic Press, 1991;20:217–304.
14. Van der Goot H, Bast A, Timmerman H. Structural requirements for histamine H_2-agonists and H_2-antagonists. In: Uvnäs B, ed Handbook of Experimental Pharmacology. Berlin:Springer Verlag, 1991;97:573–748.
15. Koper JG, Van der Vliet A, Van der Goot H, Timmerman H. New selective histamine H_1-agonists. Pharmac Weekbl Sci 1990;12:236–239.

16. Zingel V, Elz S, Schunack W. 2-Phenylhistamines with high histamine H_1 agonistic activity. Eur J Med Chem 1990;25:673–680.

17. White TE, Dickenson JM, Hill SJ. Histamine H_1-receptor mediated inositol phospholipid hydrolysis in DDT1 MF-2 cells: agonist and antagonist properties. Br J Pharmacol 1993;108:196–203.

18. Leurs R, Smit MJ, Tensen CP, Ter Laak AM, Timmerman H. Site-directed mutagenesis of the histamine H_1-receptor reveals the selective interaction of asparagine[207] with subclasses of H_1-receptor agonists. Biochem Biophys Res Commun 1994;201:295–301.

19. Ruat M, Bouthenet ML, Schwartz JC, Ganellin CR. Histamine H_1-receptor in heart: unique electrophoretic mobility and autoradiographic localization. J Neurochem 1990;55:379–385.

20. Yamashita M, Ito S, Sugama K, Fukui H, Smith B, Nakanishi K, Wada H. Biochemical characterization of histamine H_1-receptors in bovine adrenal medulla. Biochem Biophys Res Commun 1991;177:1233–1239.

21. Maloteaux J, Gossuin A, Waterkeyn C, Laduron P. Trapping of labeled ligands in intact cells: a pitfall in binding studies. Biochem Pharmacol 1983;32:2543–2548.

22. Carswell H, Nahorski SR. Distribution and characteristics of histamine H_1-receptors in guinea-pig airways identified by [^{3}H]-mepyramine. Eur J Pharmacol 1982;81:301–307.

23. Leurs R, Bast A, Timmerman H. High affinity, saturable [^{3}H]-mepyramine binding sites on rat liver plasma membranes do not represent histamine H_1-receptors. Biochem Pharmacol 1989;38:2175–2180.

24. Körner M, Bouthenet ML, Ganellin CR, Garbarg M, Gros C, Ife RJ, Sales N, Schwartz JC. [^{125}I]-iodobolpyramine, a highly sensitive probe for histamine H_1-receptors in guinea-pig brain. Eur J Pharmacol 1986;120:151–160.

25. Bouthenet ML, Ruat M, Sales N, Schwartz JC. A detailed mapping of histamine H_1-receptors in guinea-pig central nervous system established by autoradiography with [^{125}I]-iodobolpyramine. Neurosci 1988;26:553–600.

26. Villemain FM, Bach JF, Chatenoud LM. Characteristics of histamine H_1-receptor binding sites on human T lymphocytes by means of [^{125}I]-iodobolpyramine. Preferential expression of H_1-receptors on CD8 T lymphocytes. J Immunol 1990;144:1449–1459.

27. Treherne JM, Young JM. [^{3}H]-(+)-N-4-methyldiphenhydramine, a quartenary radioligand for the histamine H_1-receptor. Br J Pharmacol 1988;94:797–810.

28. Arrang JM, Schwartz JC, Schunack W. Stereoselectivity of the histamine H_3-presynaptic autoreceptor. Eur J Pharmacol 1985;117:109–114.

29. Elz S, Gerhard G, Schunack W. Histamine analogues. No 32. Synthesis and pharmacology of sopromidine, a potent and stereoselective isomer of the achiral H_2-agonist impromidine. Eur J Med Chem 1989;24:259–262.

30. Haaksma EEJ, Donne-op den Kelder GM, Timmerman H, Weinstein H. Theoretical analysis of the activity of dimaprit derivatives on the H_2-receptor. In: Timmerman H, Van der Goot H, eds. New Perspectives in Histamine Research. Basel:Birkhauser Verlag, 1991:315–324.

31. Eriks JC, Van der Goot H, Timmerman H. New activation model for the histamine H_2-receptor, explaining the activity of the different classes of histamine H_2-receptor agonists. Mol Pharmacol 1993;44:886–894.

32. Eriks JC, Sterk GJ, Van der Aar EM, Van Acker SABE, Van der Goot H, Timmerman H. 4- or 5-(ω-aminoalkyl)thiazoles and derivatives; new selective H_2-receptor agonists. In: Timmerman H, Van der Goot H, eds. New Perspectives in Histamine Research. Basel:Birkhauser Verlag, 1991: 301–314.

33. Garbarg M, Traiffort E, Ruat M, Arrang JM, Schwartz JC. Reversible labeling of H_1-, H_2-, H_3-receptors. In: Schwartz JC. Haas H, eds. The Histamine Receptor. New York:Wiley-Liss, 1992:73–96.

34. Ruat M, Traiffort E, Bouthenet ML, Schwartz JC, Hirschfeld J, Buschauer A, Schunack W. Reversible and irreversible labeling and autoradiographic localization of the cerebral histamine H_2-receptor using [^{125}I]-iodinated probes. Proc Natl Acad Sci 1990;87:1658–1662.

35. Traiffort E, Ruat M, Arrang JM, Leurs R, Piomelli D, Schwartz JC. Expression of a cloned rat histamine H_2-receptor mediating inhibition of arachidonate release and activation of cAMP accumulation. Proc Natl Acad Sci 1992;89:2649–2653.

36. Leurs R, Smit MJ, Menge WMBP, Timmerman H. Pharmacological characterization of the human histamine H_2-receptor stably expressed in Chinese Hamster Ovary cells. Br J Pharmacol 1994;112:847–854.

37. Arrang JM, Garbarg M, Lancelot JC, Lecomte JM, Pollard H, Robba M, Schunack W, Schwartz JC. Highly potent and selective ligands for histamine H_3-receptors. Nature 1987;327:117–123.

38. Lipp R, Arrang JM, Buschmann J, Garbarg M, Luger P, Schunack W, Schwartz JC. Novel chiral H_3-receptor agonists. In: Timmerman H, Van der Goot H, eds. New Perspectives in Histamine Research. Basel:Birkhauser Verlag, 1991:277–282.

39. Howson W, Parsons ME, Ravel P, Swayne GTG. Two novel and potent and selective histamine H_3-receptor agonists. Bioorg Med Chem Lett 1992; 2:77–79.

40. Van der Goot H, Schepers MJP, Sterk GJ, Timmerman H. Isothiourea analogues of histamine as potent agonists or antagonists of the histamine H_3-receptor. Eur J Med Chem 1992;27:511–517.

41. Garbarg M, Arrang JM, Rouleau A, Ligneau X, Dam Trung Toung M, Schwartz JC, Ganellin CR. S-[2-(4-imidazolyl)ethyl]isothiourea, a highly specific and potent histamine H_3-receptor agonist. J Pharmacol Exp Ther 1992;263:304–310.

42. Vollinga RJ, de Koning JP, Jansen FP, Leurs R, Menge WMPB, Timmerman H. A new potent and selective histamine H_3-receptor agonist, 4-(1H-imidazol-4-ylmethyl)piperidine. J Med Chem 1994;37:332–333.

43. Barnes JC, Brown JD, Clarke NP, Clapham J, Evans DJ, O'Shaughnessy CT. Pharmacological activity of VUF 9153, an isothiourea histamine H_3-receptor antagonist. Eur J Pharmacol 1993;250:147–152.

44. Yokohama H, Onodera K, Maeyama K, Sakurai E, Iinuma K, Leurs R,

Timmerman H, Watanabe T. Clobenpropit (VUF-9153), a new histamine H$_3$-receptor antagonist, inhibits electrically-induced convulsions in mice. Eur J Pharmacol 1994;260:23–28.

45. Korte A, Myers J, Shih NY, Egan RW, Clark MA. Characterization and tissue distribution of H$_3$-receptors in guinea-pigs by N$^\alpha$-methylhistamine. Biochem Biophys Res Commun 1990;168:979–986.

46. West RE, Zwieg A, Shih NY, Siegel MI, Egan RW, Clark MA. Identification of two H$_3$-histamine receptor subtypes. Mol Pharmacol 1990;38: 610–613.

47. Arrang JM, Roy J, Morgat JL, Schunack W, Schwartz JC. Histamine H$_3$-receptor binding sites in rat brain membranes: modulation by guanine nucleotides and divalent ions. Eur J Pharmacol 1990;188:219–227.

48. West RE, Zwieg A, Granzow RT, Siegel MI, Egan RW. Biexponential kinetics of (R)-α-methylhistamine binding to the rat brain H$_3$-histamine receptor. J Neurochem 1990;55:1612–1616.

49. Menge WMPB, Van der Goot H, Timmerman H, Eersels JLH, Herscheid JDM. Synthesis of S-[3-(4(5)-imidazolyl)propyl]-N-[2-(4-{^{125}I}-iodophenyl)-ethyl] isothioureum hydrogen sulfate (^{125}I-iodophenpropit) a new probe for histamine H$_3$-receptor binding sites. J Labelled Comp Radiopharm 1992; 31:781–786.

50. Jansen FP, Rademaker B, Bast A, Timmerman H. The first radiolabeled H$_3$-antagonist [^{125}I]-iodophenpropit; saturable and reversible binding towards rat cortex membranes. Eur J Pharmacol 1992;217:203–205.

51. Jansen FP, Wu TS, Voss HP, Steinbusch HWM, Vollinga R, Rademaker B, Bast A, Timmerman H. Characterization of the binding of the first selective radiolabeled H$_3$-antagonist [^{125}I]-iodophenpropit to rat brain. Br J Pharmacol 1994;113:355–362.

52. Ruat M, Schwartz JC. Photoaffinity labeling and electrophoretic identification of the H$_1$-receptor: comparison of several brain regions and animal species. J Neurochem 1989;53:335–339.

53. Mitsuhashi M, Payan DG. Solubilization and characterization of the pyrilamine-binding protein from cultured smooth muscle cells. Mol Pharmacol 1989;35:751–759.

54. Yamashita M, Fukui H, Sugama K, Horio Y, Ito S, Mizuguchi H, Wada H. Expression cloning of a cDNA encoding the bovine histamine H$_1$-receptor. Proc Natl Acad Sci 1991;88:11515–11519.

55. Garbarg M, Yeramian E, Körner M, Schwartz JC. Biochemical studies of cerebral H$_1$-receptors. In: Ganellin CR, Schwartz JC, eds. Frontiers in Histamine Research. Oxford:Pergamon Press, 1985:9–25.

56. Fujimoto K, Horio Y, Sugama K, Ito S, Liu YQ, Fukui H. Genomic cloning of the rat histamine H$_1$-receptor. Biochem Biophys Res Commun 1993;190: 294–301.

57. Traiffort E, Leurs R, Arrang JM, Tardivel-Lacombe J, Diaz J, Schwartz JC, Ruat M. Guinea-pig histamine H$_1$-receptor: I Gene cloning, characterization and tissue expression revealed by *in situ* hybridization. J Neurochem 1994;62:507–518.

58. Chowdhury BA, Kaliner MA, Strader CM. Cloning of a gene encoding the human H_1-receptor. Soc Neurosci Abstr 1993;19:84.

59. De Backer MD, Gommeren W, Moereels H, Nobels G, Van Gompel P, Leysen JE, Luyten WHML. Genomic cloning, heterologous expression and pharmacological characterization of a human histamine H_1-receptor. Biochem Biophys Res Commun 1993;197:1601–1608.

60. Savarese TM, Fraser CM. *In vitro* mutagenesis and the search for structure-function relationships among G protein-coupled receptors. Biochem J 1992; 283:1–19.

61. Timmerman H. Cloning of the H_1-histamine receptor. TiPS 1992;13:6–7.

62. Reyl-Desmars F, Cherifi Y, Le Romancer M, Pigeon C, Le Roux S, Lewin MJM. Solubilization and purification of the H_2-histamine receptor from the human tumoral gastric cells HGT-1. CR Acad Sci Paris 1991;312:221–224.

63. Gantz I, Schaffer M, DelValle J, Logsdon C, Campbell V, Uhler M, Yamada T. Molecular cloning of a gene encoding the histamine H_2-receptor. Proc Natl Acad Sci 1991;88:429–433.

64. Gantz I, Munzert G, Tashiro T, Schaffer M, Wang L, DelValle J, Yamada T. Molecular cloning of the human histamine H_2-receptor. Biochem Biophys Res Commun 1991;178:1386–1392.

65. Ruat M, Traiffort E, Arrang JM, Leurs R, Schwartz JC. Cloning and tissue expression of a rat histamine H_2-receptor gene. Biochem Biophys Res Commun 1991;179:1470–1478.

66. Gantz I, DelValle J, Wang L-D, Tashiro T, Munzert G, Guo Y-J, Konda Y, Yamada T. Molecular basis for the interaction of histamine with the histamine H_2-receptor. J Biol Chem 1992;267:20840–20843.

67. Zweig A, Siegel MI, Egan RW, Clark MA, Shorr RGL, West Jr, RE. Characterization of a digitonin-solubilized bovine brain H_3 histamine receptor coupled to a guanine nucleotide-binding protein. J Neurochem 1992;59:1661–1666.

68. Cherifi Y, Pigeon C, Le Romancer M, Bado A, Reyl-Desmars F, Lewin MJM. Purification of a histamine H_3-receptor negatively coupled to phosphoinositide turnover in the human gastric cell line HGT1. J Biol Chem 1992;267:25315–25320.

69. Michel RH. Inositol phospholipids and cell surface receptor function. Biochim Biophys Acta 1975;41:81–147.

70. Berridge MJ. Phosphoinositides and cell signalling. In: Fidia Research Foundation Neuroscience Research Award Lectures. New York: Raven Press, 1992;6:5–45.

71. Claro E, Garcia A, Picacoste F. Carbachol and histamine stimulation of guanine-nucleotide-dependent phosphoinositide hydrolysis in rat brain cortical membranes. Biochem J 1989;261:29–35.

72. Donaldson J, Hill SJ. Histamine-induced hydrolysis of polyphosphoinositides in guinea-pig ileum and brain. Eur J Pharmacol 1986;124:255–265.

73. Orellano S, Solski PA, Brown JH. Guanosine-5'-O-thiotriphosphate-depen-

dent inositoltriphosphate formation in membranes is inhibited by phorboles-ter and protein kinase C. J Biol Chem 1987;262:1638–1643.

74. Sakuma I, Gross SS, Levi R. Positive inotropic effect of histamine on guinea-pig left atrium: H_1-receptor induced stimulation of phosphoinositide turnover. J Pharmacol Exp Ther 1988;247:466–472.

75. Daum PR, Downes CP, Young JM. Histamine-induced inositol phospholipid breakdown mirrors H_1-receptor density in brain. Eur J Pharmacol 1993;87: 497–498.

76. Carswell H, Young JM. Regional variation in the characteristics of hista-mine H_1-agonist-mediated breakdown of inositol phospholipids in guinea-pig brain. Br J Pharmacol 1986;89:809–817.

77. Bailey SJ, Lippe IT, Holzer P. Effect of the tachykinin antagonist, [D-pro^4, D-Trp^{7,9,10}]substance P-(4-11), on tachykinin- and histamine-induced inositol phosphate generation in intestinal smooth muscle. Arch Pharmacol 1987;335:296–300.

78. Claro E, Garcia A, Picacosta F. Histamine-stimulated phosphoinositide hy-drolysis in developing rat brain. Mol Pharmacol 1987;32:384–390.

79. Claro E, Garcia A, Picacoste F. Carbachol and histamine stimulation of guanine-nucleotide-dependent phosphoinositide hydrolysis in rat brain cor-tical membranes. Biochem J 1989;261:29–35.

80. Tilly BC, Lambrechts AC, Tertoolen LGJ, De Laat SW, Molenaar WH. Regulation of phosphoinositide hydrolysis induced by histamine and gua-nine nucleotides in human HeLa carcinoma cells. FEBS Lett 1990;265: 80–84.

81. Leurs R, Traiffort E, Arrang JM, Tardivel-Lacombe J, Ruat M, Schwartz JC. Guinea-pig histamine H_1-receptor: II Stable expression in Chinese ham-ster ovary cells reveals the interaction with three major signal transduction pathways. J Neurochem 1994;62:519–527.

82. Johnston CL, Johnson CG, Bazan E, Garver D, Gruenstein E, Ahluwalia M. Histamine receptors in human fibroblasts: inositol phosphates, calcium, and cell growth. Am J Physiol 1990;258:C533–543.

83. Matsumoto T, Kanaide H, Nishimura J, Shogakiuchi Kobayashi S, Naka-mura M. Histamine activates H_1-receptors to induce cytosolic free calcium transients in cultured vascular smooth muscle cells. Biochem Biophys Res Commun 1986;135:172–177.

84. Rotrosen D, Gallin JI. Histamine type I receptor occupancy increases endo-thelial cytosolic calcium, reduces F-actin, and promotes albumin diffusion across cultured endothelial monolayers. J Cell Biol 1986;103:2379–2387.

85. Brown RD, Prendville P, Cain C. α_1-adrenergic and H_1-histamine receptor control of intracellular calcium in a muscle cell line: the influence of prior agonist exposure on receptor responsiveness. Mol Pharmacol 1986;29: 531–539.

86. Jacob R, Merrit JE, Hallam TJ, Rink TJ. Repetitive spikes in cytoplasmatic calcium evoked by histamine in human endothelial cells. Nature 1988;335: 40–45.

87. Kotlikoff MI, Murray RK, Reynolds EE. Histamine-induced calcium re-

lease and phorbol antagonism in cultured airway smooth muscle. Am J Physiol 1987;253:C561–566.

88. McDonough PM, Eubanks JH, Brown JH. Desensitization and recovery of muscarinic and histaminergic calcium mobilization in 1321N1 astrocytoma cells. Biochem J 1988;249:135–141.

89. Oakes SG, Iaizzo PA, Richelson E, Powis G. Histamine-induced intracellular free calcium, inositol phosphates and electrical changes in murine N1E-115 neuroblastoma cells. J Pharmacol Exp Ther 1988;247:114–121.

90. Tilly BC, Tertoolen LGJ, Lambrechts AC, Remoire R, De Laat SW, Molenaar WH. Histamine H_1-receptor-mediated phosphoinositide hydrolysis, calcium signalling and membrane potential oscillations in human HeLa carcinoma cells. Biochem J 1990;226:235–243.

91. Cockcroft S, Thomas GMH. Inositol-lipid-specific phospholipase C isoenzymes and their differential regulation by receptors. Biochem J 1992;288: 1–14.

92. Alexander SPH, Hill SJ, Kendall DA. Differential effects of elevated calcium ion concentration on inositol phospholipid responses in mouse and rat cerebral cortical slices. Biochem Pharmacol 1992;40:1793–1799.

93. Baird JG, Chilvers ER, Kennedy ED, Nahorski SR. Changes in extracellular calcium within the physiological range influence receptor-mediated inositol phosphate responses in brain and tracheal smooth muscle slices. Arch Pharmacol 1989;339:247–251.

94. Satoh H, Inui J. Endothelial cell-dependent relaxation and contraction induced by histamine in the isolated guinea-pig pulmonary artery. Eur J Pharmacol 1984;97:321–324.

95. Schoeffter P, Godfraind T. Characterization of histamine-induced contraction in rat isolated aorta. Eur J Pharmacol 1991;197:193–200.

96. Toda N. Mechanism of histamine actions in human coronary arteries. Circ Res 1987;61:280–286.

97. Van den Voorde J, Leusen I. Effect of histamine on aorta preparations of different species. Arch Int Pharmacodyn 1984;268:95–105.

98. Schmidt HHHW, Zernikow B, Baeblich S, Böhme E. Basal and stimulated formation and release of L-arginine-derived nitrogen oxides from cultured endothelial cells. J Pharmacol Exp Ther 1990;254:591–597.

99. Casale TB, Rodbard D, Kaliner M. Characterization of histamine H_1-receptors on human peripheral lung. Biochem Pharmacol 1985;34:3285–3292.

100. Duncan PG, Brink C, Adolphson RL, Douglas JS. Cyclic nucleotides and contraction/relaxation in airway muscle: H_1- and H_2-agonists and antagonists. J Pharm Exp Ther 1980;215:434–442.

101. Hattori Y, Sakuma I, Kanno M. Differential effects of histamine mediated by histamine H_1- and H_2-receptors on contractility, spontaneous rate and cyclic nucleotides in the rabbit heart. Eur J Pharmacol 1988;153:221–229.

102. Leurs R, Brozius MM, Jansen W, Bast A, Timmerman H. Histamine H_1-receptor-mediated cyclic GMP production in guinea-pig lung tissue is an L-arginine-dependent process. Biochem Pharmacol 1991;42:271–277.

103. Sertl K, Casale TB, Wescott SL, Kaliner MA. Immunohistochemical locali-

zation of histamine-stimulated increases in cyclic GMP in guinea-pig lung. Am Rev Respir Dis 1987;135:456–462.

104. Yuan Y, Granger HJ, Zawieja DC, Defily DV, Chilian WM. Histamine increases venular permeability via a phospholipase C-NO synthase-guanylate cyclase cascade. Am J Physiol 1993;264:H1734–H1739.

105. Lewis Baenziger N, Force LE, Becherer PR. Histamine stimulates prostacyclin synthesis in cultured human umbilical vein endothelial cells. Biochem Biophys Res Commun 1980;92:1435–1440.

106. Resink TJ, Grigorian GY, Moldabaeva AK, Danilov SM, Buhler FR. Histamine-induced phosphoinositide metabolism in cultured human umbilical vein endothelial cells. Biochem Biophys Res Commun 1987;144:438–446.

107. Murayama T, Kajiyama Y, Nomura Y. Histamine-stimulated and GTP-binding proteins-mediated phospholipase A_2 activation in rabbit platelets. J Biol Chem 1990;265:4290–4295.

108. Al-Gadi M, Hill SJ. The role of calcium in the cyclic AMP response to histamine in rabbit cerebral cortical slices. Br J Pharmacol 1987;91:213–222.

109. Donaldson J, Brown AM, Hill SJ. Temporal changes in the calcium dependence of the histamine H_1-receptor-stimulation of the cyclic AMP accumulation in guinea-pig cerebral cortex. Br J Pharmacol 1989;98:1365–1375.

110. Garbarg M, Schwartz JC. Synergism between histamine H_1- and H_2-receptors in the cyclic AMP response in guinea-pig brain slices: effects of phorbol esters and calcium. Mol Pharmacol 1988;33:38–43.

111. Magistrati PJ, Schorderet M. Norepinephrine and histamine potentiate the increase in adenosine 3′, 5′-monophosphate elicited by vasoactive intestinal polypeptide in mouse cerebral cortical slices: mediation by α_1-adrenergic and H_1-histaminergic receptors. J Neurochem 1985;5:362–368.

112. Marley PD, Thomson KA, Jachno K, Johnston MJ. Histamine-induced increases in cyclic AMP levels in bovine adrenal medullary cells. Br J Pharmacol 1991;104:839–846.

113. Hill SJ, Donaldson J. The H_1 receptor and inositol phospholipid hydrolysis. In: Schwartz JC, Haas H, eds. The Histamine Receptor. New York: Wiley-Liss, 1992:109–128.

114. Hegstrand LR, Kanof PD, Greengard P. Histamine-sensitive adenylate cyclase in mammalian brain. Nature 1976;260:163–165.

115. Ozawa K, Segawa T. Histamine increases phospholipid methylation and H_2-receptor-adenylate cyclase coupling in rat brain. J Neurochem 1988;50:551–1558.

116. Bristow MR, Cubicciotti R, Ginsburg R, Stinson EB, Johnson CL. Histamine-mediated adenylate cyclase stimulation in human myocardium. Mol Pharmacol 1982;21:671–679.

117. Sarem-Aslani A, Ratge D, Bigge HH, Walker S, Klotz U, Wisser H. A method to assess the histamine-induced cyclic AMP production in isolated gastric mucosal cells from human biopsies. Pharmacology 1991;42:327–332.

118. Agullo L, Picatoste F, Garcia A. Histamine stimulation of cyclic AMP accumulation in astrocyte-enriched and neuronal primary cultures from rat brain. J Neurochem 1990;55:1592–1598.

119. Foreman JC, Norris DB, Rising TJ, Webber SE. A study of the H_2-receptor for stimulating adenylate cyclase in homogenates of guinea-pig lung parenchyma. Br J Pharmacol 1986;87:37–44.

120. Hattori Y, Sakuma I, Kanno M. Differential effects of histamine mediated by H_1- and H_2-receptors on contractility, spontaneous rate and cyclic nucleotides in the rabbit heart. Eur J Pharmacol 1988;153:221–229.

121. Ozawa K, Fujishima Y, Segawa T. Histamine H_2-receptor in atrium: signal transduction and response. In: Timmerman H, Van der Goot H, eds. New Perspectives in Histamine Research. Basel:Birkhauser Verlag, 1991: 265–300.

122. Johnson CL. Histamine receptors and cyclic nucleotides. In: Schwartz JC, Haas H, eds. The Histamine Receptor. New York: Wiley-Liss, 1992: 129–143.

123. Ozawa K, Nomura Y, Segawa T. Histamine acting on H_2-receptors stimulates phospholipid methylation in synaptic membranes of rat brain. J Neurochem. 1987;48:1392–1398.

124. Hirata F, Axelrod J. Phospholipid methylation and biological transmission. Science 1980;209:1082–1090.

125. Haas HL, Wolf P, Palacois JM, Garbarg M, Barbin G, Schwartz JC. Hypersensitivity to histamine in the guinea-pig brain: microiontophoretic and biochemical studies. Brain Res 1978;156:275–291.

126. Mitsuhashi M, Mitshuhashi T, Payan DG. Multiple signaling pathways of histamine H_2-receptors. J Biol Chem 1989;264:8356–18362.

127. Gespach C, Saal F, Cost H, Abita J. Identification and characterization of surface receptors for histamine in the human promyelocytic leukemia cell line HL-60. Mol Pharmacol 1982;22:547–553.

128. Delvalle J, Wang L, Gantz I, Yamada T. Characterization of H_2-histamine receptor: Linkage to both adenylate cyclase and $[Ca^{2+}]_i$ signalling systems. Am J Physiol 1992;263:G967–G972.

129. Dickenson JM, White TE, Hill SJ. The effects of elevated cyclic AMP levels on histamine H_1-receptor-stimulated inositol phospholipid hydrolysis and calcium mobilization in the smooth muscle cell line DDT1 MF-2. Biochem J 1993;292:409–417.

130. Hall IP, Hill SJ. β_2-adrenoceptor stimulation inhibits histamine-stimulated inositol phospholipid hydrolysis in bovine tracheal smooth muscle. Br J Pharmacol 1988;95:1204–1212.

131. Hall IP, Donaldson J, Hill SJ. Inhibition of histamine-stimulated phospholipid hydrolysis by agents which increase cAMP levels in bovine tracheal smooth muscle. Br J Pharmacol 1989;97:603–613.

132. Cumming P, Shaw C, Vincent SR. High affinity histamine binding site is the H_3-receptor: characterization and autoradiographic localization in rat brain. Synapse 1991;8:144–151.

133. Clark MA, Korte A, Egan RW. Guanine nucleotides and pertussis toxin reduce the affinity of histamine H_3-receptors on AtT-20 cells. Agents Actions 1993;40:129–134.

134. Nozaki M, Sperelakis N. Pertussis toxin effects on transmitter release from perivascular nerve terminals. Am J Physiol 1989;256:H455–H459.
135. Oike M, Kitamura K, Kuriyama H. Histamine H_3-receptor activation augments voltage-dependent calcium current via GTP hydrolysis in rabbit sapheneous artery. J Physiol 199;448:133–152.
136. Litosch I, Sulkholutskaya I, Weng C. G protein-mediated inhibition of phospholipase C activity in a solubilized membrane preparation. J Biol Chem 1993;268:8692–8697.
137. Van der Werf JF, Bast A, Bijloo GJ, Van der Vliet A, Timmerman H. HA autoreceptor assay with superfused slices of rat brain cortex and electrical stimulation. Eur J Pharmacol 1987;138:199–206.
138. Van der Vliet A, Van der Werf JF, Bast A, Timmerman H. Frequency-dependent autoinhibition of histamine release from rat cortical slices: a possible role for H_3-receptor reserve. J Pharm Pharmacol 1988;40:577–579.
139. Arrang JM, Garbarg M, Schwartz JC. Autoregulation of histamine release in brain by presynaptic H_3-receptors. Neuroscience 1985;15:553–562.
140. Chernevskaya NI, Obukhov AG, Krishtal OA. NMDA receptor agonists selectively block N-type calcium channels in hippocampal neurons. Nature 1991;349:418–420.
141. Takemura M, Kishino J, Yamatodani A, Wada H. Inhibition of histamine release from rat hypothalamic slices by ω-conotoxin GVIA but not by nilvadipine, a dihydropyridine derivative. Brain Res 1989;496:351–356.
142. Yang OZ, Hatton GI. H_3-histamine receptors activate K^+ channels to hyperpolarize magnocellular histaminergic neurons of rat posterior hypothalamus. Soc Neurosci Abstr 1991;17:409.
143. Garbarg M, Trung Tuong MD, Gros C, Schwartz JC. Effects of histamine H_3-receptor ligands on various biochemical indices of histaminergic neuron activity in rat brain. Eur J Pharmacol 1989;164:1–11.
144. Schlicker E, Fink K, Molderings G, Göthert M. Presynaptic histamine H_3-receptors on noradrenergic and serotoninergic neurons. J Neurochem 1991;57:S37.
145. Ea Kim L, Javellaud J, Oudart N. Endothelium-dependent relaxation of rabbit middle cerebral artery to a histamine H_3-agonist is reduced by inhibitors of nitric oxide and prostacyclin synthesis. Br J Pharmacol 1992;105:103–106.
146. Shimokawa H, Flavahan NA, Lorenz RR, Vanhoutte PM. Prostacyclin releases endothelium-derived relaxing factor and potentiates its action in coronary arteries of the pig. Br J Pharmacol 1988;95:1197–1203.

2

Molecular Identification of the Histamine H_1-Receptor in Humans

Badrul Alam Chowdhury
University of Tennessee College of Medicine,
Memphis, Tennessee

Michael A. Kaliner
Institute for Asthma and Allergy, Washington Hospital Center,
Washington, D.C.

I. INTRODUCTION

Histamine is a major mediator of the immediate allergic response in humans and was the first chemical substance of mast cell origin to be identified. Extracellular histamine has a wide range of biological actions mediated through activation of specific cell surface receptors. Three histamine receptor subtypes have been identified pharmacologically, designated H_1-, H_2-, and H_3-receptors (reviewed in Ref. 1). Acting through H_1-receptors in peripheral tissues, histamine causes increased vascular permeability, pruritus, contraction of smooth muscles of the respiratory and gastrointestinal tracts, and release of inflammatory mediators and recruitment of inflammatory cells. In the central nervous system, histamine acts as a neurotransmitter through the H_1-receptor. Antagonists of the histamine H_1-receptor are useful in the treatment of allergic diseases and are among the most widely used medications in the world (2). Historically, study of the histamine H_1-receptor employed a biochemical approach in which the receptor was merely considered to be a protein to be labeled, purified, and characterized. Recently, these studies have been supplemented with binding and pharmacological studies in intact tissues and in some cell lines and the molecular genetic approach has been applied to the study of the H_1-receptor. The gene for the H_1-receptor has now

33

been cloned from the human (3–5) and other (6–9) species, and has been expressed in heterologous cell lines. It is expected that, as with other receptors, further applications of molecular biological techniques will contribute greatly to the study of the H_1-receptor.

II. BACKGROUND

Early studies showed that agonists acting at the histamine H_1-receptor activate phospholipase C with the subsequent secretion of 1,4,5-inositol triphosphate and an increase in intracellular Ca^{2+} concentrations (reviewed in Ref. 10). Those studies were done on brain and other tissues known to express relatively large amounts of the receptor and in some cell lines that express the receptor. They were limited by lack of knowledge of the primary structure of the H_1-receptor and the inherent difficulty in studying intact tissues. The application of molecular biological techniques has had a profound impact on the study of various biologically important molecules, including receptors. Genes for various receptors have recently been cloned and sequenced. Their amino acid structure has been determined, and for many receptors the gene has been stably expressed in heterologous cells. Cloning, sequence analysis, and pharmacological characterization of receptor genes have revealed the existence of receptor gene superfamilies. The largest receptor family includes several hundred receptors that mediate signal transduction through guanine nucleotide regulatory proteins (G proteins). Among this family are receptors for diverse ligands such as small amine neurotransmitters, small peptides, large glycoprotein hormones, odorants, and light. The high degrees of sequence identity at the DNA level among G protein-coupled receptors has allowed for the isolation of many new receptor genes with cross-hybridization experiments. Histamine receptors belong to this family of G protein-coupled receptors.

G protein-mediated transmembrane signaling pathways have generated much attention because of the diverse physiological and pharmacological events modulated by these mechanisms. G proteins are heterotrimeric proteins composed of α, β, and γ subunits. They are members of a large gene superfamily (11, 12). In the basal state, the G protein oligomer exists in a complex with GDP. The rate of GDP dissociation from the G protein is very slow. On agonist binding, G protein-coupled receptors undergo some conformational changes that facilitate an exchange of GTP for bound GDP at a site within the α subunit of the G protein. The binding of GTP to the α subunit of the G protein promotes dissociation of this subunit from the β and γ subunits. GTP-bound α subunit is responsible for modulating the second messenger systems intracellularly, including adenylate cyclase, phospholipase, cyclic GMP phosphodiesterase, and ion channels. These lead to metabolic and/or ionic changes within the cells. The reaction is terminated by hydrolysis of bound GTP by a GTPase intrinsic to the α subunit and subsequent reassociation of the α subunit to the β and γ subunits with return to basal state of the receptor. The receptor acts in this system as a catalyst for the activation of G proteins by the agonist specific for the receptor. Analysis of the primary structure of all G protein-coupled receptors characterized so far shows some typical features. The receptor is contained in a single polypeptide chain, mostly encoded by genes that lack introns, and shows the existence of seven putative trans-membrane spanning domains in α-helical configuration (Fig. 1).

For the histamine H₁-receptor, earlier studies suggested that the recep-tor was associated with a G protein, and that agonist stimulation led to

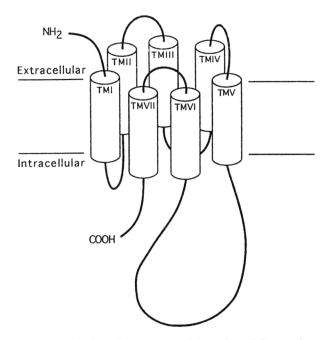

Figure 1   Model of the structural domains of G protein-coupled receptors. The transmembrane domains (TM) are shown as cylinders perpendicular to the plane of plasma membrane. The seven transmembrane domains are proposed to traverse the membrane in an α-helical configuration with the three alternating extracellular and three intracellular loops connecting to the transmembrane region. The amino-terminal of the protein is situated extracellularly and the carboxy-terminal is situ-ated intracellularly.

membrane phosphoinositide hydrolysis with mobilization of intracellular calcium. The calcium-mobilizing property of the receptor was exploited in the initial cloning of the histamine H_1-receptor cDNA from bovine adrenal medulla, using techniques of molecular cloning in an expression system and electrophysiological assay in *Xenopus* oocytes (6). The identity of the cDNA clone was confirmed by transient expression in heterologous cells showing saturable binding with radiolabeled mepyramine and stereoselective displacement of binding by different antagonists. On amino acid analysis, the typical features of a G protein-coupled receptor were seen with the presence of seven putative transmembrane domains. Later, employing cross-hybridization techniques, the genes for the rat and the guinea pig histamine H_1-receptor were cloned (7–9). Because of the physiological and pharmacological importance of histamine H_1-receptor, we began a series of experiments to isolate the genomic clone of human H_1-receptor. We and others recently reported the genomic cloning of the human H_1-receptor (3–5). We also expressed the gene permanently in Chinese hamster ovary (CHO) cells for further characterization of the receptor (13). In subsequent sections of this review, we describe the cloning, chromosomal localization, and characterization of the human histamine H_1-receptor.

III. CLONING OF THE HUMAN HISTAMINE H_1-RECEPTOR GENE

We employed the technique of polymerase chain reaction (PCR) and subsequent library screening to isolate the human histamine H_1-receptor gene. Multiple oligonucleotide primers complementary to the bovine H_1-histamine receptor cDNA sequence were synthesized, and human genomic DNA was used as the template for the PCR reaction. One pair of primers (sense: 5'-GTGG TCCTGAGCACCATGTCCTTG-3' and antisense: 5'-GTCTTTTGGCTTCCTTTTCAGAA CCTC-3') amplified a 705-bp DNA fragment extending from the amino acid #33 (transmembrane helix I) to amino acid #267 (third intracellular loop) in the bovine sequence. The DNA fragment was cloned and sequenced. Over this region, the PCR fragment displayed over 85% identity with the bovine H_1-histamine receptor at the nucleotide and amino acid level, indicating that the DNA fragment most likely represents the human histamine H_1-gene. This DNA fragment was then labeled with radiolabeled nucleotide [^{32}P]-dCTP by random priming for screening a human genomic library. The library screened was prepared from blood leukocyte DNA, which was partially digested with the restriction endonuclease Sau3AI, run over a sucrose column to separate out DNA fragments of about 14 kb, and cloned in a phage cloning vector EMBL3 SP6/T7. The enzyme Sau3AI cuts DNA at many sites into relatively small fragments. Thus the larger DNA fragments recovered after partial digestion most likely will cover overlapping areas of any portion of genomic DNA. This technique is widely used for library construction and screening (14).

With the labeled PCR-amplified DNA fragments, three clones were identified on screening four genome equivalents of the human genomic library. These clones were plaque purified and characterized by restriction mapping. Smaller restriction fragments were identified by Southern hybridization containing the human histamine H$_1$-receptor gene. A 4.5-kb PstI fragment that hybridized with the probe was used for sequence analysis. A total of 2 kb was sequenced. This fragment contained the whole open reading frame of the human histamine H$_1$-receptor gene. There were no introns within the coding region of the gene. The nucleotide and deduced amino acid sequence of the human histamine H$_1$-receptor is shown in Figure 2. Simultaneously two other groups also cloned the human histamine H$_1$-receptor (4, 5). For hybridization, both groups used the whole bovine histamine H$_1$-receptor as a probe. Sequence of the open reading frame of our gene is identical to the others, indicating that all groups identified the same gene.

To look for genes closely related to the human histamine H$_1$-receptor, we did Southern hybridization on total genomic DNA digested with different restriction endonucleases. A 1472-bp DNA fragment containing the human histamine H$_1$-receptor gene was used as the hybridization probe. A single band of the expected size as predicted from restriction analysis of the clone containing the human H$_1$-receptor gene was seen. This observation indicates that no closely related gene exists in the genome. Similar results were also obtained on guinea pig genomic DNA analysis with the cloned guinea pig histamine H$_1$-receptor (9). Taken together, these studies do not support earlier biochemical studies that suggested the existence of more than one isoform of the histamine H$_1$-receptor (15).

IV. TISSUE DISTRIBUTION OF THE HUMAN HISTAMINE H$_1$-RECEPTOR mRNA

Northern hybridization studies of mRNA from different human tissues, with an H$_1$-receptor probe, show abundant transcripts in the placenta, lung, brain, skeletal muscle, and kidneys. No message was seen in the heart, liver, and pancreas (4). Of all the tissues, the placenta has the largest amount of message (Fig. 3). This observation possibly reflects on abundance of blood vessels, which are known to express large amounts

```
-142  TTGATCACCCCCAACAGCATACAACTCGAGTCTGATGAACATCATCGCTACTAAGTGGCCACTCATCACCCAAGTCTCTGACCTTACTTTTCTCTCTTTTCTCCCAGGGAGTGAGCCATAACTGGCGGCTGCTCTTGGGCCA

   1  ATG AGC CTC CCC AAT TCC TCC TGC TTA GAA GAC AAG ATG TGT GAG GGC AAC AAG ACC ACT ATG GCC AGC CCC CAG CTG ATG CCC CTG GTG GTC GTG CTC AGC ACT
   1  Met Ser Leu Pro Asn Ser Ser Cys Leu Glu Asp Lys Met Cys Glu Gly Asn Lys Thr Thr Met Ala Ser Pro Gln Leu Met Pro Leu Val Val Val Leu Ser Thr

 109  ATC TGC TTG GTC ACA GGG CTC AAC CTG GTG TAT GTG GTA CGG AGT GAG AAG GAG CTG CGG AAC CTC TAC ATC GTC AGC CTC TCG GTG GCG
  37  Ile Cys Leu Val Thr Gly Leu Asn Leu Val Tyr Val Val Arg Ser Glu Lys Glu Leu Arg Asn Leu Tyr Ile Val Ser Leu Ser Val Ala

 217  GAC TTG ATC GTG GGT GCC GTC GTC ATG CCT ATG AAC ATC CTC TAC CTG CTC ATG TCC AAG TGG TCA CTG GGC CGT CCT CTC TGC CTT TTT TGG CTT TCC ATG GAC TAT
  73  Asp Leu Ile Val Gly Ala Val Val Met Pro Met Asn Ile Leu Tyr Leu Leu Met Ser Lys Trp Ser Leu Gly Arg Pro Leu Cys Leu Phe Trp Leu Ser Met Asp Tyr

 325  GTG GCC AGC ACA GCG TCC ATT TTC AGT GTG TTC ATT CTC TGC ATT GAT CGC TAC CTC AGG TCC GTC ATT CCC CTC AGG TAC CTT AAG TAT CGT ACC AAG ACC CGA GCC
 109  Val Ala Ser Thr Ala Ser Ile Phe Ser Val Phe Ile Leu Cys Ile Asp Arg Tyr Leu Arg Ser Val Ile Pro Leu Arg Tyr Leu Lys Tyr Arg Thr Lys Thr Arg Ala

 433  TCG GCC ACC ATT CTG GGG GCC TGG TTT CTC TCT TTC CTG GTT ATT CCC GTT TGG CTG GGC TGG AAT CAC TCA CAG ATG GAG CGA GTG CGC GAC GAG GAC AAG TGT
 145  Ser Ala Thr Ile Leu Gly Ala Trp Phe Leu Ser Phe Leu Val Ile Pro Val Trp Leu Gly Trp Asn His Ser Gln Met Glu Arg Val Arg Asp Glu Asp Lys Cys

 541  GAG ACA GAC TTC TAT GAT GTC ACT TGG TTC AAG GTC ATG ACT GCC ATC ATC AAC TTC TAC CTG CCC ACC CTG CTC ATG CTG TGG TTC TAT GCC AAG ATC TAC AAG GCC
 181  Glu Thr Asp Phe Tyr Asp Val Thr Trp Phe Lys Val Met Thr Ala Ile Ile Asn Phe Tyr Leu Pro Thr Leu Leu Met Leu Trp Phe Tyr Ala Lys Ile Tyr Lys Ala

 649  GTA CGA CAA CAC TGC CAG ACC CGG CAG ATC CTC CGG GAG CTC ATC AGG AGC TTC CCT TCC GAG TCA TTC TCC CAG ATT AAG AAG CCA GAA GAT GGG GCC AAG AAA CCA GGG
 217  Val arg Gln His Cys Gln Thr Arg Gln Ile Leu Arg Glu Leu Ile Arg Ser Phe Pro Ser Glu Ser Phe Ser Gln Ile Lys Lys Pro Glu Asp Gly Ala Lys Lys Pro Gly
```

```
757  AAG GAG TCT CCC TGG GAG GTT CTG AAA AGG AAG CCA AAA GAT GGT GGT TCT GTC TTG AAG TCA CCA TCC CAA ACC CCC AAG GAG ATG AAA TCC CCA GTT GTC
253  Lys Glu Ser Pro Trp Glu Val Leu Lys Arg Lys Pro Lys Asp Gly Gly Ser Val Leu Lys Ser Pro Ser Gln Thr Pro Lys Glu Met Lys Ser Pro Val Val

865  TTC AGC CAA GAG GAT GAT AGA GAA GTA GAC AAA CTC TAC TGC TTT CCA CTT GAT ATT GTG CAC ATG CAG GCT AGC AGG GAG AGT GCA GGG AGT GTA GCC GTC
289  Phe Ser Gln Glu Asp Asp Arg Glu Val Asp Lys Leu Tyr Cys Phe Pro Leu Asp Ile Val His Met Gln Ala Ala Glu Ser Arg Asp Tyr Val Ala Val

973  AAC CGG AGC AGC CAT GGC CTC AAG GCA GAT GAG GGC ACA CAT GGG GCC AGC GAG ATA TCA GAG GAT CAG ATG TTA GGT GAT AGC TCC TTC TCT CGA
325  Asn Arg Ser His Gly Leu Lys Ala Asp Glu Gly Thr His Gly Ala Ser Glu Ile Ser Glu Asp Gln Met Leu Gly Asp Ser Phe Ser Arg

1081 ACG GAC TCA GAT ACC ACA GAG GCA CCA GGC AAA TTG AGG AGT GGG TCT AAC ACA GGC GAT TAC ATC AAG TTT ACT TGG AAG AGG CTC CGC TCG
361  Thr Asp Ser Asp Thr Thr Glu Ala Pro Gly Lys Leu Arg Ser Gly Ser Asn Thr Gly Asp Tyr Ile Lys Phe Thr Trp Lys Arg Leu Arg Ser

1189 CAT TCA AGA CAG TAT GTA TCT GGG TTG CAC ATG GGC AAA CAG TTG GGT TTT ATC ATG GCA GCC TTC ATC CTC TGC TGG ATC CCT TAT TTC
397  His Ser Arg Gln Tyr Val Ser Gly Leu His Met Gly Lys Gln Leu Gly Phe Ile Met Ala Ala Phe Ile Leu Cys Trp Ile Pro Tyr Phe

1297 ATC TTC TTC ATG GTC ATT GCC TTC TGC ATT GAA CAT TTG TGC AAC ATT TGG AGA ATT CTG CTG TGG GGA TCA GGT TAC ATC AAC CCC CTC ATC TAC
433  Ile Phe Phe Met Val Ile Ala Phe Cys Ile Glu His Leu His Tyr Arg Asn Ser Cys Lys Asn Phe Trp Leu Gly Tyr Ile Asn Ser Thr Leu Asn Pro Leu Ile Tyr

1405 CCC TTG TGC AAT GAG AAC TTC AAG AAG ACA TTC AAG AGA ATT CTG CAT ATT CGC TCC TAA  GGGAGGCCTCTGAGGGGATGCAACAAAATGATCCTTATGATGTCCAACAAGGAAATAGAGGACG
469  Pro Leu Cys Asn Glu Asn Phe Lys Lys Thr Phe Lys Arg Ile Leu His Ile Arg Ser <0>

1528 AAGGCCTCGTCGTTGCCAGGCAGGCCACCTGGCCTTCTCTGAATCCAAACCACAGTCTTAGGGCGTTAGTTTGGAAAGTTCTTAGGCACCATAGAAGAACACAGCAGTGGCGTGATCAGCAGAGATTGAACTTTGAGGA
```

Figure 2 Nucleotide and deduced amino acid sequence of the human histamine H₁-receptor gene. Sequences of both strands of DNA were determined. The nucleotide residues are numbered from the 5' to 3' direction. The first nucleotide encodes the initiation methionine. The deduced amino acid sequence is shown below the nucleotide sequence. The seven putative transmembrane domains, as defined by hydropathy analysis, are underlined.

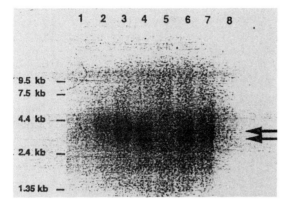

Figure 3 Northern blot analysis of mRNA from various human tissues with human H_1-receptor probe. Lanes contain mRNA from heart (lane 1), brain (lane 2), placenta (lane 3), lung (lane 4), liver (lane 5), skeletal muscle (lane 6), kidney (lane 7), and pancreas (lane 8). Positions of size markers are on the left; the arrows on the right indicate the position of the main transcripts hybridizing to the probe (adapted from Ref. 4).

of histamine H_1-receptor. On northern analysis of rat and bovine tissues, similar tissue expressions were seen (6, 7). In rats, abundant message was seen in lung, brain, heart, and small intestine. In bovine tissue, abundant message was seen in lung, small intestine, adrenal medulla, and uterus; and rare message was seen in cerebral cortex and heart. Taken together, the tissue distribution is similar to the distribution reported earlier using binding and pharmacological studies. Tissue-specific distribution of guinea pig histamine H_1-receptor in brain has recently been reported (9). The highest expression is seen in the hippocampus, cerebellum, brainstem, thalamus, and substantia nigra. Moderate levels of the message are seen in the cerebral cortex, striatum, hypothalamus, olfactory tubercles, olfactory bulb, and pituitary gland. Similar data on tissue-specific distribution or cellular localization of the human histamine H_1-receptor message is not yet available.

The mRNA transcript containing the human histamine H_1-receptor message was contained in two different-sized bands of 3.0 kb and 3.5 kb (Fig. 3). In the brain, the message was contained only in 3.5-kb mRNA transcript, whereas in peripheral tissues, such as placenta, lung, skeletal muscle, and kidneys, the message was contained equally in the 3.0- and 3.5-kb bands (4). In the guinea pig also, two different-sized mRNA transcripts of 3.3 kb and 3.7 kb with different tissue distribution were seen (9). The explanation of this observation is not clear at present. It is interesting that the mRNA sizes of the gene are different in the central nervous system as compared to the peripheral tissues. Histamine, through the H_1-receptor, is known to subserve a different function in the brain where it functions as a neurotransmitter, as opposed to the peripheral tissues where it acts mainly as a mediator of inflammation. Existence of more than one isoform of the histamine H_1-receptor is unlikely, particularly in view of the observation that on Southern hybridization with genomic DNA only one-sized band is seen. The differing mRNA sizes may correspond to the use of distinct start sites in the 5' region of the gene, or through alternate splicing of the gene through introns in the untranslated region of the gene, or through use of different polyadenylation sites. Further studies to characterize the 5' and 3' untranslated regions of the gene are required to explore these various possibilities.

To date very few studies have been done to characterize the 5' region and the promoter of the G protein-coupled receptor. Those receptors which have been studied include the dopamine D_{1A}-receptor (16), the β_2-adrenergic receptor (17), the α_{2A}-adrenergic receptor (18), and the substance P-receptor (19). The dopamine D_{1A}-receptor gene has housekeeping-gene-type promoters, while others have promoters and transcription factor binding sites like other regulated eukaryotic genes. For the human histamine H_1-receptor gene, the promoter and other regulatory elements have not yet been characterized. Sequence analysis of the 5' untranslated region shows eight potential TATA boxes, two CACCC sequences, and many consensus sequences as putative binding sites for transcriptional factors AP1, AP2, NF-GMb, Oct-1, and glucocorticoid response element (4). The 5' flanking region of the gene thus has the characteristics of a regulated gene. The 3' untranslated region has four potential polyadenylation signal sequence (AAUAA). The area also has the consensus sequences (AUCU) which are known to be involved in agonist-mediated changes in mRNA stability for other G protein-coupled receptors (20, 21). Study of the regulatory elements of histamine receptor gene and message stability will show the functional importance of these putative binding sites and consensus sequences.

V. STRUCTURE OF THE HUMAN HISTAMINE H_1-RECEPTOR

The complete nucleotide sequence of the human histamine H_1-receptor coding region contains a single open reading frame of 1461 bp which predicts a protein of 487 amino acid residues (Fig. 2). Like many other G protein-coupled receptors, there is no intron within the open reading frame. The calculated molecular weight of the protein is 55,781. Hydropathy analysis of the protein predicts seven hydrophobic domains of 20 amino acid residues each, which are characteristic of G protein-coupled receptors. The tertiary structure of the receptor protein can be predicted based on analogy with bacteriorhodopsin, a light-activated proton pump, whose three-dimensional structure was deduced from electron microscopy (22). The seven hydrophobic domains are presumed to be arranged perpendicularly on the cell membrane lipid bilayer, and the six hydrophilic regions that connect the hydrophobic domains form alternating extracellular and intracellular loops. Based on this model, the amino terminus of the protein is extracellular and the carboxy terminus is intracellular.

The sequence of the human histamine H_1-receptor is well conserved across species. At the amino acid level, the receptor displays 82, 78, and 71% identity with the bovine, rat, and guinea pig histamine H_1-receptors, respectively (Fig. 4). The identity is high in the putative membrane-spanning domains, but much lower in the extra- and intracellular domains. In particular, in the amino terminus, and the third intracellular domain which is presumed to be involved in receptor G protein interactions, the identity is only about 50%. The amino terminus domain of the receptor is short, which is typical of a G protein-coupled receptor whose natural ligand is a biogenic amine, and the third intracellular loop is large, which is characteristic of receptors that mediate signal transduction through hydrolysis of membrane phosphoinositide. In contrast, the histamine H_2-receptor, which has also been cloned (23, 24), has a short third intracellular loop which is typical for a receptor that induces cAMP turnover. Sequence comparison of the histamine H_1-receptor to other members of the same family show the highest match to the muscarinic acetylcholine receptor. The overall sequence similarity is about 45%. This observation may explain the antimuscarinic properties of older H_1-receptor antagonists. The match of the histamine H_1-receptor to the H_2-receptor is no greater than to other G protein-linked receptors, which explains the lack of cross-reactivity between H_1- and H_2-receptor antagonists.

Several amino acid residues are conserved among the histamine H_1-receptors from different species and also among other receptors from the same family. These conserved sequences may be functionally important. At the amino terminal end there are two conserved asparagine residues. Both (Asn^5 and Asn^{18}) are in the consensus sequence (Asn-X-Ser/Thr) for N-glycosylation. All G protein-coupled receptors have this consensus sequence. For histamine, like other biogenic amine receptors, it is likely that N-linked glycosylation is not crucial for ligand binding, but is essential for the subcellular distribution and trafficking of the receptor to the cell

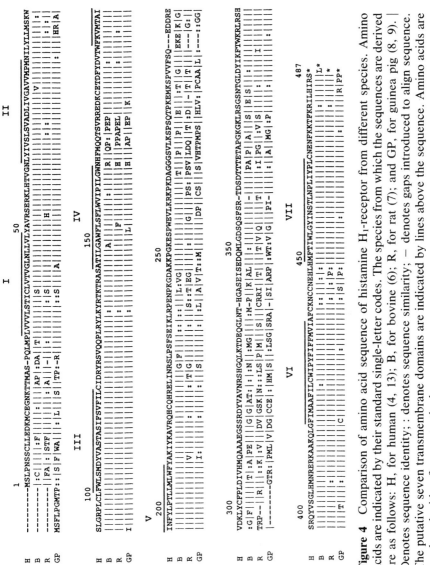

Figure 4 Comparison of amino acid sequence of histamine H₁-receptor from different species. Amino acids are indicated by their standard single-letter codes. The species from which the sequences are derived are as follows: H, for human (4, 13); B, for bovine (6); R, for rat (7); and GP, for guinea pig (8, 9). Denotes sequence identity; : denotes sequence similarity; − denotes gaps introduced to align sequence. The putative seven transmembrane domains are indicated by lines above the sequence. Amino acids are numbered with reference to the human sequence.

surface. For the β-adrenergic receptor, treatment of the receptor with endoglycosidase to remove carbohydrate moieties has no effect on ligand binding or coupling (25, 26). Also, treatment with tunicamycin, which inhibits N-linked glycosylation, has no effect on ligand binding (27). However, mutant β-adrenergic receptors lacking consensus glycosylation sites do not traffic correctly through the cell to the cell surface (28).

The first and second extracellular loops of the histamine H_1-receptor contain two cysteine residues (Cys^{100} and Cys^{180}). These residues are conserved among all G protein-coupled receptors and are implicated in the formation of disulfide bonds that maintain the tertiary structure of the receptor. Although ligand binding is not directly affected by these residues, structural integrity of the receptor is essential for the binding. Studies on mutated receptors with substitution of one or both the cysteine residues show reduced ligand binding to β_2-adrenergic receptors, muscarinic acetylcholine receptors, and rhodopsin (29–31). Previous studies on the guinea pig brain H_1-receptor by disufide bonds-disrupting thiol reagents support the notion that the two cysteine residues are critical for maintaining the affinity for histamine binding (32).

The critical amino acid residues of the histamine H_1-receptor binding to histamine are possibly the aspartic acid in the third transmembrane domain (Asp^{107}), and threonine and asparagine in the fifth transmembrane domain (Thr^{194}, Asn^{198}) (Fig. 5). The Asp residue in third transmembrane domain is conserved in all monoaminergic G protein-coupled receptors. This moiety is likely to salt-link the positively charged ammonium group of histamine. The histamine H_2-receptor has an aspartic acid residue (Asp^{98}) at the same location. The crucial difference between the histamine H_1-receptor and the H_2-receptor is the replacement of Thr^{194} and Asn^{198} in the fifth transmembrane domain of the H_1-receptor by Asp^{186} and Thr^{190} in the same locations of the H_2-receptor. This pair of amino acid residues separated by three amino acids is conserved across all species and is likely to bind the imidazole ring of the histamine molecule. By mutation of these two amino acid residues in the histamine H_2-receptor, it has been shown that the Asp^{186} defines H_2 selectivity and the Thr^{190} establishes the kinetics of histamine binding but is not essential for H_2 selectivity (33). The imidazole ring of histamine possesses two nitrogen molecules for possible proton attachment. The uncharged forms of histamine can exist in two tautomeric forms of the imidazole ring–the N^{π}-H tautomer (the proton on the N adjacent) or N^{τ}-H tautomer (the proton on the N distal). From a pharmacological perspective, a tautomeric equilibrium is required for H_2-receptor activation by an agonist, with one of the two crucial fifth transmembrane amino acid residues acting as the proton-donating and the other as the proton-accepting group. The ionization state of carboxy

Transmembrane 3   Transmembrane 5

H₁R	RPLCLF‾WLSMD‾YVASTASIFSVFILCIDRY	WFKVMTAIINFYLPTLLMLWFYA
H₂R	KVFCNIYTSLDVMLCTASILNLFMISLDRY	VYGLVDGLVTFYLPLLIMCITYY
β₁AR	SFFCELWTSVDVLCVTASIETLCVIALDRY	AYAIASSVVSFYVPLCIMAFVYL
β₂AR	NFWCEFWTSIDVLCVYASIETLCVIAVDRY	AYAIASSIVSFYVPLVIMVFVYS
α₂ₐAR	KTWCEIYLALDVLFCTSSIVHLCAISLDRY	WYVISSCIGSFFAPCLIMILVYV
D₁DR	GSFCNIWVAFDIMCSTASILNLCVISVDRY	TYAISSSVISFYIPVAIMIVTYT
5HT₁ₐR	QVTCDLFIALDVLCVTSSILHLCAIALDRY	GYTIYSTFGAFYIPLLLMLVLYG
m₁mAChR	TLACDLWLALDYVASNASVMNLLLISFDRY	IITFGTAMAAFYLPVTVMCTLYW
m₂mAChR	PVVCDLWLALDYVVSNASVMNLLIISFDRY	AVTFGTAIAAFYLPVIIMTVLYW

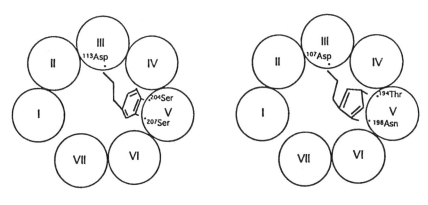

β₂-Adrenergic Receptor   H₁-Histamine Receptor

Figure 5 Amino acid alignment of third and fifth transmembrane domain of different G protein-coupled receptor and putative ligand binding sites. In the top panel, alignment of deduced amino acid sequences of the third and fifth transmembrane domains and adjacent area is shown. The transmembrane domains in the H₁-receptor sequence are identified by a line above the sequence. Standard single-letter codes for amino acids are used. The residues shown in bold are presumed to bond with the ligand. The considered receptors are as follows: H₁R, histamine H₁-receptor (4, 13); H₂R, histamine H₂-receptor (24); β₁AR, β₁-adrenergic receptor (60); β₂AR, β₂-adrenergic receptor (61); α₂ₐAR, α₂ₐ-adrenergic receptor (62); D₁DR, D₁-dopamine receptor (63); 5HT₁ₐR, 5HT₁ₐ-receptor (64); M₁MAChR, M₁-muscarinic acetylcholine receptor (65); M₂MAChR, M₂-muscarinic acetylcholine receptor (66). All receptor sequences are from humans. In the bottom panel, a model showing the binding site of the H₁-receptor as compared to β₂-adrenergic receptor is shown. The binding model for the β₂-adrenergic receptor is adapted (67). The view of the receptors is from the extracellular face of the plasma membrane. Seven transmembrane helices are numbered I through VII. The location of the conserved residues implicated in binding is indicated.

residues is unknown, but if it is uncharged, the threonine would be the proton acceptor (34, 35). For histamine H_1-receptor stimulation by an agonist, the critical requirement will be a proton-donating amino acid moiety. It is likely that the Thr^{194} or Asn^{198} will constitute the binding site of the N^{π} atom of histamine.

Similar conserved amino acid residues involved in ligand binding have been identified for other G protein-coupled receptors. For biogenic amine receptors that bind catecholamines, there are a conserved pair of Ser moieties separated by two amino acids, but not in those receptors whose endogenous ligands lack a catechol moiety. The location and separation of the critical amino acid residues that form hydrogen bonding may depend on the structure of the ring, like imidazole for histamine or catechol for catecholamines. In Figure 5, sequence alignments of the two domains involved in ligand binding from representative receptors from different classes (histamine, catecholamine, noncatecholamine) are shown. A putative model of binding for histamine as it compares to that of adrenergic receptor binding is also shown. Mutagenesis studies of these crucial amino acid residues of the histamine H_1 receptor will clarify the role of these residues as has been done for many other G protein-coupled receptors (36). The binding site for H_1-antagonists may be the same or adjacent amino acid residues. Occupancy of these sites would prevent access of histamine to its binding sites. Construction of chimeric histamine H_1-receptors and mutagenesis study will possibly define the binding site of H_1 antagonists, which are important therapeutic molecules. A site-directed mutagenesis study on the cloned guinea pig H_1-receptor has recently been reported (37). By mutating the threonine and asparagine residues in the fifth transmembrane domain (Thr^{203}, Asn^{207}) to nonfunctional alanine, it was shown that the Asn^{207} residue is involved in the binding of the N^{τ}-nitrogen atom of histamine and other analogues.

The histamine H_1-receptor has two aspartic acid residues (Asp^{73} and Asp^{124}), one located in the Leu-X-X-X-Asp-Leu motif in the second transmembrane domain and the other in anionic cationic pair (Asp^{123} and Arg^{124}) in the Asp-Arg-Tyr (D-R-Y) motif in the proximal portion of the second intracellular loop near the third transmembrane domain (Fig. 5). Most of the G protein-coupled receptors have these two conserved aspartic acid residues. The importance of these two Asp residues has been well documented for the β_2-adrenergic receptor, m_1 muscarinic acetylcholine receptor, and α_{2A}-adrenergic receptor (38–40). The Asp residue in the second transmembrane domain has been shown to be involved in guanine-sensitive high-affinity agonist binding and second messenger activation. The Asp residue in the proximal portion of the second intracellular loop appears to influence receptor/G protein coupling.

The histamine H_1-receptor also contains several Ser and Thr residues in the intracellular domains and the carboxy terminus. These are possible phosphorylation sites for various protein kinases. The third intracellular loop is particularly rich in these residues (Fig. 2).

VI. CHROMOSOMAL LOCALIZATION OF THE HUMAN H_1-RECEPTOR GENE

The 4.5 PstI fragment that contained the human histamine H_1-receptor gene was used to probe metaphase chromosomes from PHA-stimulation human peripheral blood leukocytes for chromosomal localization using the fluorescence in situ hybridization method. Specific labeling of the distal short arm of chromosome 3 was seen. On banding, the label was located 11% of the distance from the p terminus to the q terminus of the chromosome, which corresponds to the area 3p24 (4, 13). The location of the map is proximal to that for the α chain of the interleukin-5 receptor and the v-raf1 oncogene homolog, as shown in Figure 6. The proximity

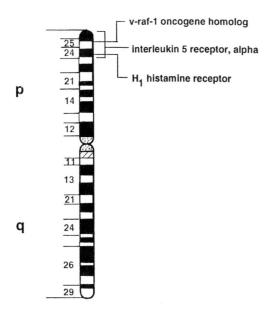

Figure 6   Chromosomal map of human H_1-receptor. Schematic representation of chromosome 3 showing the location of histamine H_1-receptor.

of the histamine H_1-receptor gene to that of the interleukin-5 gene is of interest, as both these molecules are important in mediating allergic inflammation. To date, no other G protein-coupled receptor genes have been mapped to this region of the human genome.

VII. HETEROLOGOUS EXPRESSION OF THE HUMAN HISTAMINE H_1-RECEPTOR

To verify the identification of our clone, and for further characterization of the receptor, the coding region of the gene was transfected into Chinese hamster ovary (CHO) cells to isolate stable cell lines. On preliminary screening of CHO cells by [^{3}H]-mepyramine binding, no endogenous histamine H_1-receptor was detected. The DNA sequence of the human histamine H_1-receptor gene contains several ATGs upstream to the presumed initiation codon. It is known that the presence of an additional ATG sequence between the promoter of the expression vector and the initiation codon in a gene can reduce the efficiency of mRNA translation due to multiple false starts. There were no unique restriction sites in the 5'-untranslated sequence of the human H_1-receptor gene that will allow easy removal of the upstream ATG sequence. We therefore digested the gene with Exonuclease III from the 5' end after protecting the 3' end. A smaller fragment containing the entire open reading frame of the histamine H_1-receptor gene plus 48 bp of the 5'- and 81 bp of the 3-untranslated DNA with no ATGs upstream of the presumed initiation codon was identified. This DNA was cloned in an expression vector and transfected into CHO cells using the calcium phosphate precipitation technique. The gene was expressed under a SV40 viral promoter. For selection, the same cell was also cotransfected with a selectable antibiotic gene (Geneticin) with the CMV viral promoter. Permanent transfectant cell lines were expanded and screened for the presence of the human histamine H_1-receptor by [^{3}H]-mepyramine binding assay. Cell lines expressing the receptor from densities ranging from 160 fmol/mg of protein to 1000 fmol/mg of protein (corresponding to 16,000 to 100,000 receptor/cell) were isolated. These cell lines were used for characterization of the human histamine H_1-receptor (13).

Other groups have also expressed the histamine H_1-receptor gene or cDNA from various species. The bovine H_1-receptor was transfected transiently in COS-7 cells for binding and competition studies (6), and permanently in CHO cells where Ca^{2+}-mobilization on agonist stimulation was demonstrated (41). The rat histamine H_1-receptor was transfected into C6 glioma cells and studied for [³H]-mepyramine binding and competition by antagonists (7). The guinea pig histamine H_1-receptor was transfected into CHO cells and characterized by binding studies and by investigation of signal transduction pathways (42).

VIII. PHARMACOLOGICAL CHARACTERIZATION OF HUMAN HISTAMINE H_1-RECEPTOR IN TRANSFECTED CELLS

Binding studies were done on CHO cells expressing the human histamine H_1-receptor using [³H]-mepyramine and triprolidine as competitive antagonists. Typical saturable binding, which is characteristic of receptor binding was seen (Fig. 7). Nonspecific binding was less than 10% of total binding. Scatchard analysis indicates that the radioligand bound to a single class of sites with an affinity of 1.0 ± 0.1 nM. The value is very similar to that of the cloned rat (7), guinea pig (8, 9), and bovine (6, 41) histamine H_1-receptor. Some cultured cell lines like the 1321N1 human astrocytoma cells and U937 human histiocytic lymphoma cells also show very similar binding kinetics (43, 44). The cloning of the histamine H_1-receptor from human and other species clearly show only one form of the receptor and a single binding site for [³H]-mepyramine. Earlier studies where only [³H]-mepyramine binding data were used to identify the presence of the histamine H_1-receptor in tissue need to be evaluated with caution. Many of those studies reported K_D values of [³H]-mepyramine binding varying over 1000-fold and more than one binding site with different kinetics was reported (45–47). Many of these studies also assumed the presence of more than one receptor isoform or more than one binding site, assumptions we now know to be incorrect. The discrepancies noted in earlier studies may be due to high nonspecific binding of [³H]-mepyramine to the glass fiber filter, tubes and other surfaces in the binding studies, and the presence of binding sites in tissues not related to the histamine H_1-receptor. Even before cloning data were available, it was realized that the low affinity binding of [³H]-mepyramine was likely an artifact (48, 49).

Subsequent to the cloning of the gene it became clear that [³H]-mepyramine binds to tissue components other than the histamine H_1-receptor. [³H]-mepyramine labels not only the histamine H_1-receptor, but also a [³H]-mepyramine-binding protein (MBP) that is related to the subfamily of debrisoquine 4-hydroxylase (cytochrome P4511D) isozymes. Binding to the MBP could be inhibited by quinine. On reexamination of [³H]-

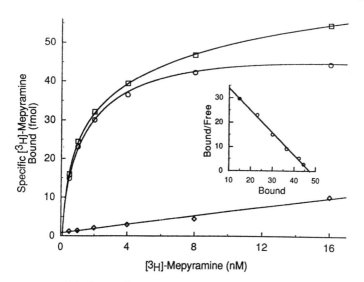

Figure 7 Binding of [³H]-mepyramine to transfected CHO cell membrane. Binding assay was done on 50-μg membrane/reaction using increasing concentration of [³H]-mepyramine in the absence (total binding) or presence (nonspecific binding) of 10-μM triprolidine in a reaction volume of 400 μL. Saturation isotherm from a representative binding assay is shown. Each point is the mean of quadruplicate determinations. Insert shows a Scatchard transformation of the data from which K_D and B_{max} values are calculated. Symbols are as follows—squares for total binding; diamonds for nonspecific binding; and circles for specific binding (adapted from Ref. 13).

mepyramine binding with quinine, it was seen that the [³H]-mepyramine binding sites in brain, lung, spleen, heart, and blood vessels were H_1-receptors, whereas most of the binding sites in the liver, kidneys, and intestine were MBP (50). Further studies using more controlled binding studies and molecular probes are needed to ascertain the true tissue and cellular distribution of the histamine H_1-receptor.

In competition studies with transfected CHO cell membranes, histamine receptor antagonists display a rank order of potency of mepyramine (H_1-receptor selective, K_i = 0.81 nM) > chlorpheniramine (H_1-receptor selective, K_i = 5.9 nM) > thioperamide (H_3-receptor selective, K_i = 100 μM) > cimetidine (H_2-receptor selective, K_i = 1.4 mM). The competition curves are shown in Figure 8. Study of cloned histamine H_1-receptors of

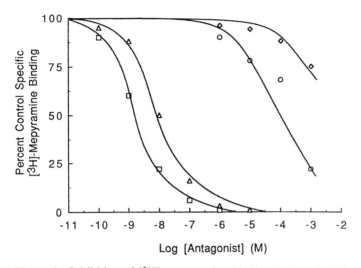

Figure 8 Inhibition of [^{3}H]-mepyramine binding to transfected CHO cell membrane with various antagonists. A series of histamine receptor antagonists was used in competition binding studies to displace [^{3}H]-mepyramine. Each reaction contained 50 μg membrane protein, 1 nM [^{3}H]-mepyramine, and increasing concentrations of mepyramine (squares), chlorpheniramine (triangles), thioperamide (circles), and cimetidine (diamonds) in a reaction volume of 400 μL. Each point is the mean of quadruplicate determinations. K_i values were obtained from the IC$_{50}$ values and the K_D of [^{3}H]-mepyramine.

other species expressed in heterologous cells also shows similar stereoselective inhibition (6, 7, 9).

Competition binding studies with histamine and [^{3}H]-mepyramine reveals the presence of high- (K_i = 5.8 μM) and low-affinity (K_i = 180 μM) binding sites for histamine. Addition of GTP results in a 10-fold shift to the right in the dose-response curve for histamine inhibition of [^{3}H]-mepyramine binding (K_i = 61 μM) and eliminates the high affinity component of binding (Fig. 9). Unlike antagonists, agonists in many ligand binding assays display complex inhibition curves and their apparent affinity is influenced by various ions or guanine nucleotides. The ability of GTP to decrease the affinity of histamine binding is consistent with a functional coupling between the expressed human histamine H_1-receptor in CHO cells and a guanine nucleotide regulatory protein. Similar binding parameters were also seen in human histiocytic lymphoma cells (44), human astrocytoma cells (43), and guinea pig cardiac tissue (15).

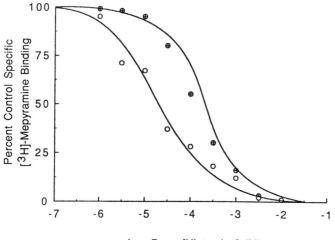

Figure 9   Guanine nucleotide-sensitive agonist binding. Histamine in the absence (open circle) or presence of 100 μM GTP (closed circle) was used in binding assay. Each reaction contained 50 μg membrane protein, 1 nM [^{3}H]-mepyramine, and increasing concentration of histamine. Each point is the mean of quadruplicate determinations.

IX. SIGNAL TRANSDUCTION PATHWAYS OF HUMAN HISTAMINE H$_1$-RECEPTOR IN TRANSFECTED CELLS

The primary intracellular signal transduction mechanism of histamine H$_1$-receptor is mediated through activation of phospholipase C, resulting in hydrolysis of membrane phosphoinositide with the generation of two intracellular second messengers 1,2-diacylglycerol (DAG) and inositol 1,4,5-triphosphate (IP3). DAG activates protein kinase C, and IP3 initiates release of intracellular calcium. The histamine H$_1$-receptor-mediated accumulation of DAG, IP3, and increase in intracellular calcium has been demonstrated in a number of tissues and cell lines (reviewed in Ref. 51). In addition, there is some evidence that histamine, through H$_1$-receptors, also secondarily potentiates cAMP turnover in cells (52). Recently, a signal transduction mechanism with cloned guinea pig histamine H$_1$-receptor cDNA was reported (42). In transfected CHO cells, agonist stimulation caused an increase in intracellular calcium, hydrolysis of membrane phosphoinositide, increases in arachidonic acid, and potentiation of forskolin-stimulated cAMP production.

To determine the signal transduction mechanisms of the human histamine H_1-receptor, we examined the effect of histamine on intracellular effector systems of transfected CHO cells (13). In cells expressing approximately 1000 fmol receptor/mg membrane protein (corresponding to about 100,000 receptors/cell), histamine produced a 2.4-fold increase in total inositol phosphate with an EC_{50} of 49.9 μM. In cells expressing approximately 500 fmol receptor/mg membrane protein (corresponding to about 50,000 receptor/cell), the maximal level of phosphoinositide hydrolysis was about 45% of that in cells expressing 1000 fmol receptor/mg protein (Fig. 10). There was no significant difference in the EC_{50} value. The decrease in the maximal level of inositol phosphate production with decrease in the density of expressed receptor, and virtually similar EC_{50} value for histamine-mediated inositol phosphate production (about 50 μM) and K_D value from binding studies (about 60 μM) suggests that there is no spare receptor reserve in transfected CHO cells.

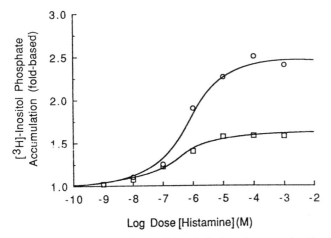

Figure 10 Histamine-induced inositol phosphate production in transfected CHO cells. Transfected CHO cells were labeled overnight with [^{3}H]-myo-inositol and then exposed to increasing concentrations of histamine for 30 min in the presence of 20 mM LiCL and 1 mM IBMX. Total inositol phosphate was extracted and quantitated for the radiolabel. Data represent the mean of a representative experiment performed in quadruplicate. The circle symbol represents data obtained from cells expressing 1000 fmol H_1-receptor/mg protein, and the square symbol represents data obtained from cells expressing 500 fmol H_1-receptor/mg protein.

Pertussis toxin had no effect on histamine-mediated inositol production, suggesting that the human histamine H_1-receptor couples to a pertussis toxin-insensitive G protein in CHO cells. Mepyramine inhibited the histamine-mediated phosphoinositide hydrolysis in a dose-dependent fashion. Chelation of extracellular calcium by EDTA completely abolished the agonist-mediated inositol phosphate production. This suggests that breakdown of phosphoinositides occurs largely as a consequence of influx of extracellular calcium. Previous studies in brain (53, 54), cultured cell lines (55), and transfected cell systems (42) have demonstrated that H_1-receptor stimulation activates phospholipase C through a pertussis toxin-insensitive G protein that is likely a member of the Gq-like family of G proteins.

Histamine had no effect on cAMP production in CHO cells transfected with the human H_1-receptor. However, when the cells were treated with a submaximal dose of forskolin—a direct stimulant of cAMP—histamine caused a dose-dependent increase of intracellular cAMP turnover with an EC_{50} of 18.4 μM. With a larger dose of forskolin, which caused a much higher level of stimulation, histamine had no further additive effect (Fig. 11). Like the histamine-mediated increase in inositol phosphate production, pertussis toxin had no effect on cAMP production in a forskolin-stimulated cell. The H_1-antagonist mepyramine blocked this cAMP pro-

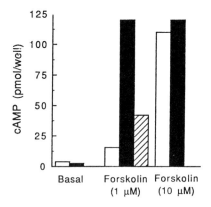

Figure 11 Histamine-induced potentiation of cAMP production in transfected CHO cells. Cells were stimulated with various concentrations of forskolin (basal = no forskolin) in the absence (open bars) or presence (solid bars) of 10^{-4} M histamine. The effect of protein kinase C inhibitor staurosporin on cAMP production by 10^{-4} M histamine in the presence of 1 μM forskolin is shown by the hatched bar.

duction in a dose-dependent fashion. The extracellular calcium chelator EDTA reduced the cAMP production by about 70%. Addition of the protein kinase C inhibitor staurosporine also reduced the ability of histamine to potentiate forskolin-stimulated cAMP turnover by about 70%. On time course study, the response to histamine occurred faster for PI hydrolysis ($t_{1/2}$ = 5 min) than for cAMP production ($t_{1/2}$ = 12 min) (13).

The mechanism of histamine H_1-receptor-mediated potentiation of cAMP accumulation is not clear, although published data suggest that cGMP is elevated (56, 57). Previous studies in animal tissues have shown that histamine acting through the H_1-receptor can augment cAMP production to stimulation of the histamine H_2-, vasoactive intestinal polypeptide-, or adenosine A_2-receptor (reviewed in Ref. 52). Employing a cloned guinea pig histamine H_1-receptor expressed in CHO cells, similar indirect augmentation of cAMP production was seen (42). A role for protein kinase C in this pathway is likely, as staurosporine, a protein kinase C inhibitor, reduced the histamine-mediated increase of forskolin-stimulated cAMP. Both the inositol phosphate and cAMP response are unaffected by pertussis toxin, are sensitive to chelation of extracellular calcium, and have similar dose-response curves. Histamine-mediated inositol phosphate release occurs earlier than cAMP potentiation. These observations suggest that the cAMP production is not directly coupled to the histamine H_1-receptor, but rather is secondary to activation of protein kinase C through the production of DAG. These data, however, do not rule out other pathways for cAMP production.

The functional properties of the human H_1-receptor expressed in CHO cells are very similar to the guinea pig H_1-receptor expressed in the same cell (42). The third intracellular loops of the receptor, which are predicted to interact with G proteins (36), display only about 50% identity between the human and guinea pig H_1-receptor (Fig. 2). However, the membrane proximal regions of the loop display about 85% identity. These data suggest that, for the histamine H_1-receptor, the membrane proximal portion of the third intracellular loop most likely interacts with G proteins as has been demonstrated for the β-adrenergic and muscarinic acetylcholine receptors (58, 59).

The histamine H_1-receptor has been linked to multiple signaling events, including PI hydrolysis, cAMP accumulation, arachidonic acid release, and calcium flux. Based on studies of transfected cell lines and tissues, we can propose a signal transduction mechanism for this receptor. On stimulation with histamine, extracellular calcium is initially mobilized and phospholipase C is activated through a calcium-dependent pathway. Membrane phosphoinositide is then hydrolyzed with generation of IP3 and DAG. IP3 results in intracellular calcium mobilization and further calcium flux. DAG activates protein kinases, including protein kinase C, which subsequently activates cAMP.

X. SUMMARY

Histamine is a major mediator of allergic reactions in humans, and antagonists of the H_1-receptor are useful in the treatment of allergic diseases. The human H_1-receptor gene was cloned from a genomic library using a PCR-amplified probe derived from the bovine H_1-histamine receptor nucleotide sequence. The human H_1-receptor gene contains no introns in the coding sequence and encodes a protein of 487 amino acids with a calculated molecular mass of 55,871 daltons. The gene maps to chromosome 3p24. The H_1-receptor has the characteristics of G protein-coupled receptors with seven transmembrane domains connected by alternating extra- and intracellular loops. The receptor was stably expressed in CHO cells for functional characterization. The transfected cells showed saturable binding toward [^{3}H]mepyramine and stereoselective antagonism in competition studies. In the transfected cells, histamine stimulation resulted in a dose-dependent increase in inositol phosphate production and potentiation of forskolin-stimulated cAMP accumulation. These responses were unaffected by pertussis toxin, suggesting the involvement of a pertussis toxin-insensitive G protein.

REFERENCES

1. Hill SJ. Distribution, properties, and functional characteristics of three classes of histamine receptor. Pharmacol Rev 1990;42:45–83.
2. Simons FER, Simons KJ. The pharmacology and use of H_1-receptor-antagonist drugs. New Engl J Med 1994;330:1663–70.
3. Chowdhury BA, Kaliner MA, Fraser CM. Cloning of a gene encoding the human H_1-histamine receptor. Abstract. J Allergy Clin Immunol 1994;93: 215.
4. Fukui H, Fujimoto K, Mizuguchi H, Sakamoto K, Horio Y, Takai S, Yamada K, Ito S. Molecular cloning of the human histamine H_1-receptor gene. Biochem Biophys Res Commun 1994;201:894–901.
5. De Backer MD, Gommeren W, Moereels H, Nobels G, Van Gompel P, Leysen JE, Luyten WHML. Genomic cloning, heterologous expression and pharmacological characterization of a human histamine H_1-receptor. Biochem Biophys Res Commun 1993;197:1601–8.
6. Yamashita M, Fukui H, Sugama K, Horio Y, Ito S, Mizuguchi H, Wada H. Expression cloning of a cDNA encoding the bovine histamine H_1-receptor. Proc Natl Acad Sci USA 1991;88:11515–9.
7. Fujimoto K, Horio Y, Sugama K, Ito S, Liu YQ, Fukui H. Genomic cloning

of the rat histamine H$_1$-receptor. Biochem Biophys Res Commun 1993;190: 294–301.

8. Horio Y, Mori Y, Higuchi I, Fujimoto K, Ito S, Fukui H. Molecular cloning of the guinea-pig histamine H$_1$-receptor gene. J Biochem (Tokyo) 1993;114: 408–14.

9. Traiffort E, Leurs R, Arrang JM, Tardivel-Lacombe J, Diaz J, Schwartz JC, Ruat M. Guinea pig histamine H$_1$-receptor. I. Gene cloning, characterization, and tissue expression revealed by in situ hybridization. J Neurochem 1994; 62:507–18.

10. Haaksma EEJ, Leurs R, Timmerman H. Histamine receptors: subclass and specific ligands. Pharmac Ther 1990;47:73–104.

11. Gilman AG. G proteins: transducers of receptor-generated signals. Ann Rev Biochem 1987;56:615–49.

12. DeVivo M, Iyengar R. G protein pathways: signal processing by effectors. Mol Cell Endocrinol 1994;100:65–70.

13. Chowdhury BA, Kaliner MA. Molecular cloning and characterization of the human H$_1$-histamine receptor gene. Allergy Clin Immunol News 1995;7: 24–28.

14. Rosenthal N. Stalking the gene–DNA libraries. New Engl J Med 1994;331: 599–600.

15. Ruat M, Bouthenet ML, Schwartz JC, Ganellin CR. Histamine H$_1$-receptor in heart: unique electrophoretic mobility and autoradiographic localization. J Neurochem 1990;55:379–385.

16. Minowa MT, Minowa T, Monsma FJ Jr, Sibley DR, Mouradian MM. Characterization of the 5′ flanking region of the human D$_{1A}$ dopamine receptor gene. Proc Natl Acad Sci USA 1992;89:3045–9.

17. Kobilka BK, Frielle T, Dohlman HG, Bolanowski MA, Dixon RA, Keller P, Caron MG, Lefkowitz RJ. Delineation of the intronless nature of the genes for the human and hamster β_2-adrenergic receptor and their putative promoter regions. J Biol Chem 1987;262:7321–7.

18. Handy DE, Gavras H. Promoter region of the human α_2A adrenergic receptor gene. J Biol Chem 1992;267:24017–22.

19. Hershey AD, Dykema PE, Krause JE. Organization, structure, and expression of the gene encoding the rat substance P receptor. J Biol Chem 1991; 266:4366–74.

20. Lee NH, Earle-Hughes J, Fraser CM. Agonist-mediated destabilization of m1 muscarinic acetylcholine receptor mRNA. Elements involved in mRNA stability are located in the 3′-untranslated region. J Biol Chem 1994;269: 4291–8.

21. Narayanan CS, Fujimoto J, Geras-Raaka E, Gershengorn MC. Regulation by thyrotropin-releasing hormone (TRH) of TRH receptor mRNA degradation in rat pituitary GH$_3$ cells. J Biol Chem 1992;267:17296–303.

22. Henderson R, Baldwin JM, Ceska TA, Zemlin F, Bechmann E, Downing KH. Model for the structure of bacteriorhodopsin based on high-resolution electron cryo-microscopy. J Mol Biol 1990;213:899–929.

23. Gantz I, Schaffer M, DelValle J, Logsdon C, Campbell V, Uhler M, Yamada T. Molecular cloning of a gene encoding the histamine H_2-receptor. Proc Natl Acad Sci USA 1991;88:429–33.

24. Gantz I, Munzert G, Tashiro T, Schaffer M, Wang L, DelValle J, Yamada T. Molecular cloning of the human histamine H_2-receptor. Biochem Biophys Res Commun 1991;178:1386–92.

25. Benovic JL, Staniszewski C, Cerione RA, Codina J, Lefkowitz RJ, Caron MG. The mammalian β-adrenergic receptor: structural and functional characterization of the carbohydrate moiety. J Recept Res 1987;7:257–81.

26. Stiles GL, Benovic JL, Caron MG, Lefkowitz RJ. Mammalian beta-adrenergic receptors. Distinct glycoprotein populations containing high mannose or complex type carbohydrate chains. J Biol Chem 1984;259:8655–63.

27. Doss RC, Kramarcy NR, Harden TK, Perkins JP. Effects of tunicamycin on the expression of β-adrenergic receptors in human astrocytoma cells during growth and recovery from agonist-induced down-regulation. Mol Pharmacol 1985;27:507–16.

28. Rands E, Candelore MR, Cheung AH, Hill WS, Strader CD, Dixon RAF. Mutational analysis of β-adrenergic receptor glycosylation. J Biol Chem 1990;265:10759–64.

29. Dixon RA, Sigal IS, Candelore MR, Register RB, Scattergood W, Rands E, Strader CD. Structural features required for ligand binding to the β-adrenergic receptor. EMBO J 1987;6:3269–75.

30. Savarese TM, Wang CD, Fraser CM. Site-directed mutagenesis of the rat m1 muscarinic acetylcholine receptor. Role of conserved cysteines in receptor function. Biol Chem 1992;267:11439–48.

31. Karnik SS, Khorana HG. Assembly of functional rhodopsin requires a disulfide bond between cysteine residues 110 and 187. J Biol Chem 1990;265:17520–4.

32. Donaldson J, Hill SJ. 1,4-Ditriothreitol-induced alteration in histamine H_1-agonist binding in guinea-pig cerebellum and cerebral cortex. Eur J Pharmacol 1986;129:25–31.

33. Gantz I, DelValle J, Wang L-D, Tashiro T, Munzert G, Guo Y-J, Konda Y, Yamada T. Molecular basis for the interaction of histamine with the histamine H_2-receptor. J Biol Chem 1992;267:20840–3.

34. Ganellin CR. Pharmacochemistry of H_1- and H_2-receptors. In: Schwartz JC and Hass HL, eds. The Histamine Receptor. New York: Wiley-Liss, 1992:1–56.

35. Timmerman H. Cloning of the H_1-histamine receptor. TiPS 1992;13:6–7.

36. Savarese TM, Fraser CM. In vitro mutagenesis and the search for structure-function relationship among G protein-coupled receptors. Biochem J 1992;283:1–19.

37. Leurs R, Smit MJ, Tensen CP, Ter Laak AM, Timmerman H. Site-directed mutagenesis of the histamine H_1-receptor reveals a selective interaction of asparagine 207 with subclasses of H_1-receptor agonists. Biochem Biophys Res Commun 1994;30:295–301.

38. Lee NH, Hu J, El-Fakahany EE. Modulation by certain conserved aspartate residues of the allosteric interaction of gallamine at the M1 muscarinic receptor. J Pharmacol Exp Ther 1992;262:312–6.

39. Neve KM, Cox BA, Henningsen RA, Spanoyannis A, Neve RL. Pivotal role for aspartate-80 in the regulation of dopamine D_2-receptor affinity for drugs and inhibition of adenylyl cyclase. Mol Pharmacol 1991;39:733–9.

40. Wang CD, Buck MA, Fraser CM. Site-directed mutagenesis of α-$_{2A}$ receptors: identification of amino acids involved in ligand binding and receptor activation by agonists. Mol Pharmacol 1991;40:168–79.

41. Iredale PA, Fukui H, Hill SJ. High, stable expression of the bovine histamine H_1 receptor coupled to [Ca^{2+}] mobilisation in CHO-K1 cells. Biochem Biophys Res Commun 1993;195:1294–300.

42. Leurs R, Traiffort E, Arrang JM, Tardivel-Lacombe J, Ruat M, Schwartz JC. Guinea pig histamine H_1 receptor. II. Stable expression in Chinese hamster ovary cells reveals the interaction with three major signal transduction pathways. J Neurochem 1994;62:519–27.

43. Nakahata N, Martin MW, Hughes A, Helper JR, Harden TK. H_1-histamine receptors on human astrocytoma cells. Mol Pharmacol 1986;29:188–95.

44. Driver AG, Kukoly CA, Bennett TE. Expression of histamine H_1-receptors on cultured histiocytic lymphoma cells. Biochem Pharmacol 1989;38:3083–91.

45. Wescott S, Kaliner M. Histamine H_1 binding site on human polymorphonuclear leukocytes. Inflammation 1983;7:291–300.

46. Casale TB, Wescott S, Rodbard D, Kaliner M. Characterization of histamine H_1-receptors on human mononuclear cells. Int J Immunopharmacol 1985;7:639–45.

47. Casale TB, Rodbard D, Kaliner M. Characterization of histamine H_1-receptors on human peripheral lung. Biochem Pharmacol 1985;34:3285–92.

48. Leurs R, Bast A, Timmerman H. High affinity, saturable [^{3}H]-mepyramine binding sites on rat liver plasma membrane do not represent histamine H_1-receptors: A warning. Biochem Pharmacol 1989;38:2175–80.

49. Bielkiewicz B, Cook D. How we isolated and characterized a novel H_1-receptor subtype in glass. Specificity in non-specific binding. TiPS 1985;6:93–4.

50. Liu Q, Horio Y, Fujimoto K, Fukui H. Does the [^{3}H]-mepyramine binding site represent the histamine H_1 receptor? Re-examination of the histamine H_1-receptor with quinine. J Pharmacol Ther 1994;268:959–64.

51. Hill SJ, Donaldson J. The H_1-receptor and inositol phospholipid hydrolysis. In: Schwartz JC and Hass HL, eds. The Histamine Receptor. New York: Wiley-Liss, 1992:109–128.

52. Johnson CL. Histamine receptors and cyclic nucleotides. In: Schwartz JC and Hass HL, eds. The Histamine Receptor. New York: Wiley-Liss, 1992:129–143.

53. Carswell H, Young JM. Regional variation in the characteristics of histamine H_1-agonist-mediated breakdown of inositol phospholipids in guinea-pig brain. Br J Pharmacol 1986;89:809–17.

54. Claro E, Garcia A, Picatoste F. Carbachol and histamine stimulation of guanine-nucleotide-dependent phosphoinositide hydrolysis in rat brain cortical membranes. Biochem J 1989;261:29–35.

55. Tohda M, Nomura Y. Effect of histamine on polyphosphoinositide metabolism in NG-108 cells. Neurochem Int 1989;14:73–8.

56. Kaliner M. Human lung tissue and anaphylaxis: the effects of histamine on the immunologic release of mediators. Am Rev Respir Dis 1978;118:1015–22.

57. Platshon LF, Kaliner M. The effects of the immunologic release of histamine upon human lung cyclic nucleotide levels and prostaglandin generation. J Clin Invest 1978;62:1113–21.

58. Hausdorff WP, Hnatowich M, O'Dowd BF, Caron MG, Lefkowitz RJ. A mutation of the β_2-adrenergic receptor impairs agonist activation of adenylyl cyclase without affecting high affinity agonist binding. Distinct molecular determinants of the receptor are involved in physical coupling to and functional activation. J Biol Chem 1990;265:1388–93.

59. Wess J, Bonner TI, Dorje F, Brann MR. Delineation of muscarinic receptor domains conferring selectivity of coupling to guanine nucleotide-binding proteins and second messengers. Mol Pharmacol 1990;38:517–23.

60. Frielle T, Collins S, Daniel KW, Caron MG, Lefkowitz RJ, Kobilka BK. Cloning of the cDNA for the human β_1-adrenergic receptor. Proc Natl Acad Sci USA 1987;84:7920–4.

61. Chung KZ, Lentes KU, Gocayne J, Fitzgerald M, Robinson D, Kerlavage AR, Fraser CM, Venter JC. Cloning and sequence analysis of the human brain β-adrenergic receptor. Evolutionary relationship to rodent and avian β-receptor and porcine muscarinic receptors. FEBS Lett 1987;211:200–6.

62. Fraser CM, Arakawa S, McCombie WR, Venter JC. Cloning, sequence analysis, and permanent expression of a human α_2-adrenergic receptor in Chinese hamster ovary cells. Evidence for independent pathways of receptor coupling to adenylate cyclase attenuation and activation. J Biol Chem 1989;264:11754–61.

63. Zhou QY, Grandy DK, Thambi L, Kushner JA, Van Tol HH, Cone R. Cloning and expression of human and rat D_1-dopamine receptors. Nature (London) 1990;347:76–80.

64. Kobilka BK, Frielle T, Collins S, Yang-Feng TL, Kobilka TS, Francke U, Lefkowitz RJ, Caron MG. An intronless gene encoding a potential member of the family of receptors coupled to guanine nucleotide regulatory proteins. Nature (London) 1987;329:75–9.

65. Peralta EG, Ashkenazi A, Winslow JW, Smith DH. Distinct primary structures, ligand-binding properties and tissue-specific expression of four human muscarinic acetylcholine receptors. EMBO J 1987;6:3923–9.

66. Peralta EG, Winslow JW, Peterson GL, Smith DH, Ashkenazi A, Ramachandran J, Schimerlik MI, Capon DJ. Primary structure and biochemical properties of an M_2-muscarinic receptor. Science 1987;236:600–5.

67. Strader CD, Sigal IS, Register RB, Candelore MR, Rands E, Dixon RAF. Identification of residues required for ligand binding to the β-adrenergic receptor. Proc Natl Acad Sci USA 1987;84:4384–8.

3

Histamine in Allergic Diseases

Martha V. White and Michael A. Kaliner
Institute for Asthma and Allergy, Washington Hospital Center, Washington, D.C.

I. INTRODUCTION

Numerous mediators, many of which are mast cell derived, have been implicated in the pathogenesis of allergic diseases. While it is clear that the clinical expression of IgE-mediated diseases depends upon the actions of multiple mediators, histamine, the earliest recognized mediator of allergy, remains a prominent contributor. Histamine was first discovered as a potent vasoactive substance by Dale and Laidlaw in 1911 (1), and later associated with tissue mast cells by Riley and West in 1953 (2). It has been studied extensively, particularly during the past two decades since its actions through H_1-, H_2-, and H_3-receptors have become clear. This review will focus upon current concepts of the biosynthesis and biodegradation of histamine, the physiology of its release from mast cells and basophils, and its effects on target tissues. The pathophysiology of the allergic response will be discussed with particular emphasis given to the role of histamine in asthma, allergic rhinitis, urticaria, and anaphylaxis.

II. HISTAMINE SYNTHESIS AND METABOLISM

Histamine, 2-(4-imidazolyl)ethylamine or 5β-amino-ethylimidazole, is formed by decarboxylation of the amino acid histidine by the pyridoxal phosphate-dependent enzyme 1-histidine decarboxylase. Most histamine is stored preformed in cytoplasmic granules of mast cells and basophils in close association with the anionic side chains of proteoglycans comprising the granule matrix. The predominant proteoglycan found varies with cell type: heparin in human connective tissue mast cells (3), chondroitin sulfates di-B and E in rodent mucosal mast cells, and chondroitin 4-sulfate in human blood basophils (4–9). The histamine content of connective tissue and mucosal mast cells is approximately 5–10 and 1–2 pg/cell, respectively. Although mast cells are the main tissue depot for histamine, the synthesis of this amine occurs in other cells as well. In rat gastric mucosa, histamine synthesis and storage occurs in two types of enterochromaffin-like cells, termed histaminocytes (10). In the dog and human mucosa, mast cells are the sole repository. In the brain, histamine is synthesized in neurons of the lateral hypothalamus projecting widely to the telencephalon, diencephalon, and lower brainstem (11). Histamine is also found within the platelets of some species.

In humans, the mast cell is found in the loose connective tissue of all organs, especially around blood vessels, nerves, and lymphatics. Mast cells are most abundant in the shock organs of allergic disease—the skin the mucosa of the upper and lower respiratory tract, the gastrointestinal tract, and the reproductive tract (12). The human heart contains large amounts of histamine, localized primarily within mast cells of the right atrium (13) and also in the walls of blood vessels. The dermis of human skin contains $3–12 \times 10^3$ mast cells/mm^3. Mast cells are found in even higher densities in the gastrointestinal mucosa, up to 20×10^3/mm^3. In the lung, they occur in concentrations of $1–7 \times 10^6$ cells/g of lung tissue (14), comprise up to 2% of alveolar cells (15), and are found in the connective tissue beneath the airway basement membrane, near the submucosal blood vessels and glands, throughout the muscle bundles, and in the interalveolar septa as well as the bronchial lumena. In the airway mucosa, mast cells represent 0.1–0.5% of all cells.

Release of mast-cell granule contents may be induced by a variety of stimuli. Cross-linkage of mast cell IgE receptors by antigen, antireceptor antibody, or lectin bound to receptor-bound IgE initiates a rapid sequence of events terminating in exocytosis. Although many of the details are yet unclear, IgE receptor aggregation is accompanied by rapid hydrolysis of membrane inositol phospholipids, resulting in the release of the two second messengers, inositol-1,4,5-trisphosphate (IP$_3$) and diacylglycerol, which activates a calcium and phospholipid-dependent protein kinase [protein kinase C (PKC)]. IP$_3$ acts through specific receptors to release calcium ions from intracellular stores (16). The identification of the biochemical components involved in signal transduction in mast cells, well characterized in other systems such as the beta-adrenergic receptor, and the intracellular targets of PKC phosphorylation and Ca^{2+} remain active areas of investigation. Events further downstream from receptor engagement include the opening of calcium channels, cleavage of arachidonic acid from phosphatidylcholine, and release of secretory granules containing preformed granule constituents such as histamine. Mast cells may also produce a number of cytokines following activation, including IL-4 and IL-5 (17). The possible significance of these potent mediators in inflammation and repair is also under active investigation at this time.

Numerous other substances also cause mast cells or basophils to release histamine (Table 1). Opioids, such as morphine, cause cutaneous mast cell degranulation through a naloxone-sensitive receptor, and it is possible that endogenous opioids, or endorphins, act similarly (18). Some of the neuropeptides are also mast cell secretagogues. The best studied, substance P (SP), induces human cutaneous mast cell degranulation at nanomolar concentrations, and at higher concentrations causes human lung mast cell degranulation. Injected intradermally, SP induces a wheal and flare response inhibitable by mast cell stabilization, histamine H_1-antagonism, or by histamine depletion with compound 48/80 (19). Depletion of neuropeptides from C fibers with capsaicin causes loss of the histamine-mediated flare, but not whealing response (20–21). These data, taken together, suggest that SP causes the release of histamine from human mast cells, that histamine is capable of causing the release of SP from sensory nerve fibers, and that both histamine and SP contribute to histamine-induced vasodilation. Histamine release in physical urticaria and angioedema caused by cold, heat, sunlight, vibration, or pressure is well described; however, the mechanism by which these physical stimuli cause mast-cell degranulation is not clear (22–26). Finally, most inflammatory cells, including lymphocytes, neutrophils, platelets, macrophages, and eosinophils, as well as endothelial cells and nasal lavage fluid produce or contain histamine-releasing factors (HRF) (27–32). This is a heterogeneous group of factors, all of which cause basophil degranulation. Several

Table 1   Mast Cell Secretagogues

IgE
C3a, C4a, C5a
Substance P
Opioids, endorphins
Physical stimuli: vibration, heat, sunlight, cold, pressure
Histamine-releasing activities from lymphocytes, neutrophils, platelets, endothelial cells, human lung macrophages, eosinophils, human nasal washings
Cytokines: IL-1, IL-3, IL-8, GM-CSF, connective tissue activating peptide III

of the cytokines, IL-1, IL-3, IL-8, and GM-CSF also cause basophil degranulation, although IL-1 is active only at high concentrations (33–40).

One of the inflammatory cell-derived HRF has been isolated and sequenced, and this factor is homologous to connective tissue–activating peptide III (CTAP III) (41). A second, IgE-dependent HRF has also been isolated and cloned (42). It appears that cytokine-induced basophil degranulation is finely controlled since IL-8, which is found in high concentration in the human airway, is a specific inhibitor of cytokine-induced basophil histamine release (44, 44a).

Certain strains of bacteria readily produce histamine by decarboxylation of histidine. Indeed, the compound was at first thought to be produced solely by bacterial action during putrefaction and was only shown to be produced by mammalian tissues in 1919, a decade after its discovery (45). Histamine is probably the cause of scombrotoxism (scombroid-fish poisoning), a clinical syndrome resulting from the ingestion of spoiled fish, usually of the families Scombridae and Scomberesocidae (46). The synthesis of histamine by bacteria suggests the possibility that inflammatory reactions surrounding areas of infection might be due in part to the formation of histamine by the invading bacteria. In one study of germ-free rats selectively infected with a histamine-producing strain of *Clostridium perfringens*, high histamine levels were found in the cecal contents, but the histamine content of the intestinal mucosa did not change (47). As noted below, the intestinal mucosa is one site where high concentrations of diamine oxidase (histaminase) have been found; presumably the enzyme functions there to prevent histamine that has been ingested or generated by the intestinal flora from penetrating the tissues.

Histamine is also produced by organisms commonly found in human airways. In a study of pathogenic and nonpathogenic respiratory bacterial species, three of the five pathogenic species studied, *Branhamella catarrhalis*, *Haemophilus parainfluenzae*, and *Pseudomonas aeruginosa* produced clinically important amounts of histamine (48). Of the nonpathogens, members of the Enterobacteriacae also produced large quantities of the amine. Thus, a neglected feature of infection or perhaps even colonization of the respiratory and gastrointestinal tracts with certain strains of bacteria may be the production of inflammatory amines by those organisms.

Morphological changes accompanying mast-cell degranulation vary depending upon the type of mast cell involved. Human skin mast cells release intact granules into the extracellular milieu. Some of the granules can subsequently be phagocytosed by connective tissue fibroblasts (49). In contrast, degranulation of human lung and nasal mucosal mast cells involves intracellular solubilization of granule contents and fusion of granular membranes with each other and with the cell membrane to form chan-

Table 2 Normal Histamine Values

Radioenzymatic and fluorometric assays:	
Plasma	200–300 pg/ml
Urine	13 ± 8 ng/ml
	14 ± 9 μg/24 h
	14 ± 12 ng/mg creatinine/ml urine
Radioimmunoassay:	
Plasma	<10 nM
	<1 ng/ml
Urine	<700 nM
	<70 ng/ml

nels to the outside through which the solubilized granule contents are extruded (50–51).

Once released, histamine diffuses rapidly into the surrounding tissues and appears in blood within 2.5 min, peaks at 5 min, and returns to baseline by 15–30 min (52). Urinary histamine elevations are more prolonged than plasma elevations. Thus, abnormalities are more easily detected with the urinary histamine assay. Of the three methods commonly employed to measure histamine, the radioenzymatic and fluorometric assays are sensitive to 100 pg/ml (normal 200–300 pg/ml) and 1–5 ng/ml, respectively. A histamine radioimmunoassay has been developed in which histamine, acetylated to a carrier molecule competes with iodinated, chemically modified histamine for recognition by a monoclonal antibody directed against the modified histamine. This assay is sensitive to 15 pg/ml and is easier to use than the two former methods (normal plasma histamine is less than 10 nM or 1 ng/ml) (53–54; Table 2).

Elevations in plasma or tissue histamine have been found following experimental provocation with physical stimuli in a variety of physical urticarias, in antigen- and exercise-induced anaphylaxis, natural and provoked bronchospasm, after intradermal skin testing, and in patients with mastocytosis. Histamine is also elevated in nasal and bronchoalveolar lavage fluids after antigen challenge. Urinary histamine is increased in mastocytosis, a subgroup of patients with idiopathic hypereosinophilic syndrome and allergic disease, some patients with Zollinger-Ellison syndrome, and occasionally in pregnancy (51).

Only 2–3% of histamine is excreted unchanged in the urine. The rest is metabolized by two major enzymatic pathways. Fifty to 70% of histamine is metabolized to N-methylhistamine by N-methyltransferase and some is further metabolized by monoamine oxidase to N-methylimidazoleacetic acid and excreted in the urine. The remaining 30–50% of histamine is metabolized to imidazole acetic acid by diamine oxidase (54) (Fig. 1).

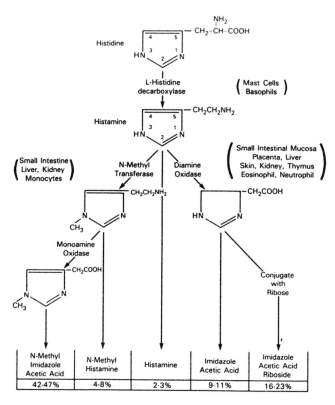

Figure 1 Synthesis and catabolism of histamine. Percent recovery of histamine and its metabolites in the urine in the 12 h after intradermally administered ^{14}C-histamine in human males. Reprinted from Ref. 51.

III. HISTAMINE RECEPTORS

The different subclasses of histamine receptors have previously been reviewed (55, 56) and are covered in Chapters 1 and 2 of this publication. The existence of more than one type of histamine receptor was suggested by Ash and Schild (57), who noted that the classic antihistamine mepyramine could block histamine-induced contractions of guinea pig ileum but not histamine-induced gastric acid secretion. The presence of a histamine H_2-receptor was later confirmed by Black (58), who introduced burimamide, the first effective H_2-antagonist. In 1983, the discovery by Arrang and co-workers (59) that histamine inhibited its own release from histaminergic nerve endings in rat cerebral cortex uncovered a third class of receptors, the H_3-receptor. The role of these histaminergic neurons is still under active investigation; however, they are thought to be involved in arousal, energy metabolism, cerebral circulation, cardiovascular reflexes, pituitary hormone release, and body temperature (60). More recently, H_3-receptors have been found in peripheral tissues (61, 62), including human bronchial smooth muscle where they mediate inhibition of cholinergic transmissions, thus leading to bronchial relaxation (63).

Histamine H_1-receptor-mediated activities are stimulated by the H_1-agonists 2-methylhistamine, 2-[2-pyridyl]ethylamine, and 2-[2-thiazolyl]-ethylamine and are inhibited by "classic" H_1-antihistamines such as diphenhydramine, chlorpheniramine, and mepyramine. Histamine H_2-receptor-mediated activities are stimulated by the H_2-agonists 4-methylhistamine and dimaprit and are inhibited by the H_2-receptor antagonists cimetidine and ranitidine as well as several newer drugs (64). Histamine H_3-receptor-mediated activities are stimulated by the H_3 chiral agonists (R)-α-methylhistamine (62) and α(R), β(S)-dimethylhistamine (65) and the nonchiral agonists imetit (66) and immepip (67). They are inhibited by the H_3-antagonist thioperamide (62). In addition, some antagonists (e.g., burimamide), formerly considered as H_2-antagonists have affinities for the H_3-receptor that are several orders of magnitude higher, and should probably be considered selective antagonists of the latter (Table 3).

Histamine receptors are distributed widely in mammalian tissues. The availability of selective ligands for H_1-receptor subtypes that can be radiolabeled to a high specific activity has made possible studies of radioligand binding to various cell types. H_1-receptors have been identified in brain,

Table 3 Histamine Agonists and Antagonists

Receptor	Agonist	Antagonist
H_1-	2-Methylhistamine 2-[2-Thiazoly]ethylamine 2[Pyridyl]ethylamine (2PEA)	Mepyramine (pyrilamine) Diphenhydramine Chlorpheniramine Promethazine Terfenadine Astemizole Loratadine Cetirizine
H_2-	4-Methylhistamine Dimaprit	Burimamide Cimetidine Ranitidine Famotidine Nizatidine
H_3-	R-α-Methylhistamine (R), (S)-Dimethylhistamine Imetit Immepip	Thioperamide

retina, adrenal medulla, liver, endothelial cells, cerebral microvasculature, lymphocytes, and smooth muscle from airway, intestinal, genitourinary, and vascular tissue (55). Striking interspecies variation in receptor distribution exists. For example, guinea pig cerebellum is richly endowed with H_1-receptors but the same tissue in humans and other species has a much lower H_1-receptor density. The human H_1-receptor has recently been cloned (68). The deduced protein contains 487 amino acids and contains 7 potential transmembrane sequences, similar to other G-protein-coupled receptors. Signal transduction by the H_1-receptor, as with mast-cell IgE-receptor-mediated exocytosis, occurs via the hydrolysis of membrane inositol phospholipids, releasing the second messengers diacylglycerol (DAG) and IP_3.

Activation of H_1-receptors causes augmentation of the cyclic adenosine 5'-monophosphate (cAMP) response generated by H_2-receptors, vasointestinal polypeptide (VIP) receptors, or adenosine A_2-receptors (55). The proposed mechanism for this augmentation includes the stimulation of adenyl cyclase by H_1-induced intracellular Ca^{2+} and maintenance of the augmentation by DAG. H_1-receptor activation also induces intracytoplasmic increases in cyclic guanosine 5'-monophosphate (cGMP) in a number of tissues including human lung, as well as guinea pig bronchial and tracheal smooth muscle (55, 69–70). In human lung, anaphylactic release of histamine induces a tenfold increase in the tissue content of cGMP, an effect mimicked by exogenous histamine and blocked by H_1-antagonists (69). These findings are concordant with two recent findings: (1) that endothelium-derived relaxing factor (EDRF), or nitric oxide, produced by the action of histamine on endothelial cell H_1-receptors, activates cytosolic guanylate cyclase in smooth muscle cells; and (2) that cGMP is probably the second messenger involved in EDRF-induced relaxation (71). Part of the function of cGMP may be to block the hydrolysis of membrane inositol lipids by inhibiting the activation of phospholipase C (72).

In contrast to H_1-receptor-mediated signaling, transmembrane signaling initiated by H_2-receptor engagement occurs via the cAMP pathway. H_2-receptor stimulation induces a rise in cAMP in a variety of tissues, including brain, gastric mucosa, adipocytes, vascular smooth muscle, basophils, and neutrophils (55). Receptor-mediated activation of adenyl cyclase, well described in the beta-adrenergic system (73, 74), is mediated by GTP-binding proteins which transduce the signal created by receptor-ligand binding to the membrane-associated enzyme. H_2-receptor stimulation has also been associated with a rise in the intracellular calcium ion concentration. The calcium ions are derived from the same intracellular source which produces the calcium rise induced by IP_3, but the mechanism of release is different. Some evidence suggests that calcium release may not be a direct consequence of the rise in cAMP (75).

The distribution and density of H_2-receptors in various tissues has proved difficult to study even when receptors are easily demonstrated by functional assays. For example, the first H_2-radioligand developed, [^{3}H]-cimetidine, was found after initial positive studies to bind nonspecifically to sites other than the H_2-receptor because it did not exhibit displacement by ranitidine or tiotidine (55). The only reliable H_2-radioligands available at present are [^{3}H]-tiotidine and its derivative [^{3}H]-ICIA 5165; for many peripheral tissues, no H_2-radioligand of sufficient specificity has yet been developed.

As mentioned previously, a new histamine receptor subtype, the H_3-receptor, was described in 1983 from work with rat brain tissue (59). Evidence is accumulating that H_3-receptors are also present in peripheral tissues. Histamine inhibits sympathetic neurotransmission in guinea pig mesenteric artery, probably by binding H_3-receptors on perivascular sympathetic nerve terminals (60). Using radioligand binding with potent and selective H_3-receptor agonists and antagonists, Arrang and co-workers (62) have found low levels of receptors in various tissues, including lung and spleen. In addition, histamine levels in tissues following in vivo exposure to the potent H_3-agonist (R)-alpha-methylhistamine were depressed some 30%, suggesting that in peripheral tissues, as in the central nervous system, histamine down-regulates its own synthesis by acting on H_3-receptors. Other than perivascular nerve terminals, the cellular locations of peripheral H_3-receptors are not presently known. Agonist/antagonist studies suggest the presence of H_3-receptors on human adenoidal mast cells; however, the data have not been confirmed (76).

Finally, evidence exists that histamine may act as an intracellular messenger in some cell types (77). A series of experiments showed that the production of intracellular histamine is necessary for human platelet aggregation induced by various external stimuli. In addition, intracellular histamine may function in the regulation of cell growth. Histamine decarboxylase activity has been noted to correlate with cell proliferation (78). Increased synthesis of histamine has been noted in various cell culture systems including lymphocytes, tumor cells, and embryological tissues (79–83).

IV. OVERVIEW OF HISTAMINE EFFECTS

As discussed earlier, the actions of histamine are mediated through its receptors (Table 4). Most of the histamine effects important in allergic diseases are mediated through the H_1-receptor. These include smooth

Table 4 Activities Mediated Through Histamine Receptors
in Humans

Receptor	Action
H_1-	Smooth muscle contraction
	Increased vascular permeability
	Increase in cGMP
	Pruritus
	Prostaglandin generation
	Decreased A-V node conduction time
	Activation of airway vagal afferent nerves
H_2-	Gastric acid secretion
	Increase in airway mucus secretion
	Increase in cAMP
	Esophageal contraction
	Inhibition of basophil histamine release
	Inhibition of neutrophil chemotaxis and enzyme release
	Stimulation of suppressor T cells
H_3-	Inhibition of sympathetic neurotransmission
	Down-regulation of histamine synthesis
H_1- + H_2-	Hypotension
	Flushing
	Headache

muscle contraction, increased vascular permeability, pruritus, prostaglandin generation, decreased A-V node conduction time with resultant tachycardia, activation of vagal reflexes, and increased cGMP. H_2-mediated effects include gastric acid secretion, increased airway mucus secretion (although in the nose this is due to a histamine H_1-mediated muscarinic reflex as described later in this chapter), increased cAMP, esophageal contraction, inhibition of basophil, but not mast cell, histamine release, inhibition of neutrophil activation and induction of suppressor T cells. Several effects are mediated through combined H_1- and H_2-receptor stimulation. These include vasodilation-related symptoms such as hypotension, flushing, and headache, as well as tachycardia stimulated indirectly through vasodilation and catecholamine secretion (84). Histamine-induced vascular permeability is an H_1-mediated effect caused by contraction of actomyosin fibers in endothelial cells of the postcapillary venules (84). Increased permeability is a feature of all allergic diseases and is therefore important in allergic rhinitis, asthma, urticaria, and anaphylaxis. The capacity for histamine to induce smooth muscle contraction is well known and, in fact, forms the basis for the histamine bioassay.

Smooth muscle contraction occurs in the airways in asthma. Histamine is the only proven mediator of pruritus, which is a prominent symptom in urticaria, anaphylaxis, eczema, and allergic rhinitis. The mechanism by which histamine mediates pruritus is indirect and involves stimulation of sensory nerve endings. Histamine-induced cutaneous vasodilation and flushing is at least partially mediated by neurohormones since capsaicin treatment inhibits the histamine-induced flare, but not the wheal. Vasodilation is a prominent component of urticaria and anaphylaxis. Mucus secretions, which are derived from a combination of vascular and glandular sources, are prominent in asthma and rhinitis. In the nose, the glandular component is indirectly mediated by histamine through a vagal reflex, since unilateral nasal histamine challenge results in contralateral glandular secretion inhibitable with atropine (85, 86). Increased mucus production is a feature of asthma, allergic rhinitis, and anaphylaxis. Hypotension and shock, seen in anaphylaxis, is secondary to increased vascular permeability and vasodilation. Tachycardia, also prominent in anaphylaxis, can be due to decreased A-V node conduction time, or occur indirectly through histamine-induced vasodilation and resultant catecholamine secretion. Finally, histamine-induced prostaglandin formation is a feature of asthma and allergic rhinitis. At least two cell types respond to histamine with prostaglandin formation (84).

The effects of histamine are complex. A discussion of the mechanism by which histamine exerts each of its major effects follows.

A. Histamine and Vasodilation

Stimulation of H_1-receptors on vascular smooth muscle produces contraction. Histamine, acting again through H_1-receptors, induces vasodilation at least in part by stimulation of specific receptors on endothelial cells with release of prostacyclin and EDRF (now recognized as nitric oxide) (56, 87–91). EDRF is also released by stimulation of endothelial cells by a number of autacoids such as bradykinin, neuropeptides such as SP, as well as acetylcholine and may be a final common pathway in the vasodilatory effects of these bioactive molecules in the evolution of an inflammatory reaction (71). Direct injection of histamine into the skin produces a flare around the injection site, which is at least partly mediated by antidromic release of SP from sensory nerve endings (vide infra) (92).

Although histamine almost invariably induces contraction of isolated mammalian smooth muscle, the injection of this compound into intact animals produces a bewildering spectrum of reactions that depend not only upon the species, but also upon the route of injection and the vascular bed involved (93). The net response is determined by variable contributions of other vasoactive mediators generated both directly and indirectly by the action of histamine on various cell types as well as a variety of neural reflexes. In skin, muscle, and splanchnic beds, the response usually consists of a drop in arterial pressure and an increase in flow, effects opposed by H_1-antagonists. High concentrations of intravascular histamine, as occurs in systemic anaphylaxis, are associated with profound decreases in blood pressure, cutaneous flushing, and gastrointestinal hypermobility. Lower concentrations may actually allow compensatory mechanisms to induce hypertension. The responses in various species are commonly bi- or triphasic.

The relative contributions of H_1- and H_2-receptors in histamine-induced hemodynamic changes have been studied extensively in vivo. The time course of receptor stimulation differs markedly; effects produced by stimulation of H_1-receptors wax and wane rapidly while those produced by H_2-receptors are slower in onset and more sustained. When the vasodilation of resistance vessels in vascular beds is measured during intravenous histamine administration in cats, H_1-receptor blockade produces displacement of the histamine dose-response curve to the right. While H_2-antagonists alone have little effect, the combination of H_1- and H_2-antagonists produces a far greater effect than either alone (94). The effect of the antagonists is time-dependent with early blood pressure changes affected by H_1 blockade and sustained changes opposed by H_2-blockade. This synergy between H_1- and H_2-antagonists has been demonstrated in other studies. In humans, pretreatment with both types of antagonists was required to significantly blunt the fall in diastolic blood pressure and pulse pressure associated with escalating infusion of low-dose histamine (95). Finally, there is some evidence that H_2-mediated vasodilation may be more important in the response of the vasculature to extraluminal histamine (96).

Because of its potent effects on the cardiovascular system, the idea was put forward many years ago that histamine might play a key role in the maintenance of normal vasomotor tone (97). This idea seems even more plausible in view of the demonstrated histamine content of heart and blood vessels and the perivascular location of tissue mast cells. However, despite the protection afforded by H_1- and H_2-antagonists against the cardiovascular effects of injected histamine, at concentrations effective for receptor blockade, no effects of these drugs have been documented on the resting cardiovascular system (98, 99) suggesting that, under normal conditions, histamine plays little or no role in the control of vasomotor tone. However, in vivo experiments demonstrate that histamine may contribute to active vasodilatation in the baroreceptor reflex (100). In canine skeletal muscle preloaded with ^{14}C-histamine, reflex dilatation results in a release of ^{14}C into the draining venous circulation (101). In addition, H_1- and H_2-antagonists inhibit reflex dilation in parallel with their inhibition of the response to exogenous histamine (102).

B. Histamine and Vascular Permeability

Edema, one cause of tissue swelling seen in areas of inflammation, is the result of the action of histamine on H_1-receptors located on endothelial cells, especially in postcapillary venules (103). Studies on intact tissues and in cultured endothelial cells demonstrate that receptor engagement results in shape change (contraction) leading to the formation of gap junctions up to 1 μm in diameter between adjacent cells and the subsequent extravasation of macromolecules. The stimulation of H_1-receptors in cultured human umbilical cells is associated with phosphatidylinositol hydrolysis and a rise in intracellular Ca^{2+} which derives initially from intracellular stores but requires extracellular Ca^{2+} for a sustained response. The requirement of endothelial cell contraction for histamine-induced permeability is supported by the ablation of the response by cytochalasin B, which inhibits actin polymerization. Agents that increase intracellular cAMP such as beta-adrenergic agonists (e.g., terbutaline), methylxanthines (e.g., theophylline), and prostaglandin E_1 all effectively reverse histamine-induced vascular permeability both in organs and in endothelial cell culture.

Clinically, edema accompanies most inflammatory reactions and is provoked as part of the response to histamine and allergen in allergic skin testing. It is seen as the familiar wheal, the third portion of the triple response described by Lewis. In accord with experimental work both in vitro and in vivo, H_1-antagonists are well known for their potent suppressive effect on these tests. While H_2-antagonists alone have not been consistently shown to affect the response, concomitant administration of H_1- and H_2-antagonists may enhance the suppression produced by H_1-antagonists alone.

C. Histamine and Neural Interactions

Interactions of histamine with the nervous system fall into the burgeoning province of psychoneuroimmunology. As discussed above, histamine is produced within the central nervous system where it acts on H_3-receptors to inhibit its own synthesis. Of greater relevance in the pathogenesis of allergic diseases, however, is its as yet poorly understood interrelationship with elements of the peripheral nervous system. It is clear that histamine influences the release and actions of neurotransmitters in a variety of organ systems, and products of the nervous system likewise influence the secretion of histamine from its major cellular location, the tissue mast cell.

The best studied in vivo models of such interactions have involved the skin and the lung. In pulmonary tissue, the calcium and cAMP second messenger systems are thought to play balanced opposing roles (104). Both histamine and acetylcholine (ACh) induce contraction of airway smooth muscle through receptors that induce PI hydrolysis and an increase in free cytoplasmic Ca^{2+}, H_1-, and M_1/M_3-receptors, respectively. Imbalances in autonomic control of airway tone may be an important factor in the development of asthma (105). Muscarinic cholinergic receptors in the lung are abundant in smooth muscle of the large airways, corresponding to the major site of cholinergic innervation (106). By contrast, the location of beta-adrenergic receptors in the lung is mainly peripheral, and direct innervation is difficult to demonstrate in most species (the dog being an exception), suggesting that these receptors are responsive mainly to circulating catecholamines under normal circumstances. In humans, bronchial smooth muscle fibers possess a relatively pure population of β_2-receptors (107).

The interaction of histamine with the parasympathetic nervous system has been well studied in canine models of asthma. Vagotomized dogs were administered histamine or antigen with or without atropine, and parasympathetic discharge, dynamic compliance, and total lung resistance were measured. Histamine caused discharge of irritant receptors, triggering both local and central vagal reflexes with resultant bronchoconstriction (108–110) The vagal reflexes were more important in low-dose compared to high-dose histamine-induced bronchoconstriction and could be inhibited by H_1- but not H_2-antagonists.

Evidence for the importance of histamine–vagus interactions in human airway constriction stems from inflammatory models of airway reactivity. When compared to healthy controls, normal subjects develop increased histamine responsiveness after upper respiratory tract infections with pathogens capable of denuding airway epithelium. This increased responsiveness can be reversed or prevented by atropine inhalation (111). In addition, evidence exists that respiratory epithelium, which is well-endowed with H_1-receptors, in a fashion analogous to vascular endothelium, releases a potent relaxing substance, and the absence of this factor with inflammatory denudation of the airway epithelium may render the underlying smooth muscle relatively more affected by bronchoconstrictors such as ACh and histamine (112).

Association of conventional neurotransmitters such as ACh with neuropeptides such as VIP within the same nerve terminals has been demonstrated. VIP and peptide histidine isoleucine have been proposed to constitute the transmitters in nonadrenergic, noncholinergic (NANC) relaxation of airway smooth muscle (106, 113). In accord with the model of opposing second messenger pathways in the control of airway tone, VIP activates adenyl cyclase and induces relaxation in airway smooth muscle. In a similar fashion to beta-adrenergic agonists, it would be expected to oppose the effects of histamine and ACh in the lung (114). In support of this concept, VIP has been shown to inhibit antigen-induced histamine release from guinea pig lung (115) and also to inhibit the bronchoconstrictive effects of a prostaglandin (PGF_2), histamine (116), and a leukotriene (LTD_4) in the guinea pig (117). In addition, VIP inhalation by some but not all human asthmatic subjects was reported to cause bronchodilatation (118–120).

A reciprocal influence of histamine and peripheral nerves was proposed as early as 1927 when Lewis proposed the axon reflex to account for the flare produced by the injection of histamine in the skin (121). The flare is ablated by interruption of the cutaneous nerve supply to the area either by nerve section or by local anesthesia (121). It is likewise diminished by pretreatment with capsaicin, which depletes sensory nerves of SP (20–21). Persuasive arguments, partly based on histologic observations, have been offered for a functional nerve-mast cell unit at least in the skin, which functions in inflammation and tissue repair (92, 122, 123). This claim has been disputed in other studies in which mast-cell degranulation was observed only after prolonged nerve stimulation (124). At the very least, strong evidence exists that mast cell and neural secretory products influence each other's secretion and actions at the tissue level (114, 125).

In the skin, the best demonstration of mast cell–neural interaction involves SP, but there are data to support the coexistence in the same sensory endings of SP, neurokinin A, and calcitonin gene-related peptide (CGRP) (92). The injection of histamine in the skin produces retrograde discharge of "C"-type fibers that subserve polymodal nociceptors and contain SP as well as other peptides. Peptides including SP, VIP, and somatostatin induce a wheal and flare reaction when injected into human skin, with SP being approximately 100 times as potent as histamine. In addition, SP, VIP, and somatostatin at micromolar concentrations all cause dose-related histamine release from isolated human skin mast cells but not from mucosal mast cells or mast cells from other tissues (e.g., intestinal serosa). The characteristics of histamine release (kinetics, relative calcium independence, lack of prostaglandin release) resemble that induced by other polycations such as compound 48/80 and poly-lysine (121, 126). The evidence suggests that, in the skin, amphiphilic polycationic peptides such as SP constitute a specialized system for the detection of injury in this most exposed of all organ systems.

D. Histamine and Prostaglandin (PG) Generation

Antigen challenge during bronchoalveolar lavage in humans (127) and sheep (128) leads to production of PGD_2, thromboxane A_2, and leukotrienes. The bronchodilatory prostanoids PGI_2 and PGE_2 are released as well. Other studies have shown lung tissue to produce 5-, 12-, and 15-

hydroxyeicosatetraenoic acids (HETEs) (129–133). All these observations are based on anaphylactic challenge of lung tissue, and the evoked prostanoids (e.g., PGD_2) have their origin at least in part from parenchymal mast cells. However, the secondary production of arachidonic acid metabolites from other cell types is likely.

Histamine-induced generation of PGs in lung has been demonstrated. Platshon and Kaliner (69) showed that antigen-induced anaphylaxis of human lung resulted in release of histamine as well as PGF_2, PGE_2, and thromboxane B_2. Release of PGF_2 could also be induced by treatment with exogenous histamine or the H_1-agonist 2-methylhistamine, but not by the H_2-agonist dimaprit. Further, the generation of PGF_2 could be blocked by pretreatment with H_1-, but not H_2-, antagonists. In a subsequent study it was found that histamine H_1-stimulation of airways led to the formation of PGE_2, whereas peripheral guinea pig lung produced PGF_2 as well as PGE_2 (134).

As noted above, the action of histamine on H_1-receptors located on endothelial cells results in the production of prostanoids including PGI_2 and PGE_2 (55, 135). Through their action on the underlying vascular smooth muscle, these arachidonic acid derivatives, in conjunction with EDRF, may be final effectors in histamine-induced vasodilatation. In support of this hypothesis, irradiation injury of human dermal explants induced by ultraviolet light is associated with liberation of both histamine and prostaglandins including PGE_2, and production of the latter is blocked by H_1-antihistamines (136).

E. Histamine and Allergic Diseases

Histamine is one of many mediators contributing to allergic diseases. Effects known to be mediated by histamine can also potentially be caused by other mast-cell-derived mediators. A systematic review of the pathophysiology of individual allergic diseases, including potential contributing mediators, will facilitate an assessment of the importance of histamine in the pathophysiology of allergic diseases.

V. ASTHMA

The cardinal features of asthma include smooth muscle spasm, mucosal edema, inflammation, and mucus secretion that can be due both to glandular secretion of mucus glycoproteins and to increased movement of interstitial fluid into the airway lumen. When one examines the mediators potentially responsible for causing the four main pathological features of asthma, one sees that two of these features, bronchospasm and mucosal edema, can be caused by H_1-receptor stimulation, while H_2- and possibly H_1-activation are probably minor causes of mucus secretion (Table 5)

Table 5 Pathological Changes in Asthma and the Mediators Possibly Responsible

Feature	Proposed mediator
Bronchospasm	Histamine (H_1-)
	Leukotrienes (LTC_4, LTD_4, LTE_4)
	Prostaglandins and thromboxane A_2
	Bradykinin
	Platelet-activating factor (PAF)
Mucosal edema	Histamine (H_1-)
	LTC_4, LTD_4, LTE_4
	PGE
	Bradykinin
	Platelet-activating factor (PAF)
Airway inflammation	Eosinophil chemotactic factor
	Neutrophil chemotactic factor
	LTB_4
	Platelet-activating factor (PAF)
Mucus secretion	Histamine (H_2—mucus glycoprotein)
	(H_1—interstitial fluid)
	LTC_4, LTD_4, LTE_4
	Prostaglandin generating factor
	HETEs

(137). Other mediators also play potentially important roles. In addition to histamine, leukotrienes, prostaglandins, bradykinin, and platelet-activating factor (PAF) can stimulate both bronchospasm and mucosal edema. Airway inflammation is not stimulated by histamine, but can be caused by inflammatory factors, such as neutrophil and eosinophil chemotactic factors, leukotriene B_4, and PAF. Mucus glycoprotein secretion can be induced by H_2-receptor activation, while increased movement of interstitial fluid into the airway lumen can be mediated by H_1-receptors. In addition, leukotrienes and prostaglandin-generating factor can mediate increased mucus secretion. Thus, histamine can cause three of the four main pathological features of asthma.

Early clinical trials employing H_1-antagonists for the treatment of asthma were disappointing. However, experiments employing the newer, nonsedating H_1-antagonists in high doses have demonstrated complete protection against histamine and dose-related partial protection (10–20%) against exercise- and antigen-induced asthma (138–140). Thus, while histamine is probably not a major mediator in asthma, it does contribute to its pathogenesis, and high-dose, nonsedating H_1-antagonists deserve further study as potential agents in the treatment of asthma. Rarely, some of the nonsedating antihistamines are associated with cardiac arrhythmias when used in high dosages. Therefore, full exploration into the use of antihistamines in asthma awaits the development of newer, safer, nonsedating antihistamines.

VI. ALLERGIC RHINITIS

Seasonal allergic rhinitis symptoms include paroxysms of sneezing, nasal and palatal pruritus, nasal congestion, and clear rhinorrhea. In severe allergic rhinitis, contiguous or adjacent mucous membranes of the eye, middle ear, or paranasal sinuses may be involved. This can lead to mucosal swelling followed by occlusion of the eustachian tubes and sinus ostia with resultant serous or purulent otitis media and/or sinusitis, respectively.

Insufflation of specific allergen into the nose of an individual with allergic rhinitis induces congestion, pruritus, sneezing, and increased secretion. Congestion is secondary to vasodilation and increased vascular permeability, while pruritus and sneezing are both caused by sensory nerve stimulation. Increased secretions are derived about 80% from the vasculature and about 20% from the glands (86).

Histamine can cause all of the pathological features of allergic rhinitis with the exception of late-phase inflammatory reactions (Table 6) (141).

Table 6 Pathological Features of Allergic Rhinitis and the Mediators Possibly Responsible

Symptoms elicited	Pathological event	Proposed mediator
Pruritus	Sensory nerve stimulation	Histamine (H_1-) Prostaglandins
Nasal congestion	Mucosal edema	Histamine (H_1-) Kinins LTC_4, LTD_4, LTE_4
Sneezing	Sensory nerve stimulation	Histamine (H_1-) LTC_4, LTD_4, LTE_4
Rhinorrhea	Increased mucus secretion	Histamine (H_1-): Direct and indirect (muscarinic discharge) LTC_4, LTD_4, LTE_4
Hyperirritability Congestion	Late-phase reaction Vasodilation	Inflammatory factors Eicosanoids Chemotactic factors (neutrophil and eosinophil)

Pruritus, which is responsible for the palatal clicking so characteristic of allergic rhinitis, is caused by stimulation of H_1-receptors on sensory nerve endings. Prostaglandins may also contribute. Mucosal edema, which is manifest as nasal congestion, can be caused by H_1 stimulation as well as eicosanoids and kinins. Sneezing, like pruritus, is a histamine H_1-mediated neural reflex and can also be mediated by eicosanoids. Nasal mucus secretion can be mediated by histamine both directly and indirectly through muscarinic discharge and by eicosanoids. Late-phase reactions, which are manifest as nasal congestion and hyperirritability, are not mediated by histamine, but rather by inflammatory factors, eicosanoids, and chemotactic factors. Thus histamine can cause four of the five pathological components of allergic rhinitis. The excellent response to H_1-antagonists experienced by most patients with allergic rhinitis suggests that histamine is a primary mediator of this disease.

VII. URTICARIA

The hallmark of urticaria is a pruritic, erythematous, raised lesion that blanches with pressure and is indicative of venous dilation and edema. The three cardinal features of urticaria are pruritus, vasodilation, and increased vascular permeability involving the superficial dermis. Histamine, acting through its H_1-receptor, can mediate all three of the pathological components of urticaria. It is the only proven mediator of pruritus and can also mediate increased vascular permeability and vasodilation. Other vasoactive mediators that might contribute to vasodilation and edema include PGD_2, LTC_4, LTD_4, PAF, and bradykinin (Table 7). PGD_2

Table 7   Pathological Changes in Urticaria and Possible Mediators

Symptoms elicited	Pathological event	Proposed mediator
Wheal	Vascular permeability	Histamine (H_1-) PGD_2 PAF Bradykinin LTC_4, LTD_4, LTE_4
Flare	Vasodilation	Histamine (H_1-) Prostaglandins PAF Bradykinin LTC_4, LTD_4, LTE_4
Pruritus	Sensory nerve stimulation	Histamine (H_1-)

is a vasodilator, while the leukotrienes C_4 and D_4 mediate increased vascular permeability (142). These eicosanoids may act synergistically in the skin (143). Release of PAF, which can mediate increased vascular permeability, has been measured during experimentally induced cold urticaria (144). Bradykinin, which is generated in the tissues after mast-cell degranulation, is also a potent vasoactive substance that induces a burning pain. The contribution of bradykinin may be more important in angioedema, which is characterized by deep dermal vasodilation and edema and burning pain. H_1-antihistamines are the mainstay of therapy for urticaria, and many patients' symptoms can be controlled with these agents alone. Thus, histamine can cause all of the pathological findings of urticaria, and the response to H_1-antagonists suggests that histamine is a primary mediator of urticaria.

VIII. ANAPHYLAXIS

Anaphylaxis is a mast-cell-mediated, life-threatening, multiorgan disorder. The major shock organs include the skin, where flushing, pruritus, urticaria, and angioedema can be noted, the cardiovascular system, where tachycardia, hypotension, shock, syncope, and arrhythmias are noted, the gastrointesinal tract, with abdominal bloating, diarrhea, and vomiting, and the respiratory tract, where symptoms range from rhinorrhea, laryngeal edema, wheezing, and bronchorrhea to asphyxiation. Other symptoms include diaphoresis and fecal or urinary incontinence. Patients complain of pruritus, especially around the face, neck and back, weakness, palpitations, bloating, crampy diarrhea, congestion, and dyspnea. They frequently have a peculiar metallic taste in their mouths as well as a feeling of impending doom.

When the pathological components of anaphylaxis are analyzed, the pattern mirrors the inherent pathology of other mast cell-mediated processes, and, in fact, all of the symptoms of anaphylaxis can be reproduced by histamine. Vascular permeability is manifest as urticaria, angioedema, and laryngeal and intestinal edema, vasodilation leads to flushing and headache, and smooth muscle contraction results in wheezing, abdominal cramping, and diarrhea. Tachycardia results in palpitations, reduced peripheral vascular resistance is responsible for syncope, and mucus secretion is manifest as rhinorrhea and bronchorrhea. Histamine, either alone or in conjunction with eicosanoids, can mediate all of the features of anaphylaxis (Table 8) (145). Local anaphylaxis is effectively treated with H_1 antagonists, but systemic anaphylaxis with respiratory and cardiovascular involvement should be treated with epinephrine to decrease mediator release and improve airway patency, followed by H_1-antagonists and, if

Table 8 Pathological Changes in Anaphylaxis and Proposed Mediators

Symptoms elicited	Pathological event	Proposed mediator
Urticaria, angioedema, laryngeal, and intestinal edema	Vascular permeability	Histamine (H_1-) eicosanoids
Flushing, headache hypotension	Vasodilation	Histamine (H_1- and H_2-) eicosanoids
Palpitations	Palpitations	Histamine (H_1-)
Rhinorrhea bronchorrhea	Mucus secretion	Histamine (H_1- and H_2-) eicosanoids

indicated, H_2-antagonists, volume expansion, pressors, and corticosteroids. Thus, histamine is a major mediator in anaphylaxis.

IX. SUMMARY

Mast cell degranulation results in the release of preformed mediators such as histamine and chemotactic factors, and in the generation of mediators such as eicosanoids and platelet-activating factor from membrane lipids and kinins by the action of mast cell enzymes on tissue precursors. These mediators work in concert to effect the pathological processes inherent in allergic disorders. Histamine can elicit all or most of the pathological processes involved in allergic rhinitis, urticaria, and anaphylaxis, and three of the four cardinal features of asthma. Further, H_1 antagonists are highly effective in the treatment of allergic rhinitis, and alone or in combination with H_2-antagonists, in urticaria and anaphylaxis. H_1-antihistamines also offer partial protection in asthma. Thus, although histamine is only one of many chemical mediators of inflammation in allergic disease, this molecule should not be overlooked when considering the cause and treatment of any mast cell-related disease.

REFERENCES

1. Dale HH, Laidlaw PP. The physiologic action of β-imidazolylethylamine. J Physiol 1911;41:318–344.
2. Riley JF, West DB. Histamine and tissue mast cells. J Physiol 1953;120:528–537.
3. Metcalfe DD, Lewis RA, Silbert JE, et al. Isolation and characterization of heparin from human lung. J Clin Invest 1979;4:1537–1543.

4. Enerback L, Kolset SO, Kusche M, et al. Glycosaminoglycans in rat mucosal mast cells. Biochem J 1985;227:661–668.
5. Ishizaka T, Conrad DH, Hugg TE, et al. Unique features of human basophilic granulocytes developed in vitro culture. Int Arch Allergy Appl Immunol 1985;77:137–143.
6. Metcalfe DD, Bland CE, Wasserman SI. Biochemical and functional characteristics of proteoglycans isolated from basophils of patients with chronic myelogenous leukemia. J Immunol 1984;130:1943–1950.
7. Razin E, Stevens RL, Akiyama F, et al. Culture from mouse bone marrow of a subclass of mast cells possessing a distinct chondroitin sulfate proteoglycan with glycosaminoglycans rich in N-acetyl galactosamine-4, 6-disulfate. J Biol Chem 1982;257:7729–7736.
8. Razin E, Stevens RL, Austen KF, et al. Cloned mouse mast cells derived from immunized lymph node cells and from foetal liver cells exhibit characteristics of bone marrow-derived mast cells containing chondroitin sulfate E proteoglycan. Immunology 1984;52:563–575.
9. Sredni B, Friedman MM, Bland CE, et al. Ultrastructural, biochemical and functional characteristics of histamine-containing cells cloned from mouse bone marrow: Tentative identification as mucosal mast cells. J Immunol 1983;131:915–922.
10. Soll AH, Lewin KJ, Beaven MA. Isolation of histamine-containing cells from rat gastric mucosa: Biochemical and morphologic differences from mast cells. Gastroenterology 1981;80:717–727.
11. Arrang JM, Garbarg M, Schwartz JC. Histamine synthesis and release in CNS: Control by autoreceptors (H_3). In: Ganellin CR, Schwartz JC, eds. Advances in the Biosciences: Frontiers in Histamine Research, A Tribute to Heinz Schild. Vol 51. New York: Pergamon Press, 1985:143.
12. Metcalfe DD. Effector cell heterogeneity in immediate hypersensitivity reactions. Clin Rev Allergy 1983;1:311–325.
13. Wolff AA, Gross SS, Levi R. Histamine receptors: Involvement in cardiac function and dysfunction. In: Settipane GA, ed. H_1 and H_2 Histamine Receptors. Providence, RI: Ocean Side Publications, 1988–1989:61–64.
14. Wasserman SI. The lung mast cell: Its physiology and potential relevance to defense of the lung. Environ Health Perspect 1980;35:153–164.
15. Fox B, Bull TB, Guz A. Mast cells in the human alveolar wall: An electron microscopic study. J Clin Pathol 1981;34:1333–1342.
16. Nahorski SR, Potter BV. Molecular recognition of inositol polyphosphates by intracellular receptors and metabolic enzymes. TiPS 1989;10:139–144.
17. Costa JJ. Mast cells and cytokines: New insights into the pathogenesis of allergic diseases. Insights Allergy 1990;5(6).
18. Casale TB, Bowman S, Kaliner M. Induction of human cutaneous mast-cell degranulation by opiates and endogenous opioid peptides: Evidence for opiate and nonopiate receptor participation. J Allergy Clin Immunol 1984; 73:775–781.

19. White MV, Kowalski ML, Kaliner MA. Mast cell secretagogues. In: Wintraub B, Tauber F, Simon AS, eds. Biochemistry of the Acute Allergic Reaction, Fifth International Symposium. New York: Alan R. Liss, Inc. 1989:83–101.
20. Lembeck F, Holzer P. Substance P as neurogenic mediator of antidromic vasodilation and neurogenic plasma extravasation. Naun Schmied Arch Pharmacol 1979;310:175–183.
21. Kiernan JA. A study of chemically-induced acute inflammation in the skin of the rat. Q J Exp Physiol 1977;62:151–161.
22. Soter NA, Wasserman S, Pathak MA, Parrish JA, Austen KF. Solar urticaria: Release of mast-cell mediators into the circulation after experimental challenge. J Invest Dermatol 1979;72:282 (Abst).
23. Kaplan AP. Urticaria and angioedema. In Middleton E, Reed CE, Ellis EF, eds. Allergy Principles and Practice. St. Louis: C.V. Mosby Co., 1983: 1341–1360.
24. Keahey TM, Indrisano J, Lavker RM, Kaliner MA. Delayed vibratory angioedema: Insights into pathophysiologic mechanisms. J Allergy Clin Immunol 1987;80:831–838.
25. Huston DP, Bressler RB, Kaliner M, Sowell LK, Baylor MW. Prevention of mast-cell degranulation by ketotifen in patients with physical urticarias. Ann Int Med 1986;104:507–510.
26. Casale TB, Keahey TM, Kaliner M. Exercise-induced anaphylactic syndromes. JAMA 1986;255:2049–2053.
26. Liu MC, Proud D, Lichtenstein LM, et al. Human lung macrophage derived histamine-releasing activity is due to IgE-dependent factors. J Immunol 1986;136:2588–2595.
27. Orchard MA, Kagey-Sobotka A, Proud D, et al. Basophil histamine release induced by a substance from stimulated human platelets. J Immunol 1986; 136:2240–2244.
28. Thueson DO, Speck LS, Lett-Brown MA, et al. Histamine-releasing activity (HRA). I. Production by mitogen- or antigen-stimulated human mononuclear cells. J Immunol 1979;626–632.
29. White MV, Baer H, Kaliner MA. Neutrophil-derived histamine-releasing activity. Ann Allergy 1985;55:273.
30. White MV, Kaliner MA. Neutrophil-induced mast cell degranulation. J Allergy Clin Immunol 1985;75:175.
31. White MV, Kaliner MA, Baer H. Stimulated neutrophils release a histamine-releasing factor. J Allergy Clin Immunol 1986;77:132.
32. Zheutlin LM, Ackerman SJ, Gleich GJ, et al. Stimulation of basophil and rat mast cell histamine release by eosinophil granule-derived cationic proteins. J Immunol 1984;133:2180–2185.
33. White MV, Yoshimura T, Hook W, et al. Neutrophil attractant/activation protein-1 (NAP-1) causes human basophil histamine release. Immunol Lett 1989;22:151–154.
34. MacDonald SM, Schleimer RP, Kagey-Sobotka A, et al. Recombinant IL-

3 induces histamine release from human basophils. J Immunol 1989;142: 3527–3532.

36. Subramanian N, Bray MA. Interleukin-1 releases histamine from human basophils and mast cells in vitro. J Immunol 1987;138:271–275.

37. Haak-Frendscho M, Arai N, Arai K, et al. Human recombinant granulocyte-macrophage colony-stimulating factor and interleukin-3 cause basophil histamine release. J Clin Invest 1988;82:17–20.

38. Haak-Frendscho M, Dinarello C, Kaplan AP. Recombinant human interleukin-1 beta causes histamine release from human basophils. J Allergy Clin Immunol 1988;82:218–223.

39. Massey WA, Randall TC, Kagey-Sobotka A, et al. Recombinant human IL-1 and IL-1β potentiate IgE-mediated histamine release from human basophils. J Immunol 1989;1875–1880.

40. Alam R, Welter JB, Forsythe PA, et al. Comparative effect of recombinant IL-1, -2, -3, -4, and -6, IFN-, granulocyte-macrophage-colony-stimulating factor, tumor necrosis factor, and histamine-releasing factors on the secretion of histamine from basophils. J Immunol 1989;142:3431–3435.

41. Baeza ML, Reddigari SR, Kornfeld D, et al. Relationship of one form of human histamine-releasing factor to connective tissue activating peptide-III. J Clin Invest 1990;85:1516–1521.

42. MacDonald S, Rafner T, Langden J, Lichtenstein LM. Molecular identification of IgE-dependent histamine-releasing factor. J Allergy Clin Immunol (in press).

43. Alam R, Welter J, Forsythe PA, et al. Detection of histamine release inhibitory factor- and histamine-releasing factor-like activities in bronchoalveolar lavage fluids. Am Rev Respir Dis 1990;141:666–671.

44. Kuna P, Reddigari SR, Kornfeld D, Kaplan AP. IL-8 inhibits histamine release from human basophils induced by histamine-releasing factors, connective tissue activating peptide III and IL-3. J Immunol 1991;147:1920–24.

45. Abel JJ, Kubota S. On the presence of histamine (β-imidazolylethyl-amine) in the hypophysis cerebri and other tissues of the body and its occurrence among the hydrolytic decomposition products of proteins. J Pharmacol Exp Therap 1919;13:243–300.

46. Morrow JD, Margolies GR, Rowland J, et al. Evidence that histamine is the causative toxin of scombroid-fish poisoning. N Engl J Med 1991;324: 716–720.

47. Beaver MH, Wostmann BS. Histamine and 5-hydroxytryptamine in the intestinal tract of germ-free animals, animals harbouring one microbial species and conventional animals. Br J Pharmacol 1962;19:385–393.

48. Devalia JL, Grady D, Harmanyeri Y, et al. Histamine synthesis by respiratory tract micro-organisms: Possible role in pathogenicity. J Clin Pathol 1989;42:516–522.

49. Friedman MM, Kaliner M. Ultrastructural changes in human skin mast cells during antigen-induced degranulation in vivo. J Allergy Clin Immunol 1988; 82:998–1005.

50. Friedman MM, Kaliner MA. In situ degranulation of human nasal mucosal mast cells: Ultrastructural features and cell-cell association. J Allergy Clin Immunol 1985;76:70–82.

51. White MV, Slater JE, Kaliner MA. Histamine and asthma. Am Rev Respir Dis 1987;135:1165–1176.

52. McBride P, Jacobs R, Bradley D, et al. Use of plasma histamine levels to monitor cutaneous mast cell degranulation. J Allergy Clin Immunol 1989; 83:374–380.

53. Histamine Radioimmunoassay Kit Manual, Immunotech International, Catalogue # 1302, Marseille, France.

53. Wasserman SI. Mediators of immediate hypersensitivity. J Allergy Clin Immunol 1983;72:101–115.

54. White MV, Kaliner MA. Regulation by histamine. In: Crystal RG, West JB, et al., eds. The Lung: Scientific Foundations, New York: Raven Press, 1991:927–939.

55. Hill SJ. Distribution, properties, and functional characteristics of three classes of histamine receptor. Pharmacol Rev 1990;42:45–83.

56. Haaksma EEJ, Leurs R, Timmerman H. Histamine receptors: Subclasses and specific ligands. Pharmac Ther 1990;47:73–104.

57. Ash ASF, Schild HO. Receptors mediating some actions of histamine. Br J Pharmacol 1966;27:427–439.

58. Black JW, Duncan WAM, Durant GJ, et al. Definition and antagonism of histamine H_2-receptors. Nature 1972;236:385–390.

59. Arrang JM, Garbarg M, Schwartz JC. Autoinhibition of brain histamine release mediated by a novel class (H_3) of histamine receptor. Nature 1983; 302:832–837.

60. Schwartz JC, Arrang JM, Garbarg M, Pollard H. Histamine H_3-receptors in the brain: Potent and selective ligands. Psychopharmacol Ser 1989;7: 10–9.

61. Ishikawa S, Sperelakis N. A novel class (H_3) of histamine receptors on perivascular nerve terminals. Nature 1987;327:158–160.

62. Arrang JM, Garbarg M, Lancelot JC, et al. Highly potent and selective ligands for histamine H_3-receptors. Nature 1987;327:117–123.

63. Ichinose M, Barnes PJ. Inhibitory histamine H_3-receptors on cholinergic nerves in human airways. Eur J Pharmacol 1989;163:383–386.

64. Feldman M, Burton ME. Histamine$_2$-receptor antagonists: Standard therapy for acid-peptic diseases. N Engl J Med 1990;323:1672–1680.

65. Lipp R, Arrang JM, Garbarg M, et al. Synthesis, absolute configuration, stereoselectivity, and receptor selectivity of $\alpha R\beta S$-$\alpha\beta$-dimethylhistamine, a novel highly potent histamine H_3-receptor agonist. J Med Chem 1992;35: 4434–4441.

66. Garbarg M, Arrang JM, Rouleau A, et al. S-[2-(4-imidazolyl)ethyl]-isothiourea, a highly specific and potent histamine H_3-receptor agonist. J Pharmacol Exp Ther 1992;263:304–310.

67. Vollinga RC, de Koning JP, Jansen FP, et al. A new potent and selective

histamine H₃-receptor agonist, 4-(1H-imidazol-4-ylmethyl)piperidine. J Med Chem 1994;37:332–333.

68. De Backer MD, Gommeren W, Moereels H, et al. Genomic cloning, heterologous expression and pharmacological characterization of a human histamine H_1-receptor. Biochem Biophys Res Commun 1993;197:1601–08.

69. Platshon LF, Kaliner MA. The effects of the immunologic release of histamine upon human lung cyclic nucleotide levels and prostaglandin generation. J Clin Invest 1978;62:1113–1121.

70. Duncan PG, Brink C, Adolphson RL, et al. Cyclic nucleotides and contraction/relaxation in airway muscle: H_1- and H_2-agonists and antagonists. J Pharmacol Exp Ther 1980;215:434–442.

71. Ignarro LJ. Endothelium-derived nitric oxide: Actions and properties. FASEB J 1989;3:31–36.

72. Hirata M, Kohse KP, Chang C, et al. Mechanism of cyclic GMP inhibition of inositol phosphate formation in rat aorta segments and cultured bovine aortic smooth muscle cells. J Biol Chem 1990;265:1268–1273.

73. Gilman AG. G proteins: Transducers of receptor-generated signals. Ann Rev Biochem 1987;56:615–649.

74. Birnbaumer L, Brown AM. G proteins and the mechanism of action of hormones, neurotransmitters, and autocrine and paracrine regulatory factors. Am Rev Respir Dis 1990;141:S106–S114.

75. Negulescu PA, Machen TE. Intracellular Ca regulation during secretogogue stimulation of the parietal cell. Am J Physiol 1988;254:C130–C140.

76. Bent S, Fehling U, Braam U, et al. The influence of H_1-, H_2- and H_3-receptors on the spontaneous and ConA-induced histamine release from human adenoidal mast cells. Agents Actions 1991;33:67–70.

77. Saxena SP, Brandes LJ, Becker AB, et al. Histamine is an intracellular messenger mediating platelet aggregation. Science 1989;243:1596–1599.

78. Kahlson G, Rosengren E, Steinhardt C. Histamine-forming capacity of multiplying cells. J Physiol 1963;169:487–498.

79. Mackay D, Marshall PB, Riley JF. Histidine decarboxylase activity in a malignant rat hepatoma. J Physiol 1960;153:31P.

80. Kahlson G, Rosengren E, Steinhardt C. Activation of histidine decarboxylase in tumour cells in mice. Nature 1962;194:380–381.

81. Kahlson G, Rosengren E, White T. The formation of histamine in the rat foetus. J Physiol 1960;151:131–138.

82. Chanda JR, Ganguly AK. Diamine oxidase activity and tissue histamine content of human skin, breast and rectal carcinoma. Cancer Lett 1987;34:207–212.

83. Oh C, Suzuki S, Nakashima I, et al. Histamine synthesis by non-mast cells through mitogen-dependent induction of histidine decarboxylase. Immunology 1988;65:143–148.

84. White MV, Slater JE, Kaliner MA. Histamine and asthma. Am Rev Respir Dis 1987;135:1165–76.

85. Raphael GD, Meredith SC, Baraniuk JN, Druce HM, Banks SM, Kaliner MA. The pathophysiology of rhinitis. II. Assessment of the sources of pro-

tein in histamine-induced nasal secretions. Am Rev Respir Dis 1989;139: 791–800.

86. Mullol J, Raphael GD, Lundgren JD, Baraniuk JN, Merida M, Shelhamer J, Kaliner M. Comparison of human nasal mucosal secretion in vivo and in vitro. J Allergy Clin Immunol 1992;89:584–92.

87. Resink TJ, Grigorian GY, Moldabaeva AK, et al. Histamine-induced phosphoinositide metabolism in cultured human umbilical vein endothelial cells: Association with thromboxane and prostacyclin release. Biochem Biophys Res Commun 1987;144:438–446.

88. Toda N. Mechanism of histamine actions in human coronary arteries. Circ Res 1987;61:280–286.

89. Palmer RMJ, Ferrige AG, Moncada S. Nitric oxide release accounts for the biological activity of endothelium-derived relaxing factor. Nature 1987; 327:524–526.

90. Myers PR, Minor RL Jr, Guerra R Jr, et al. Vasorelaxant properties of the endothelium-derived relaxing factor more closely resemble S-nitrosocysteine than nitric oxide. Nature 1990;345:161–163.

91. Wei EP, Kontos HA. H_2O_2 and endothelium-dependent cerebral arteriolar dilation: Implications for the identity of endothelium-derived relaxing factor generated by acetylcholine. Hypertension 1990;16:162–169.

92. Church MK, Lowman MA, Rees PH, et al. Mast cells, neuropeptides, and inflammation. Agents Actions 1989;27:8–16.

93. Altura BM, Halevy S. Recent developments of the histamine problem. In: Rocha e Silva M, ed. Handbook of Experimental Pharmacology, Histamine and Antihistaminics II, vol XVIII/2. New York: Springer-Verlag, 1978: 1–39.

94. Levi R, Owen DAA, Trzeciakowski J. Actions of histamine on the heart and vasculature. In: Ganellin CR, Parsons ME, eds. Pharmacology of Histamine Receptors, Bristol: John Wright & Sons (Printing) Ltd, The Stonebridge Press, 1982:236–297.

95. Kaliner M, Sigler R, Summers R, et al. Effects of infused histamine: Analysis of the effects of H_1- and H_2-histamine receptor antagonists on cardiovascular and pulmonary responses. J Allergy Clin Immunol 1981;68:365–371.

96. Wahl M, Kuschinsky W. The dilating effect of histamine on pial arteries of cats and its mediation by H_2-receptors. Circ Res 1979;44:161–165.

97. Dale HH. Antihistamine substances. Br Med J 1948;ii:281–283.

98. Black JW, Owen DAA, Parsons ME. An analysis of the depressor responses to histamine in the cat and dog: Involvement of both H_1- and H_2-receptors. Br J Pharmacol 1975;54:319–324.

99. Tucker A, Weir EK, Reeves JT, et al. Histamine H_1- and H_2-receptors in pulmonary and systemic vasculature of the dog. Am J Physiol 1975;229: 1008–1013.

100. Brody MJ. Histaminergic and cholinergic vasodilator systems. In: Vanhoutte PM, Leusen I, eds. Mechanisms of Vasodilatation, Satellite Symp, 27th Int Congr Physiol Sci., Basel: Karger, 1978.

101. Rengo F, Trimarco B, Chiariello M, et al. Relation between cholinergic and

histaminergic components in reflex vasodilatation in the dog. Am J Physiol 1978;234:H305–H311.

102. Powell JR, Brody MJ. Participation of H_1- and H_2-histamine receptors in physiological vasodilator responses. Am J Physiol 1976;231:1002–1009.

103. Grega GA. Contractile elements in endothelial cells as potential targets for drug action. Trends Pharmacol Sci 1986;7:452–457.

104. Rasmussen H, Kelley G, Douglas JS. Interactions between Ca^{2+} and cAMP messenger systems in regulation of airway smooth muscle contraction. Am J Physiol 1990;258:L279–L288.

105. Lemanske RF, Kaliner MA. Autonomic nervous system abnormalities and asthma. Am Rev Respir Dis 1990;141:S157–S161.

106. Douglas JS. Receptors on target cells: Receptors on airway smooth muscle. Am Rev Respir Dis 1990;141:S123–S126.

107. Goldie RG. Receptors in asthmatic airways. Am Rev Respir Dis 1990;141: S151–S156.

108. DeKock MA, Nadel JA, Zwi S, et al. New method for perfusing bronchial arteries: Histamine bronchoconstriction and apnea. J Appl Physiol 1966; 21:185–194.

109. Gold WM, Kessler GF, Yu DY. Role of vagus nerves in experimental asthma in dogs. J Appl Physiol 1972;33:719–725.

110. Shore SA, Bai TR, Wang CG, et al. Central and local cholinergic components of histamine-induced bronchoconstriction in dogs. J Appl Physiol 1985;58:443–451.

111. Empey DW, Laitinen LA, Jacobs L, et al. Mechanisms of bronchial hyperreactivity in normal subjects after upper respiratory tract infection. Am Rev Respir Dis 1976;113:131–139.

112. Vanhoutte PM. Epithelium-derived relaxing factor(s) and bronchial reactivity. J Allergy Clin Immunol 1989;83:855–61.

113. Dey RD, Shannon WA, Said SI. Localization of VIP-immunoreactive nerves in airways and pulmonary vessels of dogs, cats, and human subjects. Cell Tissue Res 1981;220:231–238.

114. Di Marzo V, Tippins JR, Morris HR. Neuropeptides and inflammatory mediators: Bidirectional regulatory mechanisms. Trends Pharmacol Sci 1989; 10:91–2.

115. Undem BJ, Dick EC, Buckner CK. Inhibition by vasoactive intestinal peptide of antigen-induced histamine release from guinea pig minced lung. Eur J Pharmacol 1983;88:247–249.

116. Said SI, Geumei A, Hara N. Bronchodilator effect of VIP in vivo: Protection against bronchoconstriction induced by histamine or prostaglandin F_2. In: Said DI, ed. Vasoactive Intestinal Peptide. New York: Raven Press, 1982: 185–192.

117. Hamasaki Y, Saga T, Mojarad M, et al. VIP counteracts leukotriene D_4-induced contractions of guinea pig trachea, lung and pulmonary artery. Trans Assoc Am Physicians 1983;96:406–411.

118. Barnes PJ, Dixon CMS. The effect of inhaled vasoactive intestinal peptide

on bronchial reactivity to histamine in humans. Am Rev Respir Dis 1984; 130:162–166.

119. Mojarad TL, Grode C, Cox C, et al. Differential responses of human asthmatics to inhaled vasoactive intestinal peptide (VIP). Am Rev Respir Dis 1985;131:A281

120. Morice AH, Unwin RJ, Sever PS. Vasoactive intestinal peptide as a bronchodilator in asthmatic subjects. Peptides 1984;5:439–440.

121. Church MK, Lowman MA, Robinson C, et al. Interaction of neuropeptides with human mast cells. Int Arch Allergy Appl Immunol 1989;88:70–78.

122. Stead RH, Bienenstock J. Cellular interactions between the immune and peripheral nervous systems: A normal role for mast cells? In: Burger MM, Sordat B, Zinkernagel RM, eds. Cell to Cell Interaction. Basel: Karger, 1990;170–187.

123. Stead RH, Perdue MH, Blennerhassett MG, et al. The innervation of mast cells. In Freier S, ed. The Neuroendocrine-Immune Network. Boca Raton: CRC Press, 1990:19–37.

124. Kowalski ML, Kaliner MA. Neurogenic inflammation, vascular permeability, and mast cells. J Immunol 1988;140:3905–3911.

125. Widdicombe JH. Airway diseases: Role of epithelium and inflammatory peptides. Am J Physiol 1989;257:L144–L146.

126. Pearce FL, Kassessinoff TA, Liu WL. Characteristics of histamine secretion induced by neuropeptides: Implications for the relevance of peptide-mast cell interactions in allergy and inflammation. Int Arch Allergy Appl Immunol 1989;88:129–131.

127. Wenzel SE, Westcott JY, Smith HR, et al. Spectrum of prostanoid release after bronchoalveolar allergen challenge in atopic asthmatics and in control groups. An alteration in the ratio of bronchoconstrictive to bronchoprotective mediators. Am Rev Respir Dis 1989;139:450–457.

128. Dworski R, Sheller JR, Wickersham, et al. Allergen-stimulated release of mediators into sheep bronchoalveolar lavage fluid. Effect of cyclooxygenase inhibition. Am Rev Respir Dis 1989;139:46–51.

129. Adkinson NF Jr, Newball HH, Findlay S, et al. Anaphylactic release of prostaglandins from human lung in vitro. Am Rev Respir Dis 1980;121: 911–920.

130. Hamberg M, Hedquist P, Radegran K. Identification of 15-hydroxy-5,8,11,13-eicosatetraenoic acid (15-HETE) as a major metabolite of arachadonic acid in human lung. Acta Physiol Scand 1980;110:219–221.

131. Robert RJ, Lewis RA., Oates JA, et al. Prostaglandin, thromboxane, and 12-hydroxy-5,8,10,14-eicosatetraenoic acid production by ionophore-stimulated rat serosal mast cells. Biochem Biophys Acta 1979;575:185–192.

132. Schulman ES, Adkinson NF Jr, Newball HH. Cyclooxygenase metabolites in human lung anaphylaxis: Airway vs. parenchyma. J Appl Physiol 1982; 53:589–595.

133. Schulman ES, Newball HH, Demers LM, et al. Anaphylactic release of

thromboxane A_2, prostaglandin D_2 and prostacyclin from human lung parenchyma. Am Rev Respir Dis 1981;124:402–406.

134. Steel L, Platshon L, Kaliner M. Prostaglandin generation by human and guinea pig lung tissue: Comparison of parenchymal and airway responses. J Allergy Clin Immunol 1979;64:287–293.

135. Cole OF, Lewis GP. Prostanoid production by rat aortic endothelial cells by bradykinin and histamine. Eur J Pharmacol 1989;169:307–312.

136. Pentland AP, Mahoney M, Jacobs SC, et al. Enhanced prostaglandin synthesis after ultraviolet injury is mediated by endogenous histamine stimulation. A mechanism for irradiation erythema. J Clin Invest 1990;86:566–574.

137. Kaliner M. Mast cell mediators and asthma. Chest 1985;87:2S–5S.

138. Holgate ST, Emanuel MB, Howarth PH. Astemizole and other H_1-antihistaminic drug treatment of asthma. J Allergy Clin Immunol 1985;76:375–380.

139. Tashkin DP, Brik A, Gong H, Jr. Cetirizine inhibition of histamine-induced bronchospasm. Ann Allergy 1987;59:49–52.

140. Rafferty P, Beasley R, Holgate ST. The contribution of histamine to immediate bronchoconstriction provoked by inhaled allergen and adenosine 5' monophosphate in atopic asthma. Am Rev Respir Dis 1987;136:369–373.

141. Druce HM, Kaliner MA. Allergic rhinitis. JAMA 1988;259:260–263.

142. Orange RP, Austen KF. Immunologic and pharmacologic receptor control of the release of chemical mediators from human lung. In: Ishizaka K, Dayton DH, Jr, eds. The Biological Role of the Immunoglobulin E System. Bethesda, Maryland: National Institutes of Health, U.S. Department of Health Education, and Welfare, 1972.

143. Soter NA, Lewis RA, Corey EJ, Austen KF. Local effects of synthetic leukotrienes (LTC_4, LTD_4, LTE_4 and LTB_4) in human skin. J Invest Dermatol 1983;80:115–120.

144. Kaplan AP. Urticaria and angioedema. In: Middleton E, Reed CE, Ellis EF, Adkinson NF, Yunginger JW, eds. Allergy, Principles and Practice, 3rd ed. St. Louis: CV Mosby Co., 1988:1377–1401.

145. Kaliner MA. Anaphylaxis. In: Lockey RE, Bukantz SC, eds. Fundamentals of Immunology and Allergy. Philadelphia: W.B. Saunders Co., 1987: 203–216.

4

H_1-Receptor Antagonists: Structure and Classification

Jean Bousquet and A.M. Campbell
Montpellier University, Montpellier, France

C.W. Canonica
Genoa University, Genoa, Italy

I. INTRODUCTION

Histamine, a ubiquitous cell-to-cell messenger, was identified in 1910 by Dale and Laidlaw (1) and has been recognized since the 1920s as a major mediator of allergic disorders such as rhinitis, asthma, urticaria, and anaphylaxis. The true mechanism of action of histamine remained unknown until 1966 when the H_1-histamine receptor was identified pharmacologically (2). Knowledge of the histaminergic system evolved with the later pharmacological identification of the H_2-receptor (3) responsible for gastric acid secretion, and the H_3-receptor (4), which apparently is represented mostly in the human central nervous system.

Histamine is released from basophils and mast cells, but the inflammatory response following an IgE-mediated allergic reaction is far more complex than previously thought, since many cell types and a wide array of inflammatory mediators (including arachidonic acid metabolites, toxic proteins, cytokines, growth factors, and neuromediators) are involved. Thus, it is difficult to explain why drugs that only act at the H_1-receptor level can be so effective in controlling symptoms induced by an allergic reaction.

The first antihistamine or H_1-antagonist was discovered by Bovet and Staub (5) at the Institut Pasteur in 1937. Although this compound was too weak and too toxic for clinical use, its discovery led to an enormous amount of research and, in 1942, to the development of phenbenzamine (Antegan®) the first H_1-antagonist used in the treatment of allergic diseases (6). Within a few years, three other H_1-antagonists were made available: pyrilamine (mepyramine) (7), diphenhydramine (8), and tripelennamine (9). Despite their pronounced side effects, these were the first really useful drugs employed for the symptomatic relief of allergic diseases and they are still in use today. During the last 15 years, pharmacological research has resulted in the development of compounds with higher potency, longer duration of action, and minimal sedative effect: the so-called new or second generation H_1-antagonists. These drugs have recently been the focus of considerable medical scientific interest, because of their multiple antiallergic properties, and also because of reports concerning rare but potentially severe cardiotoxic effects, and a paper describing tumor-promoting effects in a rodent model. These topics, safety in particular, are the source of much current interest because of the widespread use of the H_1-antagonists in current medical practice.

II. CHEMISTRY OF H_1-ANTAGONISTS

Histamine consists of a single heterocyclic ring (imidazole) connected directly to the ethylamine group, the unsubstituted amino terminal. H_1-antagonists bear much less structural resemblance to histamine than do H_2-antagonists. H_1-antagonists are nitrogenous bases containing an aliphatic side chain that shares with histamine the common core structure of a substituted ethylamine (10). The ethylamine side chain is essential to H_1-antagonism. A methyl group in the 2 position increases, and the same group in the 4 position decreases, the relative effectiveness of a compound for H_1-receptors in contrast with H_2-receptors (3). The side chain is attached to one, or more often two, cyclic or heterocyclic rings that may be pyridine, piperidine, pyrrolidine, piperazine, phenothiazine, or even imidazole (AR_1, AR_2). These rings are connected via a nitrogen, carbon, or oxygen linkage (X) to the ethylamine group. Unlike that of histamine, the nitrogen of this ethylamine group is tertiary with two substituents (R_1 and R_2) (Fig. 1). A study of the structure-activity relationship indicates that not all compounds with an aliphatic side chain have antihistamine activity (11). Three features of H_1-antagonists are of importance: (1) the presence of multiple aromatic or heterocyclic rings and alkyl substituents in these antagonists results in their lipophilic properties in contrast to histamine and H_2-antagonists that are hydrophilic compounds; (2) the basicity of the nitrogen group since it was shown that the nearer the pKa is

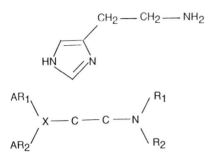

Figure 1 Structural comparison between histamine (top) and the general formula for H₁-antagonists (bottom). (From Ref. 10.)

to 8.6 the more potent it is; (3) the nature of the linkage atom X which is used to categorize the numerous H₁-antagonists into six major classes (Fig. 2). The geometries of H₁-antagonists have been studied by x-ray crystallography and data have been compared with those obtained by minimizing the conformational energy of the molecules according to a simplified model of force field. Both approaches have produced similar results, and indicate unique stereochemical requirements for optimum H₁-antihistaminic activity (12).

Some new generation H₁-antagonists present a chemical structure that may differ quite considerably from older H₁-antagonists, thereby explaining some of the properties of the new compounds (Fig. 3). Astemizole, levocabastine, and terfenadine contain a piperidine ring but they cannot be directly ascribed to this class (13). Loratadine is related to azatadine, cetirizine is a metabolite of hydroxyzine, and acrivastine is structurally related to triprolidine.

An additional group of compounds includes azelastine, ketotifen, and oxatomide. Although most H₁-antagonists possess antiallergic properties, these compounds have been directly developed as being both H₁-antagonists and inhibitors of mediator release from inflammatory cells.

Tricyclic antidepressants have a common three-ring molecular core and are classified by the type of amine on the side chain into tertiary (doxepin, amitriptyline, and trimipramine) and secondary (protriptyline, nortriptyline, and desipramine) amine tricyclics (14). These compounds were initially developed for possible use as H₁-antagonists but they were found to be potent antidepressants and were subsequently used for the treatment of depression (15). However, they possess a potent inhibitory activity on histamine-induced skin tests (16).

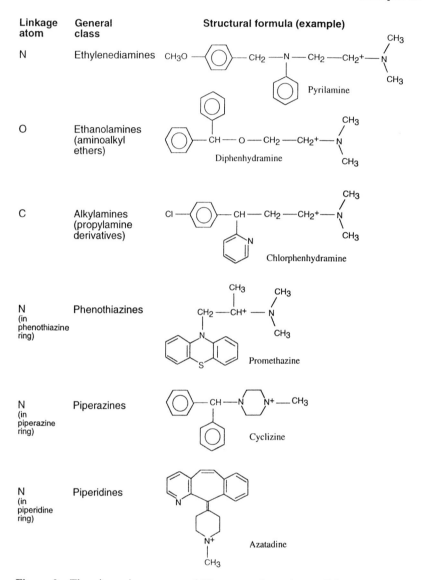

Linkage atom	General class	Structural formula (example)
N	Ethylenediamines	Pyrilamine
O	Ethanolamines (aminoalkyl ethers)	Diphenhydramine
C	Alkylamines (propylamine derivatives)	Chlorphenhydramine
N (in phenothiazine ring)	Phenothiazines	Promethazine
N (in piperazine ring)	Piperazines	Cyclizine
N (in piperidine ring)	Piperidines	Azatadine

Figure 2 The six major groups of H_1-antagonists: the traditional classification system. Chlorphenhydramine = chlorpheniramine. Pyrilamine = mepyramine.

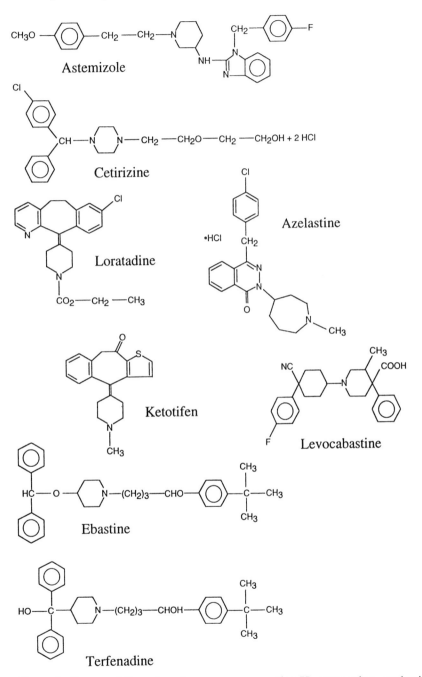

Figure 3 Structural formulas of some new generation H₁-antagonists, azelastine and ketotifen.

III. PHARMACOLOGICAL EFFECTS OF H₁-ANTAGONISTS

A. Effects Related to H₁-Mediated Responses

The pharmacological effects of H_1-antagonists derive primarily from inhibition of histamine action at H_1-receptors. The major effects of H_1-receptors that are inhibited by H_1-antagonists are smooth muscle contraction, increase in vascular permeability, possibly by direct action on endothelial cells (17), nerve stimulation, and pruritus. H_1-antagonists act by binding to H_1-histamine receptors. However, unlike histamine, binding of the antagonists to the receptors does not elicit a tissue response. The presence of H_1-antagonists at the receptor prevents histamine from binding and histaminic responses are therefore blocked. The older H_1-antagonists act competitively with histamine at the receptor level and their binding is readily reversible. They can be displaced by high levels of histamine or when they dissociate from the receptor. Doses of H_1-antagonists must therefore be high enough to compete effectively with histamine for the H_1-receptor. However, the incidence and severity of side effects increases with dose, thus limiting the use of older H_1-antagonists (18). Moreover, many of these medications have a short duration of action. Most second-generation H_1-receptor antagonists bind in a noncompetitive manner with the H_1-receptor (19, 20). Many of these new compounds dissociate slowly from the histamine H_1-receptor, are not displaced easily by histamine from the receptor site, and have a longer duration of action.

Objective dose-response studies of H_1-receptor blockade in the skin have been performed using suppression of the histamine-induced wheal and flare reactions. Results with second-generation H_1-antagonists indicate a dose-dependent increase in histamine skin test suppression. Few dose-response studies of H_1-antagonists in allergic rhinitis or chronic urticaria have been published. Recommended doses for H_1-antagonists appear to be optimal with regard to safety and should not be exceeded, as the incidence of adverse effects may increase when higher doses are used. Also, even if higher doses are given, complete symptom relief may not occur in all patients since total H_1-blockade still leaves the effects of other chemical mediators of inflammation unopposed (21).

B. Antiallergic Activity

Histamine is not the only mediator released during allergic reactions. The rank order of relative H_1-antagonism was studied by Simons et al. using skin tests with histamine and the order from the most effective to the least effective was found to be: cetirizine, 10 mg; terfenadine, 120 mg; terfenadine, 60 mg; loratadine, 10 mg; astemizole, 10 mg; chlorphenira-mine, 4 mg; and placebo (22). However, when these drugs are compared in controlled trials, it is usually impossible to differentiate their clinical efficacy. This suggests that they are clinically active because of other properties in addition to H$_1$-blocking activity or, alternatively, that incom-plete H$_1$-blockade is sufficient for clinical efficacy. Moreover, the block-ade of the release of histamine by a synthesis inhibitor was unable to suppress symptoms completely during nasal challenge (23). Thus, it ap-pears that drugs acting upon the symptoms of the allergic reaction may have additive properties to H$_1$-blockade. Over the past 15 years, it has become clear that most H$_1$-antagonists had such antiallergic properties besides H$_1$-blockade (24). These properties differ depending on the mole-cule and are fully described in Chapters 5 and 6 of this book. In vitro, high concentrations of H$_1$-antagonists are able to block mediator release from basophils and human mast cells (25–27) by mechanisms that are not yet completely understood (28). These antiallergic effects can also be seen in vivo in skin, nasal, lung, and ocular challenge studies. Cetirizine, at least in the skin, reduces eosinophil chemotaxis at currently prescribed doses (29, 30). Using nasal challenge with allergen, it has been observed that azatadine, loratadine, and terfenadine reduce histamine, PGD$_2$, and kinin release during challenge (31–34). Azelastine (35) and cetirizine (34) decreased sulfidopeptide leukotriene release but the effects of ketotifen were rather disapointing in this particular model, since mediator release was not blocked as expected. Finally, none of the H$_1$-antagonists tested (including cetirizine) were able to decrease eosinophil cationic protein release during the late-phase reaction (36). Moreover, terfenadine, cetiri-zine, and loratadine decrease the expression of intercellular adhesion mol-ecule-1 (ICAM-1) in cells from conjunctival or nasal secretions during allergen challenge (37, 38). The extent of these antiallergic effects is not completely understood, yet these studies have led to the concept of antial-lergic drugs with H$_1$-blocking properties (24). However, it would be pre-mature to attempt to reclassify the H$_1$-antagonists according to their antial-lergic properties because these properties have not been fully investigated and their relative contribution to the overall therapeutic effectiveness of each H$_1$-receptor antagonist is unknown (39).

Due to their variable H$_1$-blocking activity and antiallergic effects and possibly due to differences in tissue deposition, H$_1$-antagonists are not equally effective on skin, nose, eye, or lung symptoms. Moreover, it ap-pears that not all H$_1$-antagonists have similar effects on patients and thus nonresponders to one drug may respond favorably to another drug (40).

C. Effects Unrelated to H_1-Blockade

Most, if not all, older H_1-antagonists possess pharmacological effects that are not related to H_1-blockade. Many H_1-antagonists block cholinergic muscarinic receptors in a dose-dependent manner. Quantitative evaluation of antimuscarinic effects of antihistamines (H_1- and H_2-receptor antagonists) was carried out using a receptor-binding assay. Mequitazine, cyproheptadine, clemastine, diphenylpyraline, promethazine, homochlorcyclidine, and alimemazine had high affinities for the muscarinic receptors (K_i = 5.0–38 nM). Another group of H_1-receptor antagonists (pyrilamine, terfenadine, metapyrilen, azelastine, hydroxyzine, and meclizine) exhibited low affinities for the muscarinic receptors (K_i = 3600–30,000 nM) (41). Based on molecular cloning studies, five different muscarinic receptor subtypes exist: M1, M2, M3, M4, and M5. Stanton et al. (42) determined the affinity and selectivity of binding for three H_1-antagonists using Chinese hamster ovarian cells (CHO-K1) transfected with genes for the human muscarinic receptor subtypes. The compounds studied showed no significant selectivity among the five cloned subtypes. The most common anticholinergic side effects at usual dosages consist of dryness of the mouth and of the mucous membranes of the nose and throat. In some susceptible individuals, urinary retention and blurred vision may occur. At higher doses, more severe anticholinergic effects may be observed.

Some H_1-antagonists have local anesthetic effects and possess membrane-stabilizing or quinidine-like effects on cardiac muscle and are responsible for a prolongation of the refractory period of the heart and torsade de pointes (43). These effects have been highlighted recently since very rarely deaths have been caused by H_1-antagonists. Cardiac effects of H_1-antagonists are reviewed in Chapter 14 of this book.

Certain H_1-antagonists, particularly promethazine, possess α-adrenergic receptor-blocking properties. Others increase adrenergic effects by a cocaine-like effect to decrease reuptake of transmitter. Other H_1-antagonists possess antiserotonin(cyproheptadine) (44) or antidopamine (phenothiazines) (45) effects.

Several, but not all, H_1-antagonists, are analgesic agents and some are analgesic adjuvants as well. Those for which this effect is reported include diphenhydramine, hydroxyzine, orphenadrine, pyrilamine, phenyltoloxamine, promethazine, methdilazine, and tripelennamine. More than one mechanism of action exists for them. There is considerable evidence suggesting that histaminergic and serotoninergic central pathways are involved in nociception and that H_1-antagonist drugs can modulate their responses. The evidence for a role for norepinephrine and dopamine and the effects of H_1-antagonists on them is less well established (46).

Histamine H_1-receptors are involved in the development of the symptoms of motion sickness, including emesis. On provocative motion stimulus, a signal for sensory conflict activates the histaminergic neuron system and the histaminergic descending impulse stimulates H_1-receptors in the emetic center of the brain stem. The histaminergic input to the emetic center via H_1-receptors is independent of dopamine D_2-receptors in the chemoreceptor trigger zone and serotonin $5HT_3$-receptors in the visceral afferents from the gastrointestinal tract, which are also involved in the emetic reflex. H_1-antagonists block emetic H_1-receptors to prevent motion sickness. Acetylcholine muscarinic receptors are involved in the generation of signals for sensory conflict. Anticholinergic drugs prevent motion sickness by modifying the neural store to facilitate habituation to provocative motion stimuli (47–49).

D. Central Nervous System Side Effects

Histamine is considered to be both a local hormone and a neurotransmitter in the central nervous system (CNS) (50), where it is synthesized by neurons and mast cells. The three types of receptors are present in the CNS but differ in their localization, biochemical machinery, function, and affinities for histamine. H_1-receptors may be visualized by autoradiography and are widespread throughout the CNS. The physiological roles of H_1-receptors in the CNS need better understanding, but it is well known that H_1-antagonists induce several effects. The most common side effects of older H_1-antagonists are sedation, CNS depression, or CNS stimulation. Sedation, ranging from mild drowsiness to deep sleep, can occur frequently, even at the usual therapeutic doses. Other symptoms of CNS depression are disturbed coordination, dizziness, lassitude, and inability to concentrate. Many factors have been implicated in the CNS side effects of H_1-antagonists (51). Among them, the lipid solubility of H_1-antagonists that enables them to cross the blood–brain barrier (52) is of critical importance. Some nonsedative H_1-antagonists do not cross this barrier because of their decreased lipophilicity, others because of electrostatic charge or small volume of distribution. Moreover, there is a highly significant correlation between the sedation caused by H_1-antagonists and the level of their binding to brain receptors (53). Nonsedative H_1-antagonists may or may not have a reduced affinity for CNS histamine receptors (54, 55). The new generation of compounds is mostly devoid of CNS side effects (56). CNS side effects of older H_1-antagonists (57) but not of newer ones (58–60) are potentiated by alcohol. A review of these effects is presented in Chapter 13.

E. Gastrointestinal Disturbances

Gastrointestinal disturbances, including nausea, vomiting, diarrhea, loss of appetite, and epigastric distress, are observed with some members of the ethylenediamine class.

F. Possible Carcinogenic and Tumor-Promoting Effects

The relationship between mast cells and cancer is not clear, but it has been proposed that histamine may reduce tumor growth (61). Moreover, some H_1-antagonists have been suspected to be carcinogenic (62), for example, methapyrilene induces DNA damage (63) and is carcinogenic in rats (64). A recent experimental study on mice conducted by Brandes et al. (65), suggested a possible tumor-promoting effect for some antihistamines. A population of mice injected with melanoma or fibrosarcoma cells were given intraperitoneal doses (comparable to the human dosages) of astemizole, loratadine, hydroxyzine, doxylamine, and cetirizine. After 21 days of treatment, the wet weights of the tumors were measured and compared to those of control groups. The increased tumor growth was greatest for astemizole and loratadine, followed by hydroxyzine, while doxylamine and cetirizine were comparable to the placebo. However, these results are not consistent with clinical experience since reports of carcinogenity or enhanced tumor growth are almost unheard of in more than 50 years of clinical trials and ubiquitous use of antihistamines. Moreover, the results obtained in rodents are not directly applicable to humans because of the experimental conditions; for example, different route of H_1-antagonist administration and the different cellular metabolic systems (66, 67).

IV. FIRST-GENERATION H_1-ANTAGONISTS

Older H_1-antagonists are usually classified into six different groups. They have many effects in addition to H_1-antagonist effects and their side effects, while often troublesome, are sometimes useful therapeutically.

Group I: Ethylenediamines include pyrilamine, antazoline, methapyrilene (68), and tripelennamine. These drugs have relatively weak CNS effects but gastrointestinal side effects are common. Methapyrilene was found to possess carcinogenic properties in rats and is no longer used clinically (69).

Group II: Ethanolamines include diphenhydramine (8, 70), bromodiphenhydramine, carbinoxamine (71), clemastine (72), doxylamine, and phenyltoloxamine (73). These drugs have significant anticholinergic side effects and frequently cause sedation or other CNS symptoms. The incidence of gastrointestinal side effects is relatively low. Diphenhydramine is still sometimes used to improve voluntary movements in patients with Parkinson's disease because of its anticholinergic effects.

Group III: Alkylamines include chlorpheniramine (74, 75), brompheniramine (76), dexbrompheniramine, dexchlorpheniramine, dimethindene (77–79), pheniramine, and triprolidine (80). These drugs cause less CNS depression than members of other groups. On the other hand, some CNS stimulation may occur. This group was considered by many physicians to have the best safety/efficacy profile before the availability of the new generation H₁-antagonists.

Group IV: Phenothiazines include promethazine, mequitazine, methdilazine, and trimeprazine (81). Sedative and anticholinergic side effects are of importance with these drugs, which are primarily used as antiemetics (82). Mequitazine is effective in the treatment of allergic rhinitis (83) but it is moderately sedative (84, 85) and possesses anticholinergic effects (86, 87).

Group V: Piperazines include cyclizine (80, 88), buclizine, chlorcyclizine, hydroxyzine (89, 90), and meclizine. These drugs induce mild-to-moderate sedation and anticholinergic side effects. Cyclizine, buclizine, and meclizine are used for treating motion sickness or vertigo. Hydroxyzine is used as an H₁-antagonist (91, 92) as well as a sedative, tranquilizer, and antiemetic.

Group VI: Piperidines include cyproheptadine (91, 92) and azatadine (93–95). These drugs possess sedative and anticholinergic side effects. Cyproheptadine is used as an appetite stimulant. Conflicting observations concerning the action of cyproheptadine on pituitary function have been reported in the literature and may be a reflection of the diversity of pharmacological actions of this drug, which include potent antiserotonin activity (96, 97).

The older H₁-antagonists should no longer be used, because of potential side effects among which sedation, impaired performance (98, 99), and anticholinergic activity are the most important. The long-term efficacy of some of these drugs is another matter of potential concern, since tachyphylaxis may occur (100–102). This phenomenon is not related to increased metabolism (100, 102), but may be related to decreased compliance (100) or increased H₁-receptor number (102).

V. SECOND-GENERATION H₁-RECEPTOR ANTAGONISTS

Second-generation H₁-antagonists have been developed over the last 15 years and can be distinguished from the older H₁-antagonists by many properties, two of which are of great importance: the kinetics of their binding to and dissociation from the H_1-receptor and their lack of appreciable CNS and anticholinergic side effects. Some of the drugs have been developed for topical use.

A. Overall Properties

Several properties should be met by the second-generation H_1-antagonists:

Pharmacological properties:
potent and noncompetitive H_1-receptor blockade
antiallergic activities
no decreased absorption when administered with foods
no known drug interaction
Side effects
no sedation
no anticholinergic effects
no weight gain
no potential for cardiac side effects
Pharmacokinetics
rapid onset of action
long duration of action, at least 24 h
administration once a day
no development of tachyphylaxis (103, 104).

B. Individual Properties

Not all of the second-generation H_1-antagonists have all of the above properties. The indication for their use differs depending on the site, severity, and duration of the allergic reaction. Even newer H_1-antagonists are in various stages of development, including: norebastine (105, 106), epinastine (107–109), and mizolastine (110, 111).

1. Acrivastine

Acrivastine is a side-chain-reduced metabolite of the antihistamine triprolidine. It is a short-acting histamine H_1-receptor antagonist with a rapid onset of action. Double-blind clinical trials have shown acrivastine (usually 8 mg 3 times daily) to be an effective antihistamine in the treatment of chronic urticaria (112) and allergic rhinitis (113). Acrivastine was found to cause less drowsiness than clemastine (114), but it seems to have some sedative effects (115) and CNS interactions with alcohol have been observed (116). Because of its rapid onset of action, acrivastine is useful for "on demand" therapy in patients with intermittent symptoms (117).

2. Astemizole

Astemizole [1-(4-fluorophenylmethyl)-N-1(2-(4-methoxyphenyl)ethyl)-4-piperidinyl-1H-benzamidazole-2-amine] is a long-acting, highly selective H$_1$-antagonist with no CNS or anticholinergic effects (118–120). Astemizole and its active metabolite, desmethylastemizole, have long elimination half-life values, permitting once-daily dosing, but skin tests with histamine can be suppressed for up to 6 weeks after cessation of astemizole treatment (121). Increased appetite and weight gain may occur. Astemizole is metabolized by liver cytochrome P$_{450}$ and drug interactions with other compounds with the same metabolism can occur. Cardiac side effects in the form of torsade de pointes have been observed occasionally (122, 123) and are dose-dependent, implying that the dosage of astemizole should not be increased above the stipulated level and that drug interactions should be carefully avoided (Chapters 7 and 14). Studies have shown that astemizole 10 mg once daily is effective in the treatment of seasonal and perennial allergic rhinitis, allergic conjunctivitis, and chronic urticaria. Astemizole may not be as effective for treatment of acute allergic symptoms because of its relatively slow onset of action (124, 125).

3. Cetirizine

Cetirizine, [2-(4-[(chlorophenyl)phenylmethyl]-1-piperazinyl)ethoxy] acetic acid dihydrochloride, a piperazine derivative and carboxylated metabolite of hydroxyzine, is a specific and long-acting histamine H$_1$-receptor antagonist. The H$_1$-blocking activity of cetirizine is extremely potent (22) and it inhibits eosinophil chemotaxis during the allergic response, at least in the skin (126–128). Cetirizine is not metabolized in the liver and no cardiac side effects have been reported. Clinical trial results indicate that cetirizine 10 mg once daily is an effective treatment for seasonal/perennial allergic rhinitis and chronic idiopathic urticaria. Cetirizine may also have a role in the treatment of certain forms of physical urticaria, atopic dermatitis, and reactions to mosquito bites (129). It is associated with a significantly lower incidence of sedation than hydroxyzine. However, when sedation was subjectively assessed, cetirizine appeared to be more sedating than a placebo, loratadine, or terfenadine in some, but not all, double-blind studies (130). In contrast, when assessed objectively in pharmacodynamic comparisons, cetirizine was rarely more sedating than a placebo or other second-generation histamine H$_1$-antagonists (131, 132).

4. Ebastine

Ebastine, 4-diphenylmethoxy-1-[3-(4-ter-butylbenzoyl)-propyl] piperidine, is a piperidine derivative. Ebastine and its active metabolite carebastine are highly potent, selective H$_1$-receptor antagonists (133) devoid of any other noticeable receptor binding. Ebastine has less affinity for central than for peripheral H_1-receptors. Administered at a dose of 10 mg once daily, ebastine was found to be effective in seasonal and perennial allergic rhinitis and chronic urticaria (134) without inducing sedation (135). Ebastine has no interaction with alcohol (136). Ebastine does not appear to possess cardiovascular side effects (137).

5. Levocabastine

Levocabastine, [3S-[1(cis)-3α,4β]]-1-[4-cyano-4(4-fluorophenyl) cyclohexyl]-3-methyl-4-phenyl-4-piperidinecarboxylic acid monochloride, is a cyclohexylpiperidine derivative shown to possess long-lasting H_1-antagonism and antiallergic properties in animals (138). It has only been developed for nasal and ocular administration due to its sedative effects. In controlled trials, levocabastine was effective and well tolerated in the treatment of allergic rhinitis and allergic conjunctivitis. Comparative studies have demonstrated that levocabastine is superior to a placebo and at least as effective as sodium cromoglycate in alleviating symptoms of seasonal allergic rhinitis and conjunctivitis. However, levocabastine appears to be less effective than topical corticosteroids with regard to relieving runny and blocked nose symptoms. The incidence of adverse effects associated with levocabastine therapy is low and is similar to that observed with a placebo or sodium cromoglycate. (139). No sedation is observed with levocabastine eye drops (140).

6. Loratadine

Loratadine (Ethyl-4 (8-chloro-5,6-dihydro-11H-benzo (5-6)-cyclo-hepta (1,2-b) pyridine 11-ylidene)-1 piperidine carboxylate) is a long-acting antihistamine that has a high selectivity for peripheral histamine H_1-receptors and lacks CNS side effects (55, 141). It is rapidly metabolized into an active metabolite, descarboethoxyloratadine. Although it is metabolized in the liver, no cardiac side effects have been reported with loratadine. Results from controlled clinical trials have shown that loratadine (10 mg once daily) is a well-tolerated and effective H_1-antagonist in seasonal and perennial allergic rhinitis and chronic urticaria. It was found to be faster acting than astemizole (142). No sedative effect of loratadine is observed at recommended doses (99, 143, 144).

7. Terfenadine

Terfenadine   (alpha-[4-(1,1-dimethylethyl)phenyl]-4-(hydroxydiphenyl-methyl)-1-piperidinebutanol) is a selective histamine H_1-receptor antagonist which is devoid of CNS and anticholinergic activity (145). Terfenadine is metabolized in the liver by cytochrome P450 and interactions with ketoconazole, itraconazole, or erythromycin have been identified (146). Cardiac side effects, including torsade de pointes are exceptionally rare (147), but are dose-dependent, implying that the dosage of terfenadine should not be increased and drug interactions carefully avoided (Chapters 7 and 14). In clinical studies, terfenadine is well tolerated and at a dose of 60 mg administered twice daily or 120 mg once daily, the drug provides effective relief of symptoms in patients with allergic rhinitis (seasonal and perennial) and allergic dermatological conditions (148). Terfenadine neither impairs psychomotor performance, adversely affects subjective feelings nor enhances the depressant effects of concomitantly administered alcohol or benzodiazepines (59, 149).

VI. H₁-RECEPTOR ANTAGONISTS USED PRIMARILY FOR THEIR ANTIALLERGIC EFFECTS

A. Azelastine

Azelastine [(4-(p-chlorobenzyl)-2-(hexahydro-1-methyl-1H-azepin-4-yl)-1-(2H)-phthalazinone hydrochloride)] is an antiallergic agent that demonstrates H₁-receptor antagonist activity and also inhibits mediator release from mast cells and cells relevant to the allergic inflammation following antigen and nonantigenic stimuli (150–157). Azelastine antagonizes histamine- and leukotriene-induced bronchospasm in animal studies and reduces airway responsiveness to inhaled antigen or distilled water and exercise challenge. In comparative studies, orally administered azelastine in doses up to 4 mg/day consistently relieved symptoms in patients with seasonal or perennial rhinitis. In addition, azelastine administered as an intranasal spray was as effective as oral terfenadine 120 mg/day and intranasal budesonide 0.4 mg/day in alleviating symptoms of rhinitis (158, 159). Azelastine is also an antiasthmatic agent but this indication needs further study to be fully appreciated. The drug is superior to a placebo and comparable to oral ketotifen 2 mg/day and sustained-release theophylline 700 mg/day when administered as a twice-daily, 4-mg oral dose. Azelastine is generally well tolerated: the most common adverse effects are altered taste perception and drowsiness (154).

B. Ketotifen

Ketotifen[(4-/1-methyl-4-piperidylidene/-4H-benzo[4,5]cyclohepta[1,2-b]thiophen-10(9H)-one hydrogen fumarate)] is an H₁-antagonist with antiallergic properties. After 6 to 12 weeks of administration, ketotifen reduces respiratory symptoms and the need for concomitant antiasthmatic drugs in patients with mild-to-moderate bronchial asthma. However, absolute improvement in lung function is generally slight. Ketotifen is modestly effective in the treatment of atopic dermatitis, seasonal or perennial rhinitis, allergic conjunctivitis, chronic and acute urticaria, and food allergy. Sedation can be troublesome in older children and adults, especially during the initial weeks of treatment. Weight gain is another notable side effect (160).

C. Oxatomide

Oxatomide, [(diphenylmethyl-4-piperazinyl-1)-3 propyl]-1,3 H benzimidazolone-2, is an orally active H_1-histamine receptor antagonist that also inhibits mediator release (161). It is mostly used in chronic urticaria. Interestingly, some patients responding to oxatomide were said to be unresponsive to previously administered antihistamines. Sedation and weight gain are common side effects (162).

VII. SUMMARY

Antihistamines or H_1-antagonists were discovered by Bovet and Staub in 1937. H_1-antagonists are nitrogenous bases containing an aliphatic side chain sharing with histamine the common core structure of a substituted ethylamine. The pharmacological effects of H_1-antagonists derive primarily from inhibition of histamine action at H_1-receptors. The major effects of histamine on H_1-receptors that are inhibited by H_1-antagonists are smooth muscle contraction, increase in vascular permeability, nerve stimulation, and pruritus. H_1-antagonists act by binding to H_1-histamine receptors. The first generation of H_1-antagonists had many side effects, including anticholinergic activity and sedation. During the last 15 years, pharmacological research has produced several compounds with high potency, long duration of action, and minimal sedative effects: the so-called newer or second-generation H_1-antagonists, as opposed to the older, or first-generation H_1-antagonists. Besides their H_1-antagonism, H_1-antihistamines possess multiple antiallergic properties, the clinical importance of which is not completely understood.

REFERENCES

1. Dale H, Laidlaw P. The physiological action of β-imidazolethilamine. J Physiol (London) 1910;41:318–44.
2. Ash ASF, Schild HO. Receptors mediating some actions of histamine. Br J Pharmacol 1966;27:427–39.
3. Black JW, Duncan WAM, Durant GJ, Ganellin CR, Parsons EM. Definition and antagonism of histamine H_2 receptors. Nature 1972;236:385–90.

4. Arrang J-M, Garbarg M, Lancelot J-C, Lecomte J-M, Pollard H, Robba M, Schunack W, Schwartz J-C. Highly potent and selective ligands for histamine H_3-receptors. Nature 1987;327:117–23.

5. Staub AM, Bovet D. Action de la thymoxyéthyl-diéthylamine (929F) et des éthers phénoliques sur le choc anaphylactique du cobaye. CRS Soc Biol 1937;125:818.

6. Halpern BN. Les antihistaminiques de synthèse: essai de chimiothérapie des états allergiques. Arch Int Pharmacodyn Ther 1942;68:339–45.

7. Bovet D, Horclois R, Walthert F. Propriétés antihistaminiques de la N-p-méthoxybenzyl-N-diméthylaminoéthyl alpha aminopyridine. CRS Soc Biol 1944;138:99–108.

8. Lowe E, MacMillan R, Katser M. The antihistamine properties of Benadryl, beta-dimethyl-aminoethyl benzhydryl ether hydrochloride. J Pharmacol Exp Ther 1946;86:229.

9. Yonkman FF, Chess D, Mathieson D, Hansen N. Pharmacodynamic studies of a new antihistamine agent, N'-pyridyl-N'-benzyl-N-dimethylethylene diamine HCl, pyribenzamine HCl. I. Effects on salivation, nictitating membrane, lachrymation, pupil and blood pressure. J Pharmacol Exp Ther 1946; 87:256–264.

10. Trzeciakowski JP, Levi R. Antihistamines. In: Middleton E, Reed CE, Ellis EF, eds. Allergy Principles and Practice, 2nd ed. St Louis: Mosby, 1983: 575–92.

11. Marshall P. Some chemical and physical properties associated with histamine antagonism. Br J Pharmacol 1955;10:270.

12. Borea PA, Bertolasi V, Gilli G. Crystallographic and conformational studies on histamine H_1-receptor antagonists. IV. On the stereochemical vector of antihistaminic activity. Arzneimittelforschung 1986;36:895–9.

13. Simons FER, Simons KJ. Antihistamines. In: Middleton E, Reed CE, Ellis EF, Adkinson NF, Yunginger JW, Busse WW, eds. Allergy Principles and Practice, 4th ed. St Louis; Mosby, 1993:856–92.

14. Richelson E. Pharmacology of antidepressants. Psychopathology 1987; 20(Suppl. 1):1–12.

15. Richelson E. Tricyclic antidepressants and histamine H_1-receptors. Mayo Clin Proc 1979;54:669–74.

16. Sullivan TJ. Pharmacologic modulation of the whealing response to histamine in human skin: identification of doxepin as a potent in vivo inhibitor. J Allergy Clin Immunol 1982;69:260–7.

17. Niimi N, Noso N, Yamamoto S. The effect of histamine on cultured endothelial cells. A study of the mechanism of increased vascular permeability. Eur J Pharmacol 1992;221:325–31.

18. Carruthers SG, Shoeman DW, Hignite CE, Azarnoff DL. Correlation between plasma diphenhydramine level and sedative and antihistamine effects. Clin Pharmacol Ther 1978;23:375–82.

19. Cheng HC, Woodward JK. A kinetic study of the antihistaminic effect of terfenadine. Arzneimittelforschung 1982;32:1160–6.

20. VanWauwe J, Awouters F, Niemegeers CJ, Janssens F, VanNueten JM,

Janssen PAJ. In vivo pharmacology of astemizole, a new type of H_1-antihis-taminic compound. Arch Int Pharmacodyn 1981;251:39–51.

21. Simons FER. H_1-receptor antagonists: does a dose-response relationship exist? Ann Allergy 1993;71:592–7.

22. Simons FER, McMillan JL, Simons KJ. A double-blind, single-dose, cross-over comparison of cetirizine, terfenadine, loratadine, astemizole, and chlorpheniramine versus placebo: suppressive effects on histamine-induced wheals and flares during 24 hours in normal subjects. J Allergy Clin Immunol 1990;86:540–7.

23. Pipkorn U, Granerus G, Proud D, et al. The effect of a histamine synthesis inhibitor on the immediate nasal allergic reaction. Allergy 1987;42:496–501.

24. Bousquet J, Campbell A, Michel F-B. Antiallergic activities of antihista-mines. In: Church MK, Rihoux J-P, eds. Therapeutic Index of Antihista-mines. Lewinston, New York: Hogrefe & Huber Publishers, 1992:57–96.

25. Temple DM, McCluskey M. Loratadine, an antihistamine, blocks antigen- and ionophore-induced leukotriene release from human lung in vitro. Pros-taglandins 1988;35:549–54.

26. Campbell AM, Chanez P, Marty-Ané C, et al. Modulation of eicosanoid and histamine release from human dispersed lung cells by terfenadine. Al-lergy 1993;48:125–9.

27. Faraj BA, Jackson RT. Effect of astemizole on antigen-mediated histamine release from the blood of patients with allergic rhinitis. Allergy 1992;47: 630–4.

28. Foreman J, Rihoux J. The antiallergic activity of H_1-histamine receptor antagonists in relation to their action on cell calcium. In: Church M, Rihoux J, eds. Therapeutic Index of Antihistamines. Lewiston, New York: Hogrefe & Huber Publishers, 1992:32–46.

29. Michel L, De-Vos C, Rihoux JP, Burtin C, Benveniste J, Dubertret L. Inhibitory effect of oral cetirizine on in vivo antigen-induced histamine and PAF-acether release and eosinophil recruitment in human skin. J Allergy Clin Immunol 1988;82:101–9.

30. Charlesworth EN, Kagey-Sobotka A, Norman PS, Lichtenstein LM. Effect of cetirizine on mast cell-mediator release and cellular traffic during the cutaneous late-phase reaction. J Allergy Clin Immunol 1989;83:905–12.

31. Bousquet J, Lebel B, Chanal I, Morel A, Michel FB. Antiallergic activity of H_1-receptor antagonists assessed by nasal challenge. J Allergy Clin Im-munol 1988;82:881–7.

32. Andersson M, Nolte H, Baumgarten C, Pipkorn U. Suppressive effect of loratadine on allergen-induced histamine release in the nose. Allergy 1991; 46:540–6.

33. Naclerio RM, Kagey-Sobotka A, Lichtenstein LM, Freidhoff L, Proud D. Terfenadine, an H_1 antihistamine, inhibits histamine release in vivo in the human. Am Rev Respir Dis 1990;142:167–71.

34. Togias AG, Proud D, Kagey-Sobotka A, Freidhoff L, Lichtenstein LM, Naclerio RM. In vivo and in vitro effects of antihistamines on mast cell mediator release: a potentially important property in the treatment of aller-gic disease. Ann Allergy 1989;63:465–9.

35. Shin MH, Baroody F, Proud D, Kagey-Sobotka A, Lichtenstein LM, Naclerio RM. The effect of azelastine on the early allergic response. Clin Exp Allergy 1992;22:289–95.

36. Klementsson H, Andersson M, Pipkorn U. Allergen-induced increase in nonspecific nasal reactivity is blocked by antihistamines without a clear-cut relationship to eosinophil influx. J Allergy Clin Immunol 1990;86:466–72.

37. Ciprandi G, Buscaglia S, Pronzato C, et al. New targets for antiallergic agents. Int Acad Biomed Drug Res 1993;6:115–27.

38. Canonica G, Ciprandi G, Buscaglia S, Pesce G, Bagnasco M. Adhesion molecules of allergic inflammation: recent insights into their functional roles. Allergy 1994;49:135–41.

39. Simons FER. The antiallergic effects of antihistamines (H_1-receptor antagonists). J Allergy Clin Immunol 1992;90:705–15.

40. Carlsen KH, Kramer J, Fagertun HE, Larsen S. Loratadine and terfenadine in perennial allergic rhinitis. Treatment of nonresponders to the one drug with the other drug. Allergy 1993;48:431–6.

41. Kubo N, Shirakawa O, Kuno T, Tanaka C. Antimuscarinic effects of antihistamines: quantitative evaluation by receptor-binding assay. Jpn J Pharmacol 1987;43:277–82.

42. Stanton T, Bolden-Watson C, Cusack B, Richelson E. Antagonism of the five cloned human muscarinic cholinergic receptors expressed in CHO-K1 cells by antidepressants and antihistaminics. Biochem Pharmacol 1993;45:2352–4.

43. Roden DM. Torsade de pointes. Clin Cardiol 1993;16:683–6.

44. Van-Nueten JM, Xhonneux R, Janssen PAJ. Preliminary data on antiserotonin effects of oxatomide, a novel anti-allergic compound. Arch Int Pharmacodyn Ther 1978;232:217–20.

45. Campbell M, Bateman DN. Pharmacokinetic optimisation of antiemetic therapy. Clin Pharmacokinet 1992;23:147–60.

46. Rumore MM, Schlichting DA. Analgesic effects of antihistaminics. Life Sci 1985;36:403–16.

47. Takeda N, Morita M, Hasegawa S, Horii A, Kubo T, Matsunaga T. Neuropharmacology of motion sickness and emesis. A review. Acta Otolaryngol Suppl Stockh 1993;501:10–5.

48. Mitchelson F. Pharmacological agents affecting emesis. A review (Part I). Drugs 1992;43:295–315.

49. Mitchelson F. Pharmacological agents affecting emesis. A review (Part II). Drugs 1992;43:443–63.

50. Timmerman H. Histamine receptors in the central nervous system. Pharm Weekbl Sci 1989;11:146–50.

51. Timmerman H. Factors involved in the incidence of central nervous system effects of H_1-blockers. In: Church MK, Rihoux JP, eds. Therapeutic Index of Antihistamines. Lewiston, New York: Hogrefe & Huber Publishers, 1992:19–31.

52. Goldberg MJ, Spector R, Chiang CK. Transport of diphenhydramine in the central nervous system. J Pharmacol Exp Ther 1987;240:717–22.

53. Schwartz JC, Barbin G, Duchemin AM, et al. Histamine receptors in the brain: characterization by binding studies and biochemical effects. Adv Biochem Psychopharmacol 1980;21:169–82.

54. Janssens MML, Howarth PH. The antihistamines of the nineties. Clin Rev Allergy 1993;11:111–53.

55. Ahn HS, Barnett A. Selective displacement of [3H]mepyramine from peripheral vs. central nervous system receptors by loratadine, a non-sedating antihistamine. Eur J Pharmacol 1986;127:153–5.

56. Gengo FM, Manning C. A review of the effects of antihistamines on mental processes related to automobile driving. J Allergy Clin Immunol 1990;86: 1034–9.

57. Burns M, Moskowitz H. Effects of diphenhydramine and alcohol on skills performance. Eur J Clin Pharmacol 1980;17:259–66.

58. Bateman DN, Chapman PH, Rawlins MD. Lack of effect of astemizole on ethanol dynamics or kinetics. Eur J Clin Pharmacol 1983;25:567–8.

59. Bhatti JZ, Hindmarch I. The effects of terfenadine with and without alcohol on an aspect of car driving performance. Clin Exp Allergy 1989;19:609–11.

60. Doms M, Vanhulle G, Baelde Y, Coulie P, Dupont P, Rihoux JP. Lack of potentiation by cetirizine of alcohol-induced psychomotor disturbances. Eur J Clin Pharmacol 1988;34:619–23.

61. Burtin C. Mast cells and tumour growth. Ann Inst Pasteur Immunol 1986; 2:289–94.

62. Schmahl D, Habs M. Drug-induced cancer. Curr Top Pathol 1980;69: 333–69.

63. Althaus FR, Lawrence SD, Sattler GL, Pitot HC. DNA damage induced by the antihistaminic drug methapyrilene hydrochloride. Mutat Res 1982; 103:213–8.

64. Habs M, Shubik P, Eisenbrand G. Carcinogenicity of methapyrilene hydrochloride, mepyramine hydrochloride, thenyldiamine hydrochloride, and pyribenzamine hydrochloride in Sprague-Dawley rats. J Cancer Res Clin Oncol 1986;111:71–4.

65. Brandes L, Warrington C, Arron RJ, et al. Enhanced cancer growth in mice administered daily human-equivalent doses of some H_1-antihistamines: predictive in vitro correlates. J Nat Cancer Inst 1994;86:770–5.

66. Weed D. Between science and technology: the case of antihistamines and cancer. J Nat Cancer Inst 1994;86:740–1.

67. FDA reviews antihistamine mouse study. FDA Talk paper, May 17 1994 1994;.

68. Calandre EP, Alferez N, Hassanein K, Azarnoff DL. Methapyrilene kinetics and dynamics. Clin Pharmacol Ther 1981;29:527–32.

69. Lijinsky W, Reuber MD, Blackwell BN. Liver tumors induced in rats by oral administration of the antihistaminic methapyrilene hydrochloride. Science 1980;209:817–9.

70. Simons KJ, Watson WTA, Martin TJ, Chen XY, Simons FER. Diphenhydramine: pharmacokinetics and pharmacodynamics in elderly adults, young adults, and children. J Clin Pharmacol 1990;30:665–71.

71. Seppala T, Nuotto E, Korttila K. Single and repeated dose comparison of three antihistamines and phenylpropanolamine: psychomotor performance and subjective appraisals of sleep. Br J Clin Pharmacol 1981;12:179–88.

72. Clemastine (Tavist)—a new antihistamine. Med Lett Drugs Ther 1979;21:24.

73. Falliers CJ, Redding MA, Katsampes CP. Inhibition of cutaneous and mucosal allergy with phenyltoloxamine. Ann Allergy 1978;41:140–4.

74. Kotzan JA, Vallner JJ, Stewart JT, et al. Bioavailability of regular and controlled-release chlorpheniramine products. J Pharm Sci 1982;71:919–23.

75. Rumore MM. Clinical pharmacokinetics of chlorpheniramine. Drug Intell Clin Pharm 1984;18:701–7.

76. Simons FER, Frith EM, Simons KJ. The pharmacokinetics and antihistaminic effects of brompheniramine. J Allergy Clin Immunol 1982;70:458–64.

77. De-Graeve J, Van-Cantfort J, Gilard P, Wermeille MM. Identification of the molecular structure of the phenolic primary metabolite of dimetindene in animals and man. Arzneimittelforschung 1989;39:551–5.

78. Bhatt AD, Vaidya AB, Sane SP, et al. Comparative effect of dimethindene maleate and chlorpheniramine maleate on histamine-induced weal and flare. J Int Med Res 1991;19:479–83.

79. Towart R, Sautel M, Moret E, Costa E, Theraulaz M, Weitsch AF. Investigation of the antihistaminic action of dimethindene maleate (Fenistil) and its optical isomers. Agents Actions Suppl 1991;33:403–8.

80. Hamilton M, Bush M, Bye C, Peck AW. A comparison of triprolidine and cyclizine on histamine (H₁) antagonism, subjective effects and performance tests in man. Br J Clin Pharmacol 1982;13:441–4.

81. McGee JL, Alexander MR. Phenothiazine analgesia—fact or fantasy? Am J Hosp Pharm 1979;36:633–40.

82. Hals PA, Hall H, Dahl SG. Muscarinic cholinergic and histamine H₁-receptor binding of phenothiazine drug metabolites. Life Sci 1988;43:405–12.

83. Skassa-Brociek W, Bousquet J, Montes F, et al. Double-blind placebo-controlled study of loratadine, mequitazine, and placebo in the symptomatic treatment of seasonal allergic rhinitis. J Allergy Clin Immunol 1988;81:725–30.

84. Hugonot L, Hugonot R, Beaumont D. A double-blind comparison of terfenadine and mequitazine in the symptomatic treatment of acute pollinosis. J Int Med Res 1986;14:124–30.

85. Nicholson AN, Stone BM. The H₁-antagonist mequitazine: studies on performance and visual function. Eur J Clin Pharmacol 1983;25:563–6.

86. Martinez-Mir I, Estan L, Rubio E, Morales-Olivas FJ. Antihistaminic and anticholinergic activities of mequitazine in comparison with clemizole. J Pharm Pharmacol 1988;40:655–6.

87. Renzetti AR, Barone D, Criscuoli M. High-affinity binding of mequitamium iodide (LG 30435) to muscarinic and histamine H₁-receptors. Eur J Pharmacol 1990;182:413–20.

88. Novack GD, Stark LG, Peterson SL. Anticonvulsant effects of benzhydryl

piperazines on maximal electroshock seizures in rats. J Pharmacol Exp Ther 1979;208:480–4.

89. Schaaf L, Hendeles L, Weinberger M. Suppression of seasonal allergic rhinitis symptoms with daily hydroxyzine. J Allergy Clin Immunol 1979; 63:129–33.

90. Simons FER, Simons KJ, Frith EM. The pharmacokinetics and antihistaminic effects of the H_1-receptor antagonist hydroxyzine. J Allergy Clin Immunol 1984;73:69–75.

91. Klein GL, Galant SP. A comparison of the antipruritic efficacy of hydroxyzine and cyproheptadine in children with atopic dermatitis. Ann Allergy 1980;44:142–5.

92. Harvey RP, Wegs J, Schocket AL. A controlled trial of therapy in chronic urticaria. J Allergy Clin Immunol 1981;68:262–6.

93. Hillas JL, Somerfield SD, Wilson JD, Aman MG. Azatadine maleate in perennial allergic rhinitis: effects on clinical symptoms and choice reaction time. Br J Clin Pharmacol 1980;10:573–7.

94. Luscombe DK, Nicholls PJ, Spencer PS. Effect of azatadine on human performance. Br J Clin Pract 1980;34:75–9.

95. Wilson JD, Hillas JL, Somerfield SD. Azatadine maleate (Zadine): evaluation in the management of allergic rhinitis. N Z Med J 1981;94:79–81.

96. Kletzky OA, Marrs RP, Nicoloff JT. Effects of cyproheptadine on insulin-induced hypoglycaemia secretion of PRL, GH and cortisol. Clin Endocrinol Oxf 1980;13:231–4.

97. Gross MD, Grekin RJ, Gniadek TC, Villareal JZ. Suppression of aldosterone by cyproheptadine in idiopathic aldosteronism. N Engl J Med 1981; 305:181–5.

98. Warren R, Simpson H, Hilchie J, Cimbura G, Lucas D, Bennett R. Drugs detected in fatally injured drivers in the province of Ontario. In: Goldberg L, ed. Alcohol, Drugs, and Traffic Safety. Stockholm: Almquist and Wiksell, 1981:203–17, Vol. 1.

99. O'Hanlon JF. Antihistamines and driving safety. Cutis 1988;42:10–3.

100. Bantz EW, Dolen WK, Chadwick EW, Nelson HS. Chronic chlorpheniramine therapy: subsensitivity, drug metabolism, and compliance. Ann Allergy 1987;59:341–6.

101. Long WF, Taylor RJ, Wagner CJ, Leavengood DC, Nelson HS. Skin test suppression by antihistamines and the development of subsensitivity. J Allergy Clin Immunol 1985;76:113–7.

102. Taylor RJ, Long WF, Nelson HS. The development of subsensitivity to chlorpheniramine. J Allergy Clin Immunol 1985;76:103–7.

103. Simons FER, Watson WTA, Simons KJ. Lack of subsensitivity to terfenadine during long-term terfenadine treatment. J Allergy Clin Immunol 1988; 82:1068–75.

104. Bousquet J, Chanal I, Skassa-Brociek W, Lemonier C, Michel FB. Lack of subsensitivity to loratadine during long-term dosing during 12 weeks. J Allergy Clin Immunol 1990;86:248–53.

105. Kamali F, Emanuel M, Rawlins MD. A double-blind placebo-controlled

dose response study of noberastine on histamine-induced weal and flare. Eur J Clin Pharmacol 1991;40:83–5.

106. Knight A, Drouin MA, Yang WH, Alexander M, Del-Carpio J, Arnott WS. Clinical evaluation of the efficacy and safety of noberastine, a new H_1-antagonist, in seasonal allergic rhinitis: a placebo-controlled, dose-response study. J Allergy Clin Immunol 1991;88:926–34.

107. Kamei C, Akagi M, Mio M, et al. Antiallergic effect of epinastine (WAL 801 CL) on immediate hypersensitivity reactions: (I). Elucidation of the mechanism for histamine release inhibition. Immunopharmacol Immunotoxicol 1992;14:191–205.

108. Kamei C, Mio M, Kitazumi K, et al. Antiallergic effect of epinastine (WAL 801 CL) on immediate hypersensitivity reactions: (II). Antagonistic effect of epinastine on chemical mediators, mainly antihistaminic and anti-PAF effects. Immunopharmacol Immunotoxicol 1992;14:207–18.

109. Schilling JC, Adamus WS, Kuthan H. Antihistaminic activity and side effect profile of epinastine and terfenadine in healthy volunteers. Int J Clin Pharmacol Ther Toxicol 1990;28:493–7.

110. Rosenzweig P, Thebault JJ, Caplain H, et al. Pharmacodynamics and pharmacokinetics of mizolastine (SL 85.0324), a new nonsedative H_1-antihistamine. Ann Allergy 1992;69:135–9.

111. Danjou P, Molinier P, Berlin I, Patat A, Rosenzweig P, Morselli PL. Assessment of the anticholinergic effect of the new antihistamine mizolastine in healthy subjects. Br J Clin Pharmacol 1992;34:328–31.

112. Gibson JR, Manna VK, Salisbury J. Acrivastine: a review of its dermatopharmacology and clinical activity. J Int Med Res 1989;17:28B–34B.

113. Bojkowski CJ, Gibbs TG, Hellstern KH, Major EWT, Mullinger B. Acrivastine in allergic rhinitis: a review of clinical experience. J Int Med Res 1989; 17:54B–68B.

114. Juhlin L, Gibson JR, Harvey SG, Huson LW. Acrivastine versus clemastine in the treatment of chronic idiopathic urticaria. A double-blind, placebo-controlled study. Int J Dermatol 1987;26:653–4.

115. Simons FER. The therapeutic index of newer H_1-receptor antagonists. Clin Exp Allergy 1994;24:707–23.

116. Cohen AF, Hamilton MJ, Peck AW. The effects of acrivastine (BW825C), diphenhydramine and terfenadine in combination with alcohol on human CNS performance. Eur J Clin Pharmacol 1987;32:279–88.

117. Brogden RN, McTavish D. Acrivastine. A review of its pharmacological properties and therapeutic efficacy in allergic rhinitis, urticaria and related disorders. Drugs 1991;41:927–40.

118. Van-Wauwe J, Awouters F, Niemegeers CJ, Janssens F, Van-Nueten JM, Janssen PA. In vivo pharmacology of astemizole, a new type of H_1-antihistaminic compound. Arch Int Pharmacodyn Ther 1981;251:39–51.

119. Laduron PM, Janssen PFM, Gommeren W, Leysen JE. In vitro and in vivo binding characteristics of a new long-acting histamine H_1 antagonist, astemizole. Mol Pharmacol 1982;21:294–300.

120. Awouters FHL, Niemegeers CJE, Janssen PAJ. Pharmacology of the specific histamine H_1-antagonist astemizole. Arzneimittelforschung 1983;33: 381–8.
121. Malo JL, Fu CL, L'Archeveque J, Ghezzo H, Cartier A. Duration of the effect of astemizole on histamine-inhalation tests. J Allergy Clin Immunol 1990;85:729–36.
122. Snook J, Boothman-Burrell D, Watkins J, Colin-Jones D. Torsade de pointes ventricular tachycardia associated with astemizole overdose. Br J Clin Pract 1988;42:257–9.
123. Simons FER, Kesselman MS, Giddins NG, Pelech AN, Simons KJ. Astemizole-induced torsade de pointes. Lancet 1988;2:624.
124. Krstenansky PM, Cluxton R Jr. Astemizole: a long-acting, nonsedating antihistamine. Drug Intell Clin Pharm 1987;21:947–53.
125. Richards DM, Brogden RN, Heel RC, Speight TM, Avery GS. Astemizole. A review of its pharmacodynamic properties and therapeutic efficacy. Drugs 1984;28:38–61.
126. Rihoux JP, Mariz S. Cetirizine. An updated review of its pharmacological properties and therapeutic efficacy. Clin Rev Allergy 1993;11:65–88.
127. Walsh GM, Moqbel R, Hartnell A, Kay AB. Effects of cetirizine on human eosinophil and neutrophil activation in vitro. Int Arch Allergy Appl Immunol 1991;95:158–62.
128. De-Vos C. H_1-antagonists and inhibitors of eosinophil accumulation. Clin Exp Allergy 1991;1:277–81.
129. Reunala T, Brummer-Korvenkontio H, Karppinen A, Coulie P, Palosuo T. Treatment of mosquito bites with cetirizine. Clin Exp Allergy 1993;23: 72–5.
130. Ramaekers JG, Uiterwijk MMC, O'Hanlon JF. Effects of loratadine and cetirizine on actual driving and psychometric test performance, and EEG during driving. Eur J Clin Pharmacol 1992;42:363–9.
131. Spencer CM, Faulds D, Peters DH. Cetirizine. A reappraisal of its pharmacological properties and therapeutic use in selected allergic disorders. Drugs 1993;46:1055–80.
132. Campoli-Richards DM, Buckley MM-T, Fitton A. Cetirizine. A review of its pharmacological properties and clinical potential in allergic rhinitis, pollen-induced asthma, and chronic urticaria. Drugs 1990;40:762–81.
133. Simons FER, Watson WTA, Simons KJ. Pharmacokinetics and pharmacodynamics of ebastine in children. J Pediatr 1993;122:641–6.
134. Luria X, Bakke O. Ebastine (LAS W-090): overview of clinical trials. Drugs Today 1992;28 (Suppl B):69–79.
135. Hopes H, Meuret GH, Ungethum W, Leopold G, Wiemann H. Placebo-controlled comparison of acute effects of ebastine and clemastine on performance and EEG. Eur J Clin Pharmacol 1992;42:55–9.
136. Mattila MJ, Kuitunen T, Pletan Y. Lack of pharmacodynamic and pharmacokinetic interactions of the antihistamine ebastine with ethanol in healthy subjects. Eur J Clin Pharmacol 1992;43:179–84.

137. Llenas J, Bou J, Massingham R. Preclinical safety studies with ebastine. II. Pharmacologic effects on the cardiovascular system. Drugs Today 1992; 28 (Suppl B):29–34.

138. Awouters F, Niemegeers CJE, Jansen T, Megens AAHP, Janssen PAJ. Levocabastine: pharmacological profile of a highly effective inhibitor of allergic reactions. Agents Actions 1992;35:12–8.

139. Dechant KL, Goa KL. Levocabastine. A review of its pharmacological properties and therapeutic potential as a topical antihistamine in allergic rhinitis and conjunctivitis. Drugs 1991;41:202–24.

140. Arriaga F, Rombaut N. Absence of central effects with levocabastine eye drops. Allergy 1990;45:552–4.

141. Roman IJ, Danzig MR. Loratadine. A review of recent findings in pharmacology, pharmacokinetics, efficacy, and safety, with a look at its use in combination with pseudoephedrine. Clin Rev Allergy 1993;11:89–110.

142. Clissold SP, Sorkin EM, Goa KL. Loratadine. A preliminary review of its pharmacodynamic properties and therapeutic efficacy. Drugs 1989;37: 42–57.

143. Roth T, Roehrs T, Koshorek G, Sicklesteel J, Zorick F. Sedative effects of antihistamines. J Allergy Clin Immunol 1987;80:94–8.

144. Van Cauwenberge P. New data on the safety of loratadine. Drug Invest 1992;4:283–91.

145. Carr AA, Meyer DR. Synthesis of terfenadine. Arzneimittelforschung 1982; 32:1157–9.

146. Pohjola-Sintonen S, Viitasalo M, Toivonen L, Neuvonen P. Itraconazole prevents terfenadine metabolism and increases risk of torsades de pointes ventricular tachycardia. Eur J Clin Pharmacol 1993;45:191–3.

147. Monahan BP, Ferguson CL, Killeavy ES, Lloyd BK, Troy J, Cantilena LR. Torsades de pointes occurring in association with terfenadine use JAMA 1990;264:2788–90.

148. McTavish D, Goa KL, Ferrill M. Terfenadine. An updated review of its pharmacological properties and therapeutic efficacy. Drugs 1990;39:552–74.

149. Sorkin EM, Heel RC. Terfenadine. A review of its pharmacodynamic properties and therapeutic efficacy. Drugs 1985;29:34–56.

150. Achterrath-Tuckermann U, Simmet T, Luck W, Szelenyi I, Peskar BA. Inhibition of cysteinyl-leukotriene production by azelastine and its biological significance. Agents Actions 1988;24:217–23.

151. Little MM, Wood DR, Casale TB. Azelastine inhibits stimulated histamine release from human lung tissue in vitro but does not alter cyclic nucleotide content. Agents Actions 1989;28:16–21.

152. Casale TB. The interaction of azelastine with human lung histamine H$_1$, beta, and muscarinic receptor-binding sites. J Allergy Clin Immunol 1989; 83:771–6.

153. Busse W, Randlev B, Sedgwick J. The effect of azelastine on neutrophil and eosinophil generation of superoxide. J Allergy Clin Immunol 1989;83: 400–5.

154. McTavish D, Sorkin EM. Azelastine. A review of its pharmacodynamic and pharmacokinetic properties, and therapeutic potential. Drugs 1989;38: 778–800.

155. Fields DA, Pillar J, Diamantis W, Perhach J Jr, Sofia RD, Chand N. Inhibition by azelastine of nonallergic histamine release from rat peritoneal mast cells. J Allergy Clin Immunol 1984;73:400–3.

156. Chand N, Pillar J, Diamantis W, Sofia RD. Inhibition of IgE-mediated allergic histamine release from rat peritoneal mast cells by azelastine and selected antiallergic drugs. Agents Actions 1985;16:318–22.

157. Chand N, Pillar J, Diamantis W, Sofia RD. Inhibition of allergic histamine release by azelastine and selected antiallergic drugs from rabbit leukocytes. Int Arch Allergy Appl Immunol 1985;77:451–5.

158. Davies RJ, Lund VJ, Harten-Ash VJ. The effect of intranasal azelastine and beclomethasone on the symptoms and signs of nasal allergy in patients with perennial allergic rhinitis. Rhinology 1993;31:159–64.

159. Gambardella R. A comparison of the efficacy of azelastine nasal spray and loratadine tablets in the treatment of seasonal allergic rhinitis. J Int Med Res 1993;21:268–75.

160. Grant SM, Goa KL, Fitton A, Sorkin EM. Ketotifen. A review of its pharmacodynamic and pharmacokinetic properties, and therapeutic use in asthma and allergic disorders. Drugs 1990;40:412–48.

161. Church MK, Gradidge CF. Oxatomide: inhibition and stimulation of histamine release from human lung and leucocytes in vitro. Agents Actions 1980; 10:4–7.

162. Richards DM, Brogden RN, Heel RC, Speight TM, Avery GS. Oxatomide. A review of its pharmacodynamic properties and therapeutic efficacy. Drugs 1984;27:210–31.

5

H_1-Receptor Antagonists: Antiallergic Effects In Vitro

Martin K. Church, Andrea D. Collinson,
and Yoshimichi Okayama
University of Southampton, Southampton, England

I. INTRODUCTION

We often view antihistamines as selective antagonists of the histamine H_1-receptor that have relatively few other actions. Nothing could be further from the truth, particularly with the older antihistamines. We should remember that when Bovet and Staub (1) initially reported on compounds that relieved the exaggerated histamine-induced proinflammatory response seen in allergic individuals, they were, in fact, looking at molecules synthesized as potential anticholinergic drugs. Also, when we look at the basic structure of H_1-receptor antagonist drugs (Fig. 1), it is immediately obvious that their molecular arrangement is common to local anaesthetics, antagonists of muscarinic cholinergic, α-adrenergic, β-adrenergic, dopamine, and 5-hydroxytryptamine receptors and calcium antagonists. Thus, many drugs that share this common backbone have diverse actions; for example, the major tranquilizer chlorpromazine and the tricyclic antidepressant amitryptylline are also potent antihistamines (2), while many antihistamines, including mepyramine, also possess marked anticholinergic activity. The development of newer antihistamines, which have less central sedative actions and fewer antagonistic effects at receptors for other amines, has led to a series of drugs that are more selective in their actions. It must be emphasized, however, that they are only *selective*, not specific,

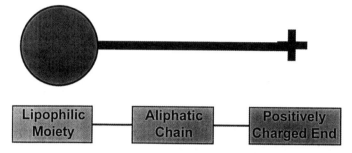

Figure 1 Generalized structure of H_1-receptor antagonist molecules. This structure is very obvious with the older H_1-antagonists but the newer ones conform less rigidly to this generalization.

as evidenced by the rare cardiotoxic effects of astemizole and terfenadine (3–6) which probably represent an effect on calcium transport across membranes. The ability of antihistamines to influence other cellular mechanisms, either by receptor effects or by direct actions on the cell membrane, does, however, endow them with the possibility of having beneficial properties that may extend their usefulness. In this chapter, we assess some of the in vitro effects of antihistamines that may contribute to their clinical profile.

II. THE MECHANISMS OF ALLERGIC DISEASE

Allergic diseases comprise a spectrum of pathological conditions that share a common initiation mechanism and have a remarkably similar cellular pathology. However, the signs and symptoms seen in individual organs vary considerably because the targets on which the inflammatory mediators act as a part of the allergic response are markedly different. For example, the primary targets involved in asthma are bronchial smooth muscle and mucous glands while the primary target producing the symptoms of rhinitis and urticaria is the blood vessels.

In a sensitized individual, the primary event following challenge of a mucosal membrane is the activation of tissue mast cells by cross-linkage of their membrane-bound IgE by allergen. In the local extracellular microenvironment this initiates the release of mediators stored preformed within the mast cell granule and mediators synthesized de novo upon cell activation. Of the preformed mediators released by exocytosis, the simple diamine, histamine, is perhaps the best characterized. Its inflammatory actions, which can be antagonized by histamine H_1-receptor antagonists, include vasodilation, plasma extravasation, and smooth muscle contraction, particularly in the bronchi and the intestine. Stimulation of phospholipid metabolism in the mast-cell membrane leads to the generation of significant quantities of prostaglandin D_2 (PGD_2), leukotriene C_4 (LTC_4), and platelet-activating factor (PAF) (7, 8). The two former products are potent contractors of bronchial smooth muscle while PAF is associated with bronchial hyperreactivity. The actions of these mediators are not prevented by the blockade of histamine H_1-receptors.

Provocation of repeated allergic reactions stimulates the development of allergic inflammation. This is initiated by the generation of proinflammatory cytokines including IL-4, IL-5, IL-6, and TNFα from mast cells (9–11), T-lymphocytes (12, 13), and possibly basophils (14). These cytokines induce the local upregulation of selectins, adhesion proteins, and chemotactic factors which leads to an initial influx of neutrophils followed by a slower, and more prolonged, accumulation of eosinophils, the latter being characteristic of allergic inflammation (15, 16). The actions of the mediators generated by these inflammatory cells are now considered to be largely responsible for the structural and functional changes seen in allergic inflammation (16). Again, these events are not influenced to any great degree by the blockade of histamine H_1-receptors.

III. INHIBITION BY ANTIHISTAMINES OF MAST-CELL AND BASOPHIL MEDIATOR RELEASE

While a specific antagonist of the histamine H_1-receptor would theoretically prevent the actions of mast-cell- and basophil-derived histamine on its target organs, it would not prevent the actions of PGD_2, LTC_4, PAF, or cytokines. Thus, a drug that prevents the generation of all mast-cell and basophil mediators rather than one that merely prevents the actions of a single mediator has obvious advantages.

The ability of antihistamines to inhibit histamine release from mast cells has been known since the early studies of Arunlakshana in 1953 (17). However, the more discerning studies of Mota and Dias de Silva in 1960 (18) using guinea pig and rat mast cells showed clearly that, not only do antihistamines have the capacity to inhibit histamine release, but at higher concentrations they may also stimulate a drug-induced release of histamine. This phenomenon of apparent dual action intrigued us and stimulated us to look more closely at the interactions between antihistamines and mast cells.

For this series of studies (19), we selected older antihistamines, and drugs used primarily for other indications but which had antihistaminic actions (Table 1). To assess their effects, we used human lung fragments and performed two experiments, the first to assess the activity of the drugs against anti-IgE-induced histamine release, and the second to examine whether or not the drugs themselves could release histamine in the absence of immunological challenge. We reported several interesting findings. First, all the drugs that we tested inhibited histamine release at lower concentrations, but at higher concentrations they induced histamine release irrespective of the presence or absence of immunological challenge. An example of the dual action of promethazine is shown in Figure 2(a). Second, there was no correlation between the concentrations of drugs that inhibited histamine release and those which released histamine (Table 1). Third, there was no correlation between either of these concentrations and the concentrations at which the drugs were effective as antagonists of the H_1-receptor effects of histamine (antihistaminic effect). Fourth, mepyramine, which is much more hydrophilic than the other drugs examined, was markedly weaker both as an inhibitor of IgE-dependent histamine release and as a histamine-releasing agent (Fig. 2b). Thus, we can

Table 1 Comparison of the Ability of Antihistamines to Block Histamine H_1-Receptors, to Inhibit Anti-IgE-Induced Histamine Release, and to Cause Histamine Release in the Absence of Challenge

	Histamine H_1-receptor blocking activity (pA_2)	Inhibition of anti-IgE-induced histamine release ($-\log IC_{50}$)	Drug-induced histamine release ($-\log RC_{50}$)
Amitriptylline	8.29*	8.35	4.67
Chlorpheniramine	8.83	7.44	3.41
Chlorpromazine	8.14	6.92	5.96
Cinnarizine	7.43	5.73	4.42
Cyclizine	7.62	5.27	4.79
Diphenhydramine	7.59	5.84	4.13
Ketotifen	9.49	7.28	4.60
Mepyramine	8.97	3.07	<3.00
Oxatomide	8.43	8.35	4.67
Phenelzine	—	6.26	4.29
Promethazine	8.84	6.47	3.96
Trimeprazine	8.11*	7.72	5.99

Asterisks indicate $-\log IC_{50}$ values as estimates of pA_2.
Source: Data from Ref. 19.

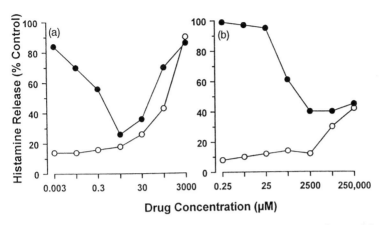

Figure 2 The effects of (a) promethazine and (b) mepyramine on histamine release from human lung fragments. The closed circles indicate the effects of the drugs against anti-IgE challenge and the open circles the effects of the drugs alone in the absence of challenge. (From Ref. 19.)

draw three conclusions from this study: (1) the abilities of an antihistamine to inhibit IgE-dependent histamine release and to release histamine in its own right are independent of any actions at the histamine H$_1$-receptor; (2) inhibition of histamine release and release of histamine probably occur by different mechanisms; (3) the property of lipophilicity is a factor in both the above effects on the mast cell but has little effect on the interaction of the H$_1$-receptor antagonists with the H$_1$-receptor.

Looking at the literature, we soon realized that our results with human lung fragments were not novel but merely confirmed and extended into human tissues previous similar experiments with rat mast cell using antihistamines, tricyclic antidepressants, neuroleptics, and local anesthetics (20–22). Perhaps the most meticulous studies were performed by Seeman (23, 24). He used a series of lipophilic cationic drugs with local anesthetic properties to advance the following mechanisms of mast-cell stabilization and labilization. All the drugs he tested are lipophilic and readily dissolved in the lipid that forms most of the cell membrane (Fig. 3). At low concentrations this had three effects. First, it caused a small expansion of the membrane, thus leading to its physical stabilization in a nonspecific manner. Second, the dissolving of the lipophilic end of the molecule in the cell membrane led to the presentation of a positive charge on the outside of the cell membrane, which competitively inhibited the binding of calcium

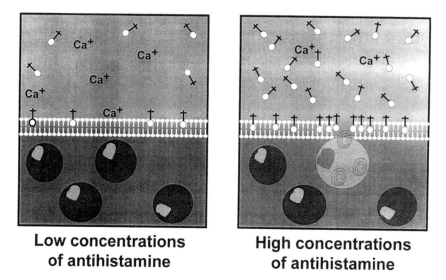

**Low concentrations   High concentrations
of antihistamine   of antihistamine**

Figure 3 Theoretical interactions of H_1-receptor antagonists with the mast-cell membrane. The lipophilic portions of H_1-antagonist molecules dissolve in the cell membrane while the cationic part, being charged and thus less lipophilic, remains on the outer surface of the membrane. This decreases the relative negative charge of the membrane, thus reducing its ability to attract and bind positively charged calcium ions. At high concentrations, the presence of the drug causes expansion and eventual instability of the membrane.

to the cell membrane (23, 25). Third, the inhibition of calcium binding led to a reduction of calcium transport through the cell membrane, thus reducing the activity of calcium-dependent enzymes, such as calmodulin (24). As the influx of calcium ions and the activation of calmodulin-dependent enzymes are crucial to propagation of the mast-cell-mediator release cascade, this would present a tenable hypothesis for a mechanism of action of antihistaminic drugs in stabilizing mast cells (26, 27).

Histamine release by antihistamines, Seeman postulates, has a slightly different mechanism (24) (Fig. 3). As the concentration of drug increases, more becomes dissolved in the membrane and greater and greater expansion of the cell membrane occurs. Eventually, at around 4% expansion, a critical state is reached when the membrane can no longer maintain its structural integrity and disruption occurs, with the consequential cytotoxic release of preformed mediators. Thus, histamine liberation by antihistamines would appear to reflect only their lipophilicity. Although these theories are now over 20 years old, they have not been superseded and many of the more recent studies would tend to support them rather than disprove them.

Moving to the newer antihistamines, in order to make one drug stand out above others there seems to be a quest to find properties additional to histamine H_1-antagonism, (e.g., to have mast-cell stabilizing activity). This has led to many studies being performed with a variety of histamine-releasing cells, such as rodent mast cells, human tissue mast cells, and peripheral blood basophils, and a variety of immunological stimuli, anti-IgE, and specific allergen, and nonimmunological stimuli, compound 48/80, substance P, concanavalin A, and calcium ionophore A23187. It has been found that histamine release from rat peritoneal mast cells, stimulated by immunological and nonimmunological stimuli, is inhibited by terfenadine (2–10 μM), astemizole (10–100 μM), and loratadine (12–13 μM), whereas only nonimmunological stimulation of histamine release was blocked by ketotifen (28–32). In contrast, ketotifen has been reported to inhibit both histamine and leukotriene release from human basophil leukocytes stimulated by both anti-IgE and low concentrations of calcium ionophore A23187 (33–35). With human lung mast cells, ketotifen has been reported to be an effective inhibitor of the release of leukotrienes but a poor inhibitor of histamine release (36). In contrast, azelastine has been reported to inhibit, by 50%, both histamine and leukotriene release from human lung mast cells at a concentration of around 3 μM (37–39). Astemizole, an effective inhibitor in rat mast cells, was reported in one study (40) to be an inconsistent inhibitor of antigen-induced histamine release from human lung mast cells, while loratadine was reported to reduce leukotriene, but not histamine, release (41). Thus, there seems to be a great deal of diversity and even confusion regarding the results of studies of antihistamines on histamine release.

In order to try to rationalize this situation, we examined the effects of three antihistamines, ketotifen, terfenadine, and cetirizine, on human dispersed lung, tonsillar and skin mast cells, all stimulated immunologically with anti-IgE (42). As outcome measures, we estimated both histamine and PGD_2 release, the former being a marker of preformed mediator release and the latter being a marker for newly generated mediator formation. The results showed variations both among drugs and among mast-cell sources. For example, terfenadine inhibited mediator release from all tissue mast cells in the 5- to 10-μM range, while it enhanced release above 10 μM in lung and skin mast cells only (Fig. 4). Ketotifen, on the other hand, was a weak inhibitor of histamine release and induced release only from skin mast cells. The third drug, cetirizine, caused some inhibition of both histamine and PGD_2 release from all tissues but did not induce

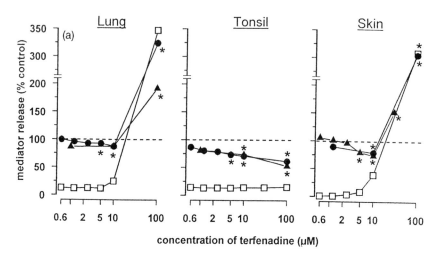

Figure 4   The effects of (a) terfenadine, (b) ketotifen, and (c) cetirizine on histamine and prostaglandin D_2 (PGD_2) release from human dispersed lung, tonsillar, and skin mast cells. The closed circles and triangles indicate the effects of the drugs against anti-IgE-induced histamine and PGD_2 release, respectively, and the open squares the ability of the drugs alone to release histamine in the absence of challenge. Asterisks indicate significant difference ($p < 0.05$) from control. (From Ref. 42.)

release at higher concentrations. Several conclusions may be drawn from this study. First, newly-generated mediator release is inhibited in parallel with histamine, thus making a mast-cell stabilizing property of an antihistamine a more attractive posulate. Second, the stimulation of PGD_2 generation at high concentrations suggests that the high dose effect is not merely cytotoxic, as previously hypothesized, but a stimulation of synthetic as well as secretory pathways. Third, mast cells derived from different tissues are likely to respond to antihistamines in quantitatively and, perhaps, qualitatively different ways. Fourth, and probably most important, the concentrations necessary to reduce mast-cell-mediator release are well above those found in the blood, and by inference in the tissues, during therapy in humans. This fourth point is common to all the studies reported previously in this review.

But can these in vitro studies be extrapolated into clinical efficacy? An old English phrase says "The proof of the pudding is in the eating." In more scientific terms, this means that the only way to find out whether or not antihistamines inhibit mast-cell-mediator release in clinical practice

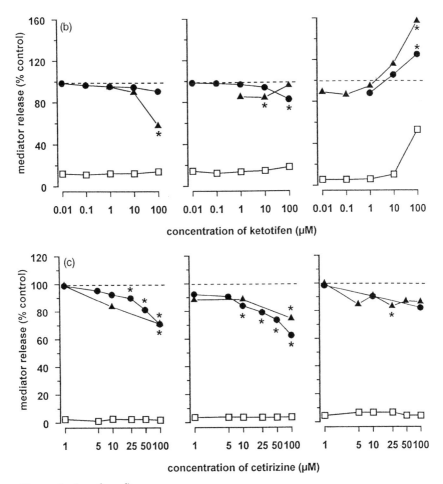

Figure 4 (continued)

is to test them in humans. An early study with astemizole in allergen provocation in asthma (43) showed that plasma histamine levels were unaffected. Although such a histamine measurement may only be considered to be indirect, it may be taken as circumstantial evidence that astemizole does not inhibit lung mast-cell histamine release in vivo. In subjects with allergic rhinitis, Naclerio and colleagues have performed a number of studies (44–46) in which they lavaged the nasal cavities following out-of-season allergen provocation. Summarizing their results, terfenadine and azatadine significantly inhibited histamine release while loratadine caused a small reduction of histamine release that was not statistically significant

(44–46). Furthermore, both terfenadine and loratadine also caused a small, but not statistically significant, reduction of PGD_2 generation (47). In contrast, cetirizine, ketotifen, azelastine, and diphenhydramine were without effect (46–50). In the skin, Hägermark and colleagues (51) found that clemastine and loratadine suppressed the flare response induced by compound 48/80 more effectively than a similar response induced by histamine. From these results they concluded that the drugs probably inhibited mast cell histamine release in addition to blocking the histamine H_1-receptor.

In contrast to these indirect assessments, we have recently undertaken a more direct assessment of antihistaminic activity in the skin using microdialysis (Malhotra et al., in preparation). Briefly, 216-μm-diameter 2-kDa microdialysis fibers were inserted into the upper layer of the dermis of the forearm in vivo and the skin was challenged by infusion of 10 mg/ml codeine to induce local mast cell mediator release. Assay of histamine in the outflow of the microdialysis tubes showed 7.6 $\pm$ 1.2 ng histamine to be released in the 20-min period following provocation with codeine in 10 subjects. Two hours after ingestion of 10 mg of either cetirizine or loratadine by the same subjects, the histamine release values were 6.3 $\pm$ 0.5 and 7.4 $\pm$ 1.1 ng/20-min observation period, respectively. These results show clearly that neither cetirizine nor loratadine, given in conventional doses, inhibited codeine-induced mast cell mediator release under these conditions.

To conclude this section on mast cell mediator release, it is clear that antihistamines have the potential to inhibit histamine release from mast cells and basophils. This effect is not mediated by an action on the H_1-receptor but by properties more closely related to lipophilicity and the ionic charge characteristics of the molecules. Thus, if pharmaceutical companies wish to pursue this aspect, perhaps development should be made in this direction at the expense of H_1-antagonism. With antihistamines currently available, it is our view that inhibition of mast cell mediator release is likely to be weak and inconsistent in clinical practice and, therefore, unlikely to contribute significantly to the efficacy of the drugs.

IV. EFFECTS OF ANTIHISTAMINES ON THE MIGRATION AND ACTIVATION OF INFLAMMATORY CELLS

As stated earlier, the accumulation and activation of inflammatory cells within the sites of allergic reactions are responsible for the chronic structural and functional changes that are associated with allergic inflammation. Thus, the ability of an antihistamine to influence these events would obviously extend its therapeutic potential. Many studies have been performed with a view to establishing an activity of antihistamines on inflammatory cells. As those who are familiar with the literature in this area will have already concluded, these studies are somewhat of a potpourri of experiments with little to relate one to another. Also, the failure to include a spectrum of drugs within many individual studies means that the reader cannot readily ascertain whether the action reported is unique to a particular antihistamine or whether it is a property of the entire drug group.

Studies on the effects of antihistamines on adhesion protein expression are relatively sparse. However, Ciprandi and colleagues (52) have demonstrated that the expression of intercellular adhesion molecule-1 (ICAM-1) (CD 54) on conjunctival epithelial cells in response to local allergen challenge in vivo is reduced by pretreatment of the subjects with cetirizine. The accumulation of inflammatory cells in the biopsies taken 6 h after challenge is reduced in parallel. The finding that loratadine also reduces inflammatory cell accumulation in this model (53) may indicate that this drug has a similar effect on adhesion protein expression, but no results on this were reported. In in vitro experiments, cetirizine has also been shown to down-regulate ICAM-1 expression in an epithelial cell line (G. W. Canonica, personal communication). In other studies, Sehmi and colleagues (54) reported that cetirizine reduces the adhesion of eosinophils to cultured human umbilical vein endothelial cells while Kyan and colleagues (55), using a similar system, found that only a high concentration of cetirizine (100 μg/ml) inhibited eosinophil adhesion stimulated by IL-1 or fMLP (formyl-methionyl-leucyl-phenylalanine). The adhesion of neutrophils to human umbilical vein endothelial cells was not inhibited even at this concentration of cetirizine. Yet another experiment designed to examine the effects of an antihistamine on cellular adhesion was conducted by Walsh and colleagues (56), who showed that cetirizine inhibited the adherence of human eosinophils to a plasma-coated glass plate with an IC_{50} of around 2×10^{-5} M.

In contrast, Gonzales and colleagues (57) reported that ketotifen at concentrations of up to 10 μg/ml failed to inhibit the adherence of neutrophils to glass. Whether these inhibitory actions of antihistamines are non-histamine receptor-mediated effects or whether they result from inhibition of H₁-receptor stimulation is not known, as histamine has been shown to augment the expression of ICAM-1 on epidermal keratinocytes induced by TNFα, an effect blocked by diphenhydramine (58). Furthermore, histamine has been shown to induce a profound increase in leukocyte rolling in perfused rat mesenteric blood vessels (59). Interestingly, this effect was enhanced by PAF. That the effect of histamine was abrogated by diphenhydramine and antibodies to P-selectin but not to CD 18 led these authors to speculate that the mechanism was an up-regulation of P-selectin expression by histamine. This observation would also explain the report by Johnson and colleagues (60) that tracheal eosinophilia in beagle dogs induced by inhalation of Ascaris antigen was reduced by astemizole, azelastine, cetirizine, and mepyramine. The positive effect obtained with mepyramine is particularly interesting because this compound is highly hydrophilic and would, therefore, be unlikely to interfere with biochemical processes occurring in the cell membrane. The possibility that antihistamines may modulate granulocyte-endothelial cell interactions would appear to be a fruitful area for further research.

Studies of inflammatory cell chemotaxis, an in vitro test designed to model some aspects of the ability of a cell to migrate toward a site of inflammation, are more numerous. An early study using human neutrophils (61) demonstrated that ketotifen strongly suppressed chemotaxis stimulated by calcium ionophore A23187, the lectin concanavalin A, and the tetrapeptide fMLP. However, the inhibitory effect against chemotaxis required higher drug concentrations than those necessary to inhibit superoxide radical production. In contrast, chemotaxis induced by opsonized zymosan was only weakly inhibitable. The ability of ketotifen to inhibit neutrophil chemotaxis and chemokinesis at concentrations of around 10 μg/ml was confirmed by Gonzales and colleagues (57). Another antihistamine, terfenadine, has also been shown to inhibit neutrophil chemotaxis induced by PAF and fMLP in vitro (62), but again concentrations higher than those expected to be found in clinical practice are required (63). Studies with azelastine on neutrophil chemotaxis are equivocal; one using chemotactic stimuli reported it to be without effect (64) while another, using conditioned medium from ethanol-stimulated hepatocytes as a model of alcoholic hepatitis found both azelastine and terfenadine to be inhibitory at concentrations as low as 0.01 μM (65). However, the observation that azelastine reduced the allergen-induced production of pleural exudate but not the accompanying neutrophilia in rats would support the view that inhibition of neutrophil migration may not occur in vivo. Finally, cetirizine has been reported by Van Epps and colleagues (66) to inhibit neutrophil chemotaxis in vitro, but again this effect required high concentrations, above 35 μg/ml, which led the authors to conclude that, with this drug also, the observation was unlikely to be of clinical relevance.

Perhaps the cell studied most with respect to chemotaxis is the eosinophil. Work on eosinophils has been stimulated partly by its importance in the pathophysiology of allergic disease and partly because of the clinical observations that cetirizine reduces eosinophil accumulation in the skin, using the skin window technique (67–71), in the nose (48), and in the airways (72). In an early study, Leprevost and colleagues (73) reported that PAF- and fMLP-induced eosinophil chemotaxis was reduced by cetirizine at therapeutic concentrations and that this property was not shared by polaramine. This was confirmed by Okada and colleagues (74), who showed that cetirizine inhibited PAF and fMLP-induced chemotaxis of human eosinophils with an IC_{50} of 0.1 μg/ml, within the therapeutic range.

Other antihistamines have also been tested for their ability to inhibit eosinophil chemotaxis. Ketotifen has been reported to reduce PAF-induced eosinophil chemotaxis in vitro at 10 μM, a lower concentration than that needed to suppress LTC_4 production (75), while in the rat in vivo, a dose of 2 mg/kg intraperitoneally reduced the pulmonary eosinophilia induced by intratracheal installation of IL-5 (76). Ketotifen has also been shown to reduce shape change, an early event preparing cells for migration, stimulated by IgG in human eosinophils (77). In the eosinophil cell EOL-1, it reduced PAF-induced actin polymerization, a suggested intracellular component of the chemotactic response (78), but in another study Numao and colleagues (79) reported that PAF-induced chemotaxis in human eosinophils was not blocked by either ketotifen or azelastine. In further experiments by Eda and colleagues, both loratadine (80) and terfenadine (63) have been shown to inhibit chemotaxis of human eosinophils induced by PAF at concentrations equivalent to or marginally above those that would be expected to be found in the blood after a single oral dose in humans. It is difficult to draw firm conclusions in the absence of comprehensive comparative trials using a representative spectrum of drugs but, with our present state of knowledge, we would tend to agree with Bernheim and colleagues (81), who concluded from their review of eosinophil chemotaxis studies in vitro and in vivo that cetirizine was inhibitory while other antihistamines were either less active or inactive.

The number of inflammatory cells present within the site of an allergic reaction is dependent on two main factors, the rate at which cells migrate into the area and the rate at which they die and are removed from the area. Dying inflammatory cells do not merely disintegrate and release their potentially harmful granule-associated mediators into the local environment. Evolution has provided a mechanism of programmed cell death, or apoptosis, whereby aging cells undergo a number of degenerative changes, including the expression of specific cell surface proteins that encourage their phagocytosis by macrophages. Apoptosis is delayed and survival increased in eosinophils by the cytokine IL-5 (82) and in mast cells by stromal cell-derived stem cell factor (SCF) (83). In contrast, corticosteroids have been shown to promote apoptosis in animal experiments in vivo (84), an action that is considered to be important in reducing leukocyte numbers at sites of inflammation. Recently, it has been demonstrated in vitro that ketotifen and theophylline may also decrease the viability of eosinophils in the presence of IL-5, although the concentrations necessary to produce this effect, 10^{-4} and 10^{-3} μM, respectively, are high (85).

As the secretion of cytokines from lymphocytes, particularly the TH-2 subset of lymphocytes, appears to be central to the establishment and maintenance of allergic inflammation (12, 13, 86), it would seem pertinent to examine the effects of antihistamines on these cells. An early experiment by Gushchin and colleagues (87) suggested that ketotifen, at concentrations of up to 50 μM, enhanced phytohemaglutinin (PHA)-stimulated lymphocyte proliferation, and only extremely high concentrations of the drug inhibited the response. In 1990, Todoroki and colleagues (88) found that azelastine had a weak suppressive effect on the induction of IL-2 responsiveness by *D. farinae* antigen in peripheral blood mononuclear cells from patients with asthma. From experiments that showed the inhibitory effect to be on adherent antigen-presenting cells rather than on lymphocytes themselves, they concluded that azelastine may have some suppressive effects on antigen processing and/or presentation. A study by Kondo and colleagues (89) on the antigen-induced proliferative response of peripheral blood mononuclear cells from children with atopic dermatitis showed an inhibitory effect of ketotifen. Although the authors suggest a direct effect on the lymphocytes, the observation that proliferation induced by PHA or tetanus toxoid was not inhibited indicates a possible effect of ketotifen on antigen presentation. This interpretation would also be consistent with the observation of Canonica and colleagues (90) that proliferation of peripheral blood mononuclear cells to PHA or antibodies to CD2, CD3, or CD28 was not prevented by therapeutic concentrations of cetirizine. Furthermore, this study showed no effect of cetirizine on the induction of ICAM-1, HLA-DR, or CD25 expression or the presence of α-1-acid glycoprotein on the lymphocyte membrane.

One positive study in this area was that conducted by Brodde and colleagues (91). They reported that the down-regulation of lymphocyte β_2-receptors (stimulation of which reduces many aspects of lymphocyte function) by exposure to terbutaline in vitro could be prevented by ketotifen. By assessment of cardiovascular parameters in human volunteers, these workers also showed that ketotifen markedly blunted terbutaline-induced down-regulation of β_2-adrenoceptors in vivo. This ability of antihistamines to prevent the down-regulation of adrenoceptors is supported by the observations that β-receptor numbers, assessed by ligand binding in lung membrane preparations, were reduced by administration of terbutaline to guinea pigs for 7 days, and that azelastine prevented this (92).

The potential of antihistamines to inhibit indices of inflammatory cell activation has been studied widely. These include the de novo generation, by membrane-associated enzymes, of proinflammatory products such as superoxide radicals (O_2^-) and the arachidonic acid products LTB_4 and LTC_4, and the release of granule-associated products, such as neutrophil elastase and eosinophil cationic protein (ECP).

An early study performed on human neutrophils (61) demonstrated that superoxide radical production stimulated by calcium ionophore A23187, concanavalin A, or fMLP was strongly suppressed by ketotifen and that this effect required lower drug concentrations than did inhibition of chemotaxis. However, superoxide radical production induced by opsonized zymosan was only weakly inhibitable. In contrast, Van Epps and co-workers (66) reported that cetirizine inhibited neutrophil superoxide radical production only at concentrations above 35 $\mu g/ml$, higher than those required for suppression of chemotaxis. Since then, many reports have confirmed the inhibitory effects of antihistamines on neutrophil superoxide radical generation using ketotifen (93–97), azelastine (64, 94–96, 98–106), oxatomide (95, 96, 104, 106) and clemastine (98, 99). Furthermore, inhibition of oxygen-free radical production has been reported for ketotifen in human alveolar macrophages (107) and for azelastine (103), oxatomide (108), loratadine (80), and cetirizine (74) in human eosinophils. Interestingly, this last study demonstrated a marked difference between eosinophils obtained from allergic and nonallergic donors, the former being sensitive to low concentrations of cetirizine (0.02–2 μM) while the latter were insensitive to the effects of the drug in this concentration range. In contrast, in other studies little effect on superoxide generation was seen with azelastine in human alveolar macrophages (109, 110) and in guinea pig eosinophils (111). The reasons for these negative results are not immediately clear, but it must be recognized that experimental conditions and differences in provoking stimuli may have a great influence on the observed result.

Stimulation of inflammatory leukocytes initiates the activation of membrane-associated phospholipases, which leads to the liberation of arachidonic acid. This is the rate-limiting step in the generation of LTB_4, a potent chemoattractant for granulocytes, and LTC_4, a potent bronchoconstrictor agent that also up-regulates the expression of adhesion proteins. Manabe and colleagues (37) reported that oxatomide inhibits LTC_4 generation from human granulocytes, the effect on eosinophils being more pronounced than the effect on neutrophils, while Nabe and co-workers showed inhibition of calcium ionophore A23187-induced LTC_4 release from human eosinophils to be inhibited by both terfenadine (112) and ketotifen (75). Using human neutrophils stimulated by A23187, Tanaguchi and colleagues have reported that azelastine and oxatomide block the liberation of arachidonic acid and the formation of LTB_4 (104), while azelastine, oxatomide, and diphenhydramine are effective inhibitors of $LTC_4/LTD_4/LTE_4$ generation

(113). One feature common to all these studies is that drug concentrations in excess of 10 μM were required for effective activity, again raising the question of clinical relevance.

Studies on the ability of antihistamines to suppress the granulocyte exocytosis and the liberation of granule-associated mediators have proved to be less fruitful. Van Epps and co-workers (66) reported that concentrations above 35 μg/ml of cetirizine were necessary to inhibit the release of neutrophil lysosomal enzymes, while Werner and colleagues (114) found that concentrations of 10–100 μM of azelastine, astemizole, and oxatomide were required to prevent fMLP-induced release of neutrophil elastase. Ketotifen and mepyramine were without effect in these experiments. Also using human neutrophils, Renesto and colleagues (115) found that concentrations of 100 μM of azelastine were necessary to inhibit cathepsin G-stimulated β-glucuronidase release. In experiments using human eosinophils, both loratadine (80) and terfenadine (63) were shown to be ineffective against PAF-induced eosinophil cationic protein (ECP) release at concentrations that markedly reduced chemotaxis and superoxide radical generation. Thus antihistamines would appear to be ineffective at inhibiting the release of preformed granulocyte mediators at clinically relevant concentrations.

Experiments on platelets have produced mixed results. Wang and co-workers (116), using rabbit platelets, reported that extremely low concentrations of ketotifen (IC_{50} ~ 33 pM) inhibited PAF-induced platelet aggregation while higher concentrations were required to inhibit aggregation induced by ADP and arachidonic acid (IC_{50} values of ~ 95 and 143 μM, respectively). The authors found the result with PAF to be particularly encouraging. However, this optimism was not fulfilled in the study by Chan and colleagues (117) who reported an IC_{50} value of around 250 μM for the inhibition by ketotifen of PAF-induced aggregation of human platelets. The observation by these workers that lignocaine and propranolol were similarly effective suggested that this may be a physical effect of the drugs at the platelet membrane. Finally, the report by Renesto and co-workers (115) that a concentration of 100 μM of azelastine was required to inhibit neutrophil-stimulated aggregation of human platelets would reinforce the view that inhibition by antihistamines of platelet aggregation is likely to be an in vitro phenomenon.

Indirect methods of assessing the effect of antihistamines on granulocyte function have also been used. In 1989, Chihara and colleagues (118) reported that the cytotoxic effects of eosinophils and the eosinophil cell line, EOL-1, against bronchial epithelial cells were reduced by oxatomide. Two years later, Walsh and co-workers (56) reported that cetirizine inhibited PAF-induced enhancement of C3b- and IgG-dependent neutrophil and eosinophil rosette formation with an IC_{50} of 2×10^{-5} M. There was a comparable inhibition of PAF-dependent enhancement of eosinophil cytotoxicity for complement-coated schistosomula of *Schistosoma mansoni*. The relative successfulness of the antihistamines in these experiments suggests that the cytotoxicity used as an indicator of activity was more likely to result from the de novo synthesis of cell-membrane-derived products than from the liberation of granule-associated proteins.

V. INVESTIGATIONS INTO THE MECHANISMS OF ACTION OF ANTIHISTAMINES ON INFLAMMATORY CELLS

Many studies have been performed with the aim of elucidating the mechanisms by which antihistamines may modulate the migration and activation of inflammatory cells. These may be arbitrarily divided into three areas, the way in which drugs associate with the cell membrane, the influence of drugs on calcium mobilization, and the inhibition of membrane-associated enzymes. To investigate the way in which drugs associate with the cell membrane, Takanaka and colleagues (98–100, 113) compared the ability of ketotifen, azelastine, clemastine, diphenhydramine, chlorpheniramine, polaramine, indomethacin, and procaine to prevent fMLP-induced oxygen-free radical production and arachidonic acid liberation from human neutrophils. Their findings that only azelastine and clemastine ($IC_{50} \sim 20$ μM) and ketotifen ($IC_{50} \sim 50$ μM) inhibited these responses and that changes in membrane potential were also reduced at these concentrations led these authors to the conclusion that these cationic drugs interacted with the membrane to inhibit membrane-associated enzymes or receptors.

Further studies by these workers (104, 119) showed that the acidic compound dodecylbenzenesulphonic acid (DBS) could displace antihistamines from the membrane, thereby reversing their inhibitory effects. This would indicate that the binding of antihistamines to membrane elements was ionic and reversible and would not be consistent with interactions with specific cell membrane receptors or binding sites. Thus, differences in the overall cationic charge of drugs, in the charge distribution within drugs, and the lipophilicity of drugs will affect their potency and efficacy as will the variable nature of cell membrane charge characteristics. Obviously the binding of antihistamines to specific histamine H₁-receptors is an exception to this rule, but we know that the inhibitory effects described above do not result from H₁-receptor blockade.

As mentioned earlier in the section on mast cells and basophils, the association of a positively charged lipophilic drug with the cell membrane will potentially inhibit, in a competitive manner, the binding of calcium to that membrane (23, 25), thereby reducing the activity of calcium-dependent enzymes, such as calmodulin (24). The possibility that antihistamines may affect inflammatory cell activation by influencing calcium mobilization has been considered. Nakamura and colleagues (120) reported that azelastine reduced the changes in intracellular calcium levels induced in guinea pig peritoneal macrophages by PAF or fMLP in the same concentration range as would be expected for inhibition of superoxide radical generation. Also, Kakuta and co-workers (107) reported that ketotifen reduces phorbol myristate acetate (PMA)-induced calcium-activated potassium conductance at the same concentrations at which it inhibits oxygen-free radical production. A quite different effect of antihistamines was proposed by Letari and colleagues (121), who found that loratadine and terfenadine increased the resting levels of intracellular calcium in rat macrophages and human platelets. They hypothesized that the discharge of intracellular calcium stores by the drugs prevents rises in intracellular calcium induced by physiological activators such as PAF and ADP. In contrast, Subramanian (111), using a series of pharmacological inhibitors and antagonists in guinea pig eosinophils, concluded that there was no functional association between LTB_4-induced calcium ion fluxes and oxygen-free radical production. Finally, Senn and colleagues (122), using rabbit cultured airway smooth muscle cells rather than inflammatory cells, demonstrated that ketotifen prevented the endothelin-1-induced rapid increase in cytosolic free calcium levels without changing resting calcium levels or inhibiting agonist-induced calcium fluxes. However, as it is known that endothelin-1-induced contraction of airways smooth muscle is mediated, at least in part, indirectly through the release of histamine and other inflammatory mediators (123–125), a histamine H_1-receptor-mediated effect cannot be ruled out in these experiments. Thus, based on the evidence accumulated so far, no clear relationship between the effects of antihistamines on calcium fluxes and the modulation of inflammatory cell or smooth muscle function has been established.

The final aspect of antihistamine activity to be considered is their influence on the activity of membrane-associated enzymes. Yoshikawa and colleagues (94), using a cell-free electron spin resonance assay, showed that azelastine and ketotifen did not inhibit oxygen-free radical production by the xanthine oxidase system while they did reduce production in whole neutrophils induced by PMA. Schmidt and colleagues (102), from studies with pharmacological inhibitors in human neutrophils and guinea pig alveolar macrophages, suggested that azelastine inhibits oxygen-free radical production by an action on protein kinase C, an enzyme dependent on the presence of free intracellular calcium for expression of its activity. Umeki (95, 106) showed that the inhibition by ketotifen, azelastine, and oxatomide of superoxide generation by human neutrophils exposed to PMA in a whole-cell system, and the activation of the cell-membrane-associated superoxide-generating enzyme, NADPH oxidase, by sodium dodecyl sulfate (SDS) in a cell-free system occurred at similar concentrations of the drugs. The studies of Hojo and colleagues (96), using the same drugs, were in general agreement with this finding. We can conclude, therefore, that inhibition of superoxide radical production by antihistamines is most likely to reflect an inhibition of NADPH oxidase. However, the relatively high drug concentrations, in the micromolar range, required to achieve this inhibition suggest that this, too, may not be clinically relevant.

VI. SUMMARY

In this review of H₁-receptor antagonists, we have highlighted several potentially important points. First, from an analysis of the basic structure of antihistaminic drugs we can predict that they may exert many biochemical effects that are completely unrelated to their ability to antagonize the interaction of histamine with the H₁-receptor. Second, the cationic amphiphilic nature of many antihistamines means that they may readily form an ionic association with cell membranes where they may discourage calcium binding and inhibit membrane-associated enzymes. Third, antihistamines, at least in vitro, are able to affect the function of most inflammatory cells. Fourth, the observations that antihistamines in general produce a similar spectrum of non-H₁-receptor-mediated inhibitory effects and yet have widely differing structures would suggest these effects to be relatively nonspecific. Fifth, with the exception of some specific examples, such as the effect of cetirizine on eosinophil accumulation, the non-H₁-receptor-mediated effects of antihistamines require higher concentrations than would be expected to occur in clinical practice. This conclusion, of course, carries the caveat that antihistamines are not preferentially concentrated in inflammatory cells in vivo, although there is no evidence to suggest that they are. Finally, given the wide spectrum of potential biological effects of the basic molecule from which all antihistamines are derived, we find it somewhat surprising that no pharmaceutical company has undertaken research into optimizing the molecule for potent inhibition of inflammatory cell activation rather than trying to demonstrate this activity in drugs specifically designed for their antagonism of histamine H₁-receptors.

ACKNOWLEDGMENTS

ADC is supported by a research studentship provided by Janssen Pharmaceuticals, UK.

REFERENCES

1. Bovet D, Staub A. Action protectrice des ethers phenoliques au cours de l'intoxication histaminique. CRS Soc Biol (Paris) 1937;124:527–49.
2. Meares RA, Mills JE, Horvath TB, Atkinson JM, Pun LQ, Rand MJ. Amitriptylline and asthma. Med J Aust 1971;2:25–8.
3. Saviuc P, Danel V, Dixmerias F. Prolonged QT interval and torsade de pointes following astemizole overdose. Clin Toxicol 1993;31:121–5.
4. Pohjola-Sintonen S, Viitasalo M, Toivonen L, Neuvonen P. Itraconazole prevents terfenadine metabolism and increases risk of torsades de pointes ventricular tachycardia. Eur J Clin Pharmacol 1993;45:191–3.
5. Lang DG, Wang CM, Wenger TL. Terfenadine alters action potentials in isolated canine Purkinje fibers more than acrivastine. J Cardiovasc Pharmacol 1993;22:438–42.
6. Honig PK, Wortham DC, Hull R, Zamani K, Smith JE, Cantilena LR. Itraconazole affects single-dose terfenadine pharmacokinetics and cardiac repolarization pharmacodynamics. J Clin Pharmacol 1993;33:1201–6.
7. Fox CC, Kagey-Sobotka A, Schleimer RP, Peters SP, MacGlashan DW, Lichtenstein LM. Mediator release from human basophils and mast cells from the lung and intestinal mucosa. Int Arch Allergy Appl Immunol 1985; 77:130–6.
8. Schleimer RP, MacGlashan DW, Peters SP, Pinckard RN, Adkinson NF, Lichtenstein LM. Characterization of inflammatory mediator release from purified human lung mast cells. Am Rev Respir Dis 1986;133:614–7.
9. Bradding P, Feather IH, Howarth PH, Mueller R, Roberts JA, Britten K, Bews JPA, Hunt TC, Okayama Y, Heusser CH, Bullock GR, Church MK, Holgate ST. Interleukin-4 is localized to and released by human mast cells. J Exp Med 1992;176:1381–6.
10. Church MK, Okayama Y, Bradding P. The role of the mast cell in acute and chronic allergic inflammation. In: Chignard M, Pretolani M, Renesto P, Vargaftig BB, eds. Cells and Cytokines in Lung Inflammation. New York Academy of Science, 1994, pp. 13–21.
11. Okayama Y, Semper A, Holgate ST, Church MK. Multiple cytokine mRNA in human mast cells stimulated via Fc∈R1. Int Arch Allergy Immunol 1995; 107:29–33.
12. Durham SR, Ying S, Varney VA, Jacobson MR, Sudderick RM, Mackay IS, Kay AB, Hamid QA. Cytokine messenger RNA expression for IL-3, IL-4, IL-5, and granulocyte/macrophage-colony-stimulating factor in the nasal mucosa after local allergen provocation: Relationship to tissue eosinophilia. J Immunol 1992;148:2390–4.

13. Robinson D, Hamid Q, Bentley A, Ying S, Kay AB, Durham SR. Activation of CD4+ T cells, increased T_{H2}-type cytokine mRNA expression, and eosinophil recruitment in bronchoalveolar lavage after allergen inhalation challenge in patients with atopic asthma. J Allergy Clin Immunol 1993;92: 313–24.

14. Gauchat J-F, Henchoz S, Mazzei G, Aubry J-P, Brunner T, Blasey H, Life P, Talabot D, Flores-Romo L, Thompson J, Kishi K, Butterfield J, Dahinden CA, Bonnefoy J-Y. Induction of human IgE synthesis in B cells by mast cells and basophils. Nature 1993;365:340–3.

15. Montefort S, Gratziou C, Goulding D, Polosa R, Haskard DO, Howarth PH, Holgate ST, Carroll MP. Bronchial biopsy evidence for leukocyte infiltration and upregulation of leukocyte endothelial cell adhesion molecules 6 hours after local allergen challenge of sensitized asthmatic airways. J Clin Invest 1994;93:1411–21.

16. Djukanovic R, Roche WR, Wilson JW, Beasley CRW, Twentyman OP, Howarth PH, Holgate ST. Mucosal inflammation in asthma. Am Rev Respir Dis 1990;142:434–57.

17. Arunlakshana O, Schild HO. Histamine release by antihistamines. J Physiol (Lond) 1953;119:47P–8P.

18. Mota I, Da Silva WD. The anti-anaphylactic and histamine releasing properties of the antihistamines: their effect on the mast cells. Br J Pharmacol 1960;15:396–404.

19. Church MK, Gradidge CF. Inhibition of histamine release from human lung in vitro by antihistamines and related drugs. Br J Pharmacol 1980;69:663–7.

20. Gushchin IS, Deryugin IL, Kaminka ME. Histamine liberating action of antihistamines on isolated rat mast cells. Bull Exp Biol Med 1978;85:352–5.

21. Frisk-Holmberg M, van der Kleijn E. The relationship between the lipophilic nature of tricyclic neuroleptics and anti-depressants and histamine release. Eur J Pharmacol 1972;18:139–47.

22. Kazimierczak W, Peret M, Maslinski C. The action of local anaesthetics on histamine release. Biochem Pharmacol 1976;25:1747–50.

23. Kwant WO, Seeman P. The displacement of membrane calcium by a local anaesthetic (chlorpromazine). Biochim Biophys Acta 1969;193:338–49.

24. Seeman P. The membrane actions of anaesthetics and tranquillizers. Pharmacol Rev 1972;24:583–655.

25. Lullmann H, Plosch H, Ziegler A. Calcium replacement by cationic amphiphillic drugs from lipid monolayers. Biochem Pharmacol 1980;29:2969–74.

26. Peachell PT, Pearce FL. Effect of calmodulin inhibitors on histamine secretion from mast cells. Agents Actions 1985;16:43–4.

27. Peachell PT, Pearce FL. Divalent cation dependence of the inhibition by phenothiazines of mediator release from mast cells. Br J Pharmacol 1989; 97:547–55.

28. Akagi M, Mio M, Tasaka K, Kinitra S. Mechanisms of histamine release inhibition induced by azelastine. Pharmacokinetics 1983;26:191–8.

29. Kreutner W, Chapman RW, Gulbenkian A, Siegel MI. Antiallergic activity of loratadine, a nonsedating antihistamine. Allergy 1987;42:57–63.

30. De Clerck F, Van Reempts J, Borgers M. Comparative effects of oxatomide on the release of histamine from rat peritoneal mast cells. Agents Actions 1981;11:184–92.

31. Chand N, Pillar J, Diamantis W, Sofia RD. Inhibition of allergic histamine release by azelastine and selected antiallergic drugs from rabbit leukocytes. Int Arch Allergy Appl Immunol 1985;77:451–5.

32. Hachisuka H, Nomura H, Sakamoto F, Mori O, Okubo K, Sasai Y. Effect of antianaphylactic agents on substance-P induced histamine release from rat peritoneal mast cells. Arch Dermatol Res 1988;280:158–62.

33. Tomioka H, Yoshida S, Tanaka M, Kumagai A. Inhibition of chemical mediator release from human leukocytes by a new antiasthma drug HC20-511 (ketotifen). Monogr Allergy 1979;14:313–7.

34. Radermecker M. Inhibition of allergen-mediated histamine release from human cells by ketotifen and oxatomide: comparison with other H_1 antihistamines. Respiration 1981;41:45–55.

35. Wilhelms OH. Inhibition profiles of picumast and ketotifen on the in vitro release of prostanoids, slow-reacting substance of anaphylaxis, histamine and enzyme from human leukocytes and rat alveolar macrophages. Int Arch Allergy Appl Immunol 1987;82:547–9.

36. Ross WJ, Harrison RG, Jolley MR, Neville MC, Todd A, Verge JP, Dawson W, Sweatman WJ. Antianaphylactic agents 1. 2-(Acylamino) oxazoles. J Med Chem 1979;22:412–7.

37. Manabe H, Ohmori K, Tomioka H, Yoshida S. Oxatomide inhibits the release of chemical mediators from human lung tissues and from granulocytes. Int Arch Allergy Appl Immunol 1988;87:91–7.

38. Josephs LK, Gregg I, Mullee MA, Holgate ST. Non-specific bronchial reactivity and its relationship to the clinical expression of asthma: a longitudinal study. Am Rev Respir Dis 1989;140:350–7.

39. Schmutzler W, Delmich K, Eichelberg D, Gluck S, Greven T, Jurgensen H, Riesener KP. The human adenoidal mast cell. Susceptibility to different secretagogues and secretion inhibitors. Int Arch Allergy Appl Immunol 1985;77:177–8.

40. Awouters FHL, Niemegeers CJE, Janssen PAJ. Pharmacology of the specific histamine H_1-antagonist astemizole. Arzneimittelforschung 1983;33:381–8.

41. Temple DM, McCluskey M. Loratadine, an antihistamine, blocks antigen- and ionophore-induced leukotriene release from human lung in vitro. Prostaglandins 1988;35:549–54.

42. Okayama Y, Church MK. Comparison of the modulatory effect of ketotifen, sodium cromoglycate, procaterol and salbutamol in human skin, lung and tonsil mast cells. Int Arch Allergy Appl Immunol 1992;97:216–25.

43. Holgate ST, Howarth PH. Astemizole, an H_1-antagonist in allergic asthma (Abstract). J Allergy Clin Immunol 1985;75(Suppl):166.

44. Naclerio RM, Kagey-Sobotka A, Lichtenstein LM, Freidhoff L, Proud D. Terfenadine, an H_1 antihistamine, inhibits histamine release in vivo in the human. Am Rev Respir Dis 1990;142:167–71.

45. Naclerio RM. The effect of antihistamines on the immediate allergic response: A comparative review. Otolaryngol Head Neck Surg 1993;108: 723–30.
46. Naclerio RM. Inhibition of mediator release during the early reaction to antigen. J Allergy Clin Immunol 1992;90:(Suppl.)715–9.
47. Naclerio RM. Effects of antihistamines on inflammatory mediators. Ann Allergy 1993;71:292–5.
48. Naclerio RM. Additional properties of cetirizine, a new H₁-antagonist. Allergy Proc 1991;12:187–91.
49. Majchel AM, Proud D, Kagey-Sobotka A, Lichtenstein LM, Naclerio RM. Ketotifen reduces sneezing but not histamine release following nasal challenge with antigen. Clin Exp Allergy 1990;20:701–5.
50. Naclerio RM, Proud D, Kagey-Sobotka A, Freidhoff L, Norman PS, Lichtenstein LM. The effect of cetirizine on early allergic response. Laryngoscope 1989;99:596–9.
51. Hägermark O, Wahlgren C-F, Giös I. Inhibitory effect of loratadine and clemastine on histamine release in human skin. Skin Pharmacol 1992;5: 93–8.
52. Ciprandi G, Buscaglia S, Pronzato C, Ricca V, Pesca GP, Villaggio B, Fiorino N, Albano M, Scordamaglia M, Bagnasco M, Canonica GW. New targets for antiallergic drugs. In: Langer SZ, Church MK, Vargaftig BB, Nicosia S, eds. New Developments in the Therapy of Allergic Disorders and Asthma. Basel: Karger, 1993, pp. 115–27.
53. Ciprandi G, Buscaglia S, Marchesi E, Danzig M, Cuss F, Canonica G-W. Protective effect of loratadine on late phase reaction induced by conjunctival provocation test. Int Arch Allergy Immunol 1993;100:185–9.
54. Sehmi R, Walsh GM, Hartnell A, Barkans J, North J, Kay AB, Moqbel R. Modulation of human eosinophil chemotaxis and adhesion by anti-allergic drugs in vitro. Pediatr Allergy Immunol 1993;4:13–8.
55. Kyan Aung U, Hallsworth M, Haskard D, De Vos C, Lee TH. The effects of cetirizine on the adhesion of human eosinophils and neutrophils to cultured human umbilical vein endothelial cells. J Allergy Clin Immunol 1992;90: 270–2.
56. Walsh GM, Moqbel R, Hartnell A, Kay AB. Effects of cetirizine on human eosinophil and neutrophil activation in vitro. Int Arch Allergy Appl Immunol 1991;95:158–62.
57. Gonzalez JA, Lorente F, Romo A, Muriel M, Palomero B, Salazar V. Action of ketotifen on different functions of neutrophil polymorphonuclear cells. Allergol Immunopathol Madr 1986;14:215–20.
58. Mitra RS, Shimizu Y, Nickoloff BJ. Histamine and cis-urocanic acid augment tumor necrosis factor-alpha-mediated induction of keratinocyte intercellular adhesion molecule-1 expression. J Cell Physiol 1993;156:348–57.
59. Kubes P, Kanwar S. Histamine induces leukocyte rolling in post-capillary venules—A P-selectin-mediated event. J Immunol 1994;152:3570–7.
60. Johnson HG, McNee ML, Nugent RA. Canine in vivo tracheal chemotaxis of eosinophils to antigen in sensitized dogs: inhibition by a steroid, a sys-

temic lazaroid U-78517F, and several topical H_1 antihistamines. Am Rev Respir Dis 1992;146:621–5.

61. Kato T, Terui T, Tagami H. Effects of HC 20-511 (ketotifen) on chemiluminescence of human neutrophils. Inflammation 1985;9:45–51.

62. Okada C, Hopp RJ, Miyagawa H, Sugiyama H, Nair NM, Bewtra AK, Townley RG. Effect of terfenadine on neutrophil and eosinophil chemotactic activities after inhalation of platelet-activating factor in vivo and on neutrophil chemotaxis in vitro. Int Arch Allergy Immunol 1992;97:181–6.

63. Eda R, Townley RG, Hopp RJ. Effect of terfenadine on human eosinophil and neutrophil chemotactic response and generation of superoxide. Ann Allergy 1994;73:154–60.

64. Akamatsu H, Miyachi Y, Asada Y, Niwa Y. Effects of azelastine on neutrophil chemotaxis, phagocytosis and oxygen radical generation. Jpn J Pharmacol 1991;57:583–9.

65. Shiratori Y, Takada H, Hai K, Kiriyama H, Mawet E, Komatsu Y, Niwa Y, Matsumura M, Shiina S, Kawase T, et al. Effect of anti-allergic agents on chemotaxis of neutrophils by stimulation of chemotactic factor released from hepatocytes exposed to ethanol. Dig Dis Sci 1994;39:1569–75.

66. Van Epps DE, Kutvirt SG, Potter JW. In vitro effects of cetirizine and histamine on human neutrophil function. Ann Allergy 1987;59:13–9.

67. Fadel R, Herpin-Richard N, Rihoux J-P, Henocq E. Inhibitory effect of cetirizine 2HC1 on eosinophil migration in vivo. Clin Allergy 1987;17:373–9.

68. Michel L, De Vos C, Rihoux J-P, Burtin C, Benveniste J, Dubertret L. Inhibitory effect of oral cetirizine on in vivo antigen induced histamine and PAF-acether release and eosinophil recruitment in human skin. J Allergy Clin Immunol 1988;82:101–9.

69. De Vos C, Joseph M, Leprevost C, Vorng H, Tomassini M, Capron M, Capron A. Inhibition of human eosinophil chemotaxis and of the IgE-dependent stimulation of human blood platelets by cetirizine. Int Arch Allergy Appl Immunol 1989;88:212–5.

70. Rihoux JP. The inhibiting effect of cetirizine 2 HC1 on eosinophil migration and its link to H_1 blockade. Agents Actions Suppl 1991;33:409–15.

71. Rihoux JP, Fadel R, Juhlin L. Platelet-activating factor-induced immediate and late cutaneous reactions. Int Arch Allergy Appl Immunol 1991;94: 299–300.

72. Redier H, Chanez P, De Vos C, Rifai N, Clauzel AM, Michel FB, Godard P. Inhibitory effect of cetirizine on the bronchial eosinophil recruitment induced by allergen inhalation challenge in allergic patients with asthma. J Allergy Clin Immunol 1992;90:215–24.

73. Leprevost C, Capron M, De Vos C, Tomassini M, Capron A. Inhibition of eosinophil chemotaxis by a new antiallergic compound (cetirizine). Int Arch Allergy Appl Immunol 1988;87:9–13.

74. Asako H, Kurose I, Wolf R, DeFrees S, Zheng Z-L, Phillips ML, Paulson JC, Granger DN. Role of H_1-receptors and P-selectin in histamine-induced leukocyte rolling and adhesion in postcapillary venules. J Clin Invest 1994; 93:1508–15.

75. Nabe M, Miyagawa H, Agrawal DK, Sugiyama H, Townley RG. The effect of ketotifen on eosinophils as measured on LTC4 release and by chemotaxis. Allergy Proc 1991;12:267–71.

76. Iwama T, Nagai H, Suda H, Tsuruoka N, Koda A. Effect of murine recombinant interleukin-5 on the cell population in guinea-pig airways. Br J Pharmacol 1992;105:19–22.

77. Kishimoto T, Sato T, Ono T, Takahashi K, Kimura I. Effect of ketotifen on the reactivity of eosinophils with the incubation of anti-IgG. Br J Clin Pract 1990;44:226–30.

78. Morita M, Tsuruta S, Mori KJ, Mayumi M, Mikawa H. Ketotifen inhibits PAF-induced actin polymerization in a human eosinophilic leukaemia cell line, EoL-1. Eur Respir J 1990;3:1173–8.

79. Numao T, Fukuda T, Akutsu I, Makino S. Effects of anti-asthmatic drugs on human eosinophil chemotaxis. Nippon Kyobu Shikkan Gakkai Zasshi 1991;29:65–71.

80. Eda R, Sugiyama H, Hopp RJ, Bewtra AK, Townley RG. Effect of loratadine on human eosinophil function in vitro. Ann Allergy 1993;71:373–8.

81. Bernheim J, Arendt C, De Vos C. Cetirizine: more than an antihistamine? Agents Actions Suppl 1991;34:269–93.

82. Stern M, Meagher L, Savill J, Haslett C. Apoptosis in human eosinophils: Programmed cell death in the eosinophil leads to phagocytosis by macrophages and is modulated by IL-5. J Immunol 1992;148:3543–9.

83. Iemura A, Tsai M, Ando A, Wershil BK, Galli SJ. The C-kit ligand, stem cell factor, promotes mast cell survival by suppressing apoptosis. Am J Pathol 1994;144:321–8.

84. Soda K, Kawabori S, Kanai N, Bienenstock J, Perdue MH. Steroid-induced depletion of mucosal mast cells and eosinophils in intestine of athymic nude rats. Int Arch Allergy Immunol 1993;101:39–46.

85. Hossain M, Okubo Y, Sekiguchi M. Effects of various drugs (staurosporine, herbimycin A, ketotifen, theophylline, FK506 and cyclosporin A) on eosinophil viability. Arerugi 1994;43:711–7.

86. Kay AB. Lymphocytes-T and their products in atopic allergy and asthma. Int Arch Allergy Appl Immunol 1991;94:189–93.

87. Gushchin IS, Serov AA, Zebrev AI, Sapozhnihov AM, Khristenko AV. Ketotifen effect on secretion of histamine from basophils and on proliferative response of mononuclear cells of human peripheral blood. Allergol Immunopathol Madr 1985;13:111–21.

88. Todoroki I, Yoshizawa I, Kawano Y, Noma T. Suppressive effect of azelastine on the induction of Df antigen-specific IL-2 responsiveness in lymphocytes from patients with bronchial asthma. Arerugi 1990;39:1509–14.

89. Kondo N, Fukutomi O, Kameyama T, Nishida T, Li GP, Agata H, Shinbara M, Shinoda S, Yano M, Orii T. Suppression of proliferative responses of lymphocytes to food antigens by an anti-allergic drug, ketotifen fumarate, in patients with food-sensitive atopic dermatitis. Int Arch Allergy Immunol 1994;103:234–8.

90. Canonica GW, Pesce G, Ruffoni S, Buscaglia S, Boero F, Jing G, Rihoux

JP, Ciprandi G. Cetirizine does not influence the immune response. Ann Allergy 1992;68:251–4.

91. Brodde OE, Petrasch S, Bauch HJ, Daul A, Gnadt M, Oefler D, Michel MC. Terbutaline-induced desensitization of beta 2-adrenoceptor in vivo function in humans: attenuation by ketotifen. J Cardiovasc Pharmacol 1992; 20:434–9.

92. Yin KS, Hayashi K, Taki F, Watanabe T, Takagi K, Satake T. Effect of azelastine on the down regulation of β-adrenoceptors in guinea pig lung. Arzneimittelforschung 1991;41:525–7.

93. Hsu K, Wang D, Shen CY, Hsu YS, Shieh SD. Protective effect of ketotifen on chemiluminescence of neutrophils. Taiwan I Hsueh Hui Tsa Chih 1988; 87:939–43.

94. Yoshikawa T, Naito Y, Takahashi S, Tanigawa T, Oyamada H, Ueda S, Takemura T, Sugino S, Kondo M. Influences of anti-allergic drugs on superoxide generation from the hypoxanthine-xanthine oxidase system or polymorphonuclear leukocytes. Arerugi 1989;38:486–92.

95. Umeki S. Biochemical study on anti-inflammatory action of anti-allergic drugs—with regard to NADPH oxidase. Arerugi 1991;40:1511–20.

96. Hojo M, Hamasaki Y, Fujita I, Koga H, Matsumoto S, Miyazaki S. Effects of anti-allergy drugs on fMet-Leu-Phe-stimulated superoxide generation in human neutrophils. Ann Allergy 1994;73:21–6.

97. Xin XH, Bian RL. Effects of ketotifen on human neutrophil respiratory burst and intracellular free calcium. Chung Kuo Yao Li Hsueh Pao 1992; 13:367–70.

98. Taniguchi K, Urakami M, Takanaka K. Effects of antiallergic agents on polymorphonuclear leukocytes. The inhibition of arachidonic acid release and superoxide production. Nippon Yakurigaku Zasshi 1987;90:97–103.

99. Taniguchi K, Urakami M, Takanaka K. Effects of various drugs on superoxide generation, arachidonic acid release and phospholipase A2 in polymorphonuclear leukocytes. Jpn J Pharmacol 1988;46:275–84.

100. Taniguchi K, Takanaka K. Inhibitions of metabolic responses of polymorphonuclear leukocytes by antiallergic drugs. J Pharmacobiodyn 1989;12: 37–42.

101. Busse W, Randlev B, Sedgwick J. The effect of azelastine on neutrophil and eosinophil generation of superoxide. J Allergy Clin Immunol 1989;83: 400–5.

102. Schmidt J, Kaufmann B, Lindstaedt R, Szelenyi I. Inhibition of chemiluminescence in granulocytes and alveolar macrophages by azelastine. Agents Actions 1990;31:229–36.

103. Kurosawa M, Hanawa K, Kobayashi S, Nakano M. Inhibitory effects of azelastine on superoxide anion generation from activated inflammatory cells measured by a simple chemiluminescence method. Arzneimittelforschung 1990;40:767–70.

104. Taniguchi K, Masuda Y, Takanaka K. Action sites of antiallergic drugs on human neutrophils. Jpn J Pharmacol 1990;52:101–8.

105. Kato M, Morikawa A, Kimura H, Shimizu T, Nakano M, Kuroume T. Effects of antiasthma drugs on superoxide anion generation from human polymorphonuclear leukocytes or hypoxanthine-xanthine oxidase system. Int Arch Allergy Appl Immunol 1991;96:128–33.

106. Umeki S. Effects of anti-allergic drugs on human neutrophil superoxide-generating NADPH oxidase. Biochem Pharmacol 1992;43:1109–17.

107. Kakuta Y, Kato T, Sasaki H, Takishima T. Effect of ketotifen on human alveolar macrophages. J Allergy Clin Immunol 1988;81:469–74.

108. Chihara J, Yasukawa A, Yamamoto T, Kurachi D, Haraguchi R, Mori T, Seguchi M, Uenishi H, Nakajima S. [The inhibitory effect of oxatomide on oxygen radical products from human eosinophils and an eosinophilic cell line.] Arerugi 1991;40:689–94.

109. Kurosawa M, Ishizuka T, Kobayashi S, Nakano M. Effects of antiallergic drugs on superoxide anion generation from activated human alveolar macrophages measured by chemiluminescence method. Arzneimittelforschung 1991;41:47–51.

110. Kurosawa M, Ishizuka T, Kobayashi S, Nakano M, Yodoi J. Effects of azelastine on superoxide anion generation from gamma interferon treated human monoblast cell line U937 measured by chemiluminescence method. Arzneimittelforschung 1991;41:43–7.

111. Subramanian N. Leukotriene B4-induced steady state calcium rise and superoxide anion generation in guinea pig eosinophils are not related events. Biochem Biophys Res Commun 1992;187:670–6.

112. Nabe M, Agrawal DK, Sarmiento EU, Townley RG. Inhibitory effect of terfenadine on mediator release from human blood basophils and eosinophils. Clin Exp Allergy 1989;19:515–20.

113. Taniguchi K, Masuda Y, Takanaka K. Inhibitory effects of histamine H$_1$-receptor blocking drugs on metabolic activations of neutrophils. J Pharmacobiodyn 1991;14:87–93.

114. Werner U, Seitz O, Szelenyi I. Stimulated elastase release from human leukocytes: influence of anti-asthmatic, anti-inflammatory and calcium antagonist drugs in vitro. Agents Actions 1993;38 Spec No:C112–4.

115. Renesto P, Balloy V, Vargaftig BB, Chignard M. Interference of anti-inflammatory and anti-asthmatic drugs with neutrophil-mediated platelet activation: singularity of azelastine. Br J Pharmacol 1991;103:1435–40.

116. Wang XD, Bian RL. [Effects of ketotifen on rabbit platelet aggregation and platelet activating factor formation from rat neutrophils.] Chung Kuo Yao Li Hsueh Pao 1990;11:524–7.

117. Chan WP, Levy JV. Effects of antiplatelet agents on platelet aggregation induced by platelet-activating factor (PAF) in human whole blood. Prostaglandins 1991;42:337–42.

118. Chihara J, Sugihara R, Kubo H, Yamamoto T, Nakano N, Uenishi H, Nakajima S. The inhibitory effect of oxatomide, an anti-allergic agent, on eosinophil-mediated and eosinophilic cell line-mediated natural cytotoxicity against bronchial epithelial cells. Arerugi 1989;38:1180–4.

119. Takanaka K, Taniguchi K, Masuda Y, O'Brien PJ. Reversible drug effects on the metabolic activation of polymorphonuclear leukocytes. Chem Biol Interact 1990;73:309–21.

120. Nakamura T, Nishizawa Y, Sato T, Yamato C. Effect of azelastine on the intracellular Ca^{2+} mobilization in guinea pig peritoneal macrophages. Eur J Pharmacol 1988;148:35–41.

121. Letari O, Miozzo A, Folco G, Belloni PA, Sala A, Rovati GE, Nicosia S. Effects of loratadine on cytosolic Ca^{2+} levels and leukotriene release—novel mechanisms of action independent of the anti-histamine activity. Eur J Pharmacol-Molec Pharm 1994;266:219–27.

122. Senn N, Jeanclos E, Garay R. Action of azelastine on intracellular Ca^{2+} in cultured airway smooth muscle. Eur J Pharmacol 1991;205:29–34.

123. Lueddeckens G, Becker K, Rappold R, Forster W. Influence of aminophylline and ketotifen in comparison to the lipoxygenase inhibitors NDGA and esculetin and the PAF antagonists WEB 2170 and BN 52021 on endothelin-1 induced vaso- and bronchoconstriction. Prostaglandins Leukot Essent Fatty Acids 1991;44:155–8.

124. Nomura A, Ninomiya H, Saotome M, Ohse H, Ishii Y, Uchida Y, Hirata F, Hasegawa S. Multiple mechanisms of bronchoconstrictive responses to endothelin-1. J Cardiovasc Pharmacol 1991;17 Suppl 7:S213–5.

125. Uchida Y, Saotome M, Nomura A, Ninomiya H, Ohse H, Hirata F, Hasegawa S. Endothelin-1-induced relaxation of guinea pig trachealis muscles. J Cardiovasc Pharmacol 1991;17 Suppl 7:S210–2.

6

H_1-Receptor Antagonists: Antiallergic Effects in Humans

Robert M. Naclerio and Fuad M. Baroody
The University of Chicago, Chicago, Illinois

I. INTRODUCTION

Dale and Laidlaw first studied the physiological effects of histamine in the early 20th century. They showed that it induced smooth muscle contraction and vasodilation (1). They made the analogy between the effect of histamine and the immediate response to an injected foreign protein in previously sensitized animals and postulated that histamine might play a role in this reaction. In 1927, Best and colleagues isolated histamine from animal tissues (2), and Lewis reported that a histamine-like substance was released from cells in the skin by the interaction of antigen and antibody (3). In 1937, Bovet and Staub discovered the first drug with antihistaminic activity and, in 1944, the first antihistamine was developed for human use (4, 5). Since then, the use of antihistamines in the treatment of allergic disease has grown rapidly. There are more than 20 different H_1-antagonists and over 100 different antihistamine-containing products currently available for clinical use. In addition, several new H_1-antihistamines are in various stages of development.

Numerous clinical trials support the efficacy of these drugs in reducing sneezing and rhinorrhea during episodes of allergic rhinitis, and their utility in treating this condition is indisputable (6). Despite the abundance of clinical data, the mechanism by which antihistamines relieve allergic symptoms is not fully understood. Their primary mechanism of action is believed to be competitive antagonism of histamine binding to cellular receptors—specifically, the H_1-receptors present on nerve endings, smooth muscle cells, and glandular cells. This notion is supported by the fact that structurally unrelated drugs antagonize the H_1-receptor. However, H_1-antagonism may not be their sole mechanism of action.

Most H_1-antihistamines also have anticholinergic, sedative, local anesthetic, and antiserotonin effects that might favorably affect the symptoms of the allergic response or contribute to side effects (6). These additional properties are not uniformly distributed among drugs classified as H_1-receptor antagonists. Azatadine, for example, is a tricyclic compound with H_1-antagonist, some anticholinergic, and almost no antiserotonin properties (7). At high concentrations (10^{-4}–10^{-2} M), it inhibits in vitro IgE-mediated histamine and leukotriene release from mast cells and basophils (8). In contrast, terfenadine is minimally anticholinergic, does not depress the central nervous system, and reduces IgE-mediated histamine release. These characteristics of H_1-antihistamines may contribute to their clinical usefulness, but the biological mechanisms of these effects are not well understood.

The terminology used to describe these other properties of H_1-receptor antagonists lacks precision. The term "anti-inflammatory" commonly appears in the literature and, when used to describe treatments for allergic rhinitis, often implies effects beyond pure H_1-receptor antagonism. However, in Stedman's medical dictionary (9), H_1-receptor antagonists are used as examples of anti-inflammatory drugs. The term antiallergic is defined as any agent or measure that prevents, inhibits, or alleviates an allergic reaction. It also often refers to additional properties such as the inhibition of mediator release from mast cells. As the number of proinflammatory substances identified in mast cells increases (mediators, interleukins), the definition of the term becomes more difficult: for example, does an antiallergic drug have to inhibit the release of all mediators or just some of them? For the purposes of this chapter, the term antiallergic will be used to describe drugs primarily classified as H_1-receptor antagonists that may have properties leading to additional beneficial effects on the allergic response.

Each section of this chapter describes a different H_1-antagonist and the studies performed using this medication. Since our understanding of the allergic response continues to evolve, most drugs have not been tested in vivo for every action. Similarly, model systems to test certain properties do not exist. A discussion of our current understanding of the pathophysiology is presented first to provide a background for the studies to follow.

II. PATHOPHYSIOLOGY OF THE ALLERGIC RESPONSE

In order to understand the potential antiallergic effects of H$_1$-antihistamines, it is necessary to have an overview of the pathophysiology of the allergic reaction (Fig. 1). Our current view of the allergic response involves a phase of sensitization followed by a period of clinical disease. Individuals with the genetic predisposition for allergic disease develop IgE antibodies to allergens after exposure. These antibodies bind to high-affinity receptors on mast cells and basophils and to low-affinity receptors on certain other cells (10–13). Upon reexposure to antigen, mast cells degranulate, releasing multiple inflammatory mediators that stimulate the nasal end organs to produce itching, sneezing, rhinorrhea, and congestion (14). This immediate response to antigen can be shown in over 90% of patients with a history of allergic rhinitis and a positive skin test (15). The exact role of individual mediators and their interactions has not been totally elucidated, but nasal provocation has provided useful information. Histamine provocation, for example, induces all the signs and symptoms produced by acute antigen provocation (16).

Acute mast-cell degranulation leading to an episode of nasal anaphylaxis does not entirely mimic clinical disease. This statement reflects differences between the clinical manifestations of disease and the acute response to experimental nasal provocation with antigen, including the longer duration of seasonal disease, differential responsiveness to systemic corticosteroids, different histologies, and the association of clinical disease with alterations in reactivity to antigen and irritants. Blackley, in 1873, described not only the immediate response to nasal antigen challenge but also the spontaneous recurrence of symptoms hours later (17). This event was subsequently termed a late-phase reaction and serves as a means of linking clinical disease with in vivo provocation (18). In approximately 50% of subjects with seasonal allergic rhinitis, spontaneous recurrence of symptoms occurs. The spontaneous local reappearance of some (including histamine), but not all, inflammatory mediators occurs in association with symptoms (19) (Fig. 2). The late reaction is suppressed by systemic administration of glucocorticosteroids, which do not affect the antecedent early reaction (20). Hence, a drug with clinical utility in the management of allergic rhinitis does not affect mast-cell degranulation.

An influx of eosinophils, neutrophils, basophils, and lymphocytes accompanies the late reaction (21, 22). These cells enter the nose and become activated. The number of subjects demonstrating a cellular influx exceeds those with late mediator release, but a tendency exists for those with late

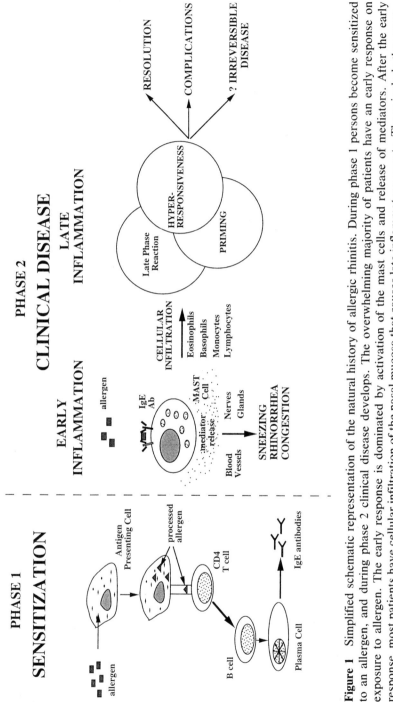

Figure 1 Simplified schematic representation of the natural history of allergic rhinitis. During phase 1 persons become sensitized to an allergen, and during phase 2 clinical disease develops. The overwhelming majority of patients have an early response on exposure to allergen. The early response is dominated by activation of the mast cells and release of mediators. After the early response, most patients have cellular infiltration of the nasal mucosa that causes late inflammatory events. These include the spontaneous recurrence of release of mediators (late-phase reaction), hyperresponsiveness to irritants, and increased responsiveness to allergen (priming). The circles indicate the heterogeneity of these late inflammatory events. The inflammation can resolve spontaneously, cause a complication, or potentially lead to an irreversible form of chronic rhinitis. (Reprinted with permission from Naclerio RM. Allergic rhinitis. N Engl J Med 1991;325:860–9.)

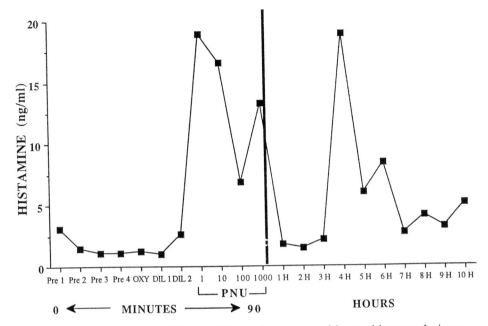

Figure 2 An example of levels of histamine recovered in nasal lavages during a late-phase reaction. The protocol is shown on the abscissa: Pre = prewashes done to bring mediator levels to a stable baseline; OXY = oxymetazoline which is used to maintain the patency of the nasal airway and facilitate lavage; DIL = diluent for the allergen extracts, 1, 10, 100, and 1000 = dose of antigen in PNU, 1h to 10 h = hours after administration of the last dose of allergen. The heavy vertical line separates measurements made approximately every 10 min from those measured at hourly intervals.

reactions to have the greatest eosinophil influx (15). The above changes also occur during seasonal exposure. Both the number of eosinophils and the level of eosinophil cationic protein, another eosinophil-derived mediator, increase during seasonal exposure and are reduced by treatment with topical steroids and immunotherapy (23, 24). Mast-cell and eosinophil progenitors increase in nasal scrapings and correlate with the severity of the disease (25). The changes in cellular constituents probably cause the changes in reactivity associated with allergic rhinitis.

About half of allergic individuals develop increased sensitivity to antigen after repeated daily exposure (i.e., priming) (15, 26). The increased sensitivity to antigen is associated with a cellular influx and an increased release of some mediators, such as histamine (21). The increased histamine appears to come from basophils, which increase in nasal secretions hours after antigen provocation in the laboratory and during seasonal exposure (21, 27). In addition, the basophils are a target for antigen during priming. Continued antigen exposure, as occurs during the season, also increases the number of mast cells in the epithelial and subepithelial layers (28).

Surprisingly, the development of a late reaction is not obligatorily linked to priming (15). Patients are equally likely to express any combination of inflammatory events. This observation points to the heterogeneity of the inflammation occurring after antigen stimulation and may explain the differential response to treatment. Unfortunately, early and late skin test reactivity, specific IgE levels, cellular influx, and basophil histamine release only weakly correlate with the inflammatory events and have minimal predictive value (15).

Increases in nasal reactivity to histamine occur following antigen provocation and may reflect increased sensitivity to irritants, such as tobacco smoke, described by symptomatic allergic patients (29). Some researchers have found similar increases in nasal reactivity during seasonal exposure, while others relate the changes in reactivity to a shift in baseline (30, 31). The increased responsiveness following antigen provocation reverses spontaneously when no further antigen exposure occurs. Although significant correlations exist between the cellular changes and measured changes in reactivity, their predictability is small (29).

Changes in reactivity describe only part of allergic inflammation. The role of many other components needs to be studied in greater detail, including the epithelial cells, neuropeptides, lymphocytes, adhesion molecules, chemoattractants, fibroblasts, and other connective tissue cells. Such knowledge should refine our concept of the allergic response. Equally important will be an appreciation of the influence of environmental factors, such as viruses and pollutants on allergic inflammation.

III. AZATADINE

Based on the observations that tricyclic antihistamines inhibit histamine release from basophils in vitro, we attempted to study this property in vivo (7). In order to accomplish this goal, we developed a nasal challenge model that permitted observation of the biochemical events after antigen provocation (32) (Table 1). Levels of histamine in nasal secretions, before and after antigen provocation, were used as a quantitative indicator of mast-cell activation.

In vitro experimentation with azatadine base, a tricyclic antihistamine of the piperidine class, showed that it inhibited histamine and leukotriene

Table 1 Effect of H_1-Antihistamines on the Response to Nasal Antigen Challenge

Antihistamine	Histamine release	Leukotriene (LTC$_4$) generation	Hyperresponsiveness to methacholine
Azatadine (intranasal)	Decreased	ND	ND
Azelastine	No change	Decreased	ND
Cetirizine	No change	Decreased	Decreased
Ketotifen	No change	ND	ND
Levocabastine (intranasal)	ND	ND	ND
Loratadine	Decreased	No change	Decreased
Terfenadine	Decreased	No change	Decreased

ND = not evaluated

release by 45 and 85%, respectively, when used at concentrations of 10^{-3} M (8). Concentrations less than 10^{-4} M had no significant effect. As these concentrations would be difficult to obtain from oral administration of the drug, we investigated topical delivery. Eight patients were given 0.5 mg of azatadine intranasally in a double-blind, cross-over study (8). Azatadine significantly suppressed sneezing ($p < 0.05$) and inhibited the elevation of levels of histamine ($p < 0.07$), N-α-tosyl-L-arginine methyl ester (TAME)-esterase activity ($p < 0.01$), and kinins ($p < 0.05$) in nasal lavage fluids collected after antigen challenges. Our observations were consistent with the notion that azatadine, given in sufficient concentration, was capable of inhibiting mast-cell histamine release in vitro and in vivo. The development of this challenge model permitted the investigation of properties of antiallergic agents including their H_1-antihistaminic activity and other anti-inflammatory and mediator suppressor characteristics.

IV. TERFENADINE

Terfenadine is one of the best-studied H_1-antihistamines. When investigating its antiallergic effects in a nasal provocation model, potential effects of the drug that might lead to spurious conclusions must be evaluated. To better understand the effect of terfenadine in allergic rhinitis, we used in vivo models to monitor symptoms and mediator release into nasal secretions following nasal challenge with either histamine or antigen (32).

Because terfenadine is a nonsedating antihistamine, we assessed whether a larger dose (300 mg orally b.i.d.) would permit additional properties to become manifest and provide additional therapeutic advantages (33). A double-blind, placebo-controlled clinical trial was performed with 12 subjects with allergic rhinitis to evaluate the effect of one-week pretreatment with terfenadine (60 mg b.i.d. or 300 mg b.i.d.) on the response to sequential nasal challenges with allergen and histamine. The response to allergen challenge was monitored by counting sneezes and quantifying the levels of histamine, kinins, TAME-esterase activity, and albumin in recovered nasal lavages.

The volunteers were challenged twice with the diluent for the antigen extract and then 3 times with increasing amounts of ragweed or grass antigen extract. Each challenge was separated by 12 min. A nasal lavage was performed 10 min after each challenge, with an additional lavage performed 20 min after the 1000 protein nitrogen units (PNU) challenge. Immediately following the last lavage after antigen provocation, we challenged the nose with three doses of histamine (0.04, 0.2, and 1.0 mg), recording the number of sneezes, and lavaging the nose 10 min after each challenge to measure the levels of albumin. The challenge with histamine served as a positive control.

Prolonged treatment with terfenadine reduced the number of sneezes following antigen challenge. Associated with clinical efficacy, terfenadine significantly reduced the levels of kinins, albumin, and TAME-esterase activity, which are, in part, indicators of vascular permeability, suggesting a major role for histamine on the nasal vasculature during the immediate allergic reaction. Of particular interest was a significant reduction ($p <$ 0.03) in the levels of histamine, suggesting an inhibition of mast-cell degranulation both by 60 and 300 mg of terfenadine (Fig. 3). The clinical importance of the inhibition of histamine release by terfenadine is unknown. As both the low and high doses decreased histamine release, to achieve this effect using a dose of terfenadine greater than that recommended by the manufacturer is neither necessary nor advisable since the incidence of adverse events increases with high doses.

Following histamine challenge, terfenadine dramatically reduced the number of sneezes and the levels of albumin, again demonstrating its potency as an H_1-antagonist. This study, like the one reported above, emphasizes the importance of the H_1-receptor in mediating sneezing and increased vascular permeability. To reproduce the results of this observation and address possible physiological effects, we tested the effect of terfenadine in another nasal challenge system.

Challenging the nose with high doses of histamine reproduces the symptoms of the allergic reaction, including sneezing, rhinorrhea, and nasal congestion (34). To objectively measure the secretory response to histamine, we used filter paper discs to challenge the nose with histamine and

HISTAMINE

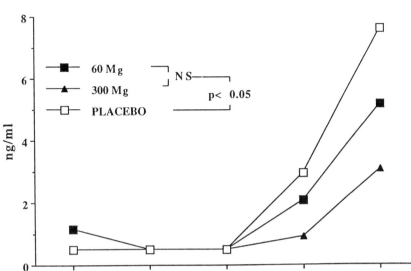

Figure 3 The effect of premedication with terfenadine, an H_1-antihistamine, on the early allergic response. Twelve subjects with allergic rhinitis were premedicated with 60 or 300 mg b.i.d. of active drug or placebo for 1 week prior to challenge. The challenge protocol is on the abscissa: DIL = diluent for the allergen extract; PNU = protein nitrogen units. The median response for the group is plotted. Statistical comparison between treatments is performed for the sum of net changes from diluent. Both treatments resulted in a significant decrease in histamine levels after allergen provocation ($p < 0.05$) and there was no significant difference between the two drug doses (NS = nonsignificant). (Reprinted with permission from Ref. 33.)

measure secretions on the ipsilateral and contralateral sides of stimulation (35). Using this technique, we then examined the effect of terfenadine on the nasal secretory response to histamine and the nasonasal reflex.

We challenged seven subjects with increasing doses of histamine with and without pretreatment with 60 mg of terfenadine (35). Filter paper challenge discs soaked with varying concentrations of histamine (0.004–3 mg) were applied for 1 min on the anterior part of the nasal septum, just posterior to the mucocutaneous junction. Thirty seconds after removal of the challenge disc, a preweighed collection disc was applied for 30 s to the same area of the mucosa to collect secretions. The recovered disc was immediately returned to its microtube, which was then sealed. Contralateral, reflex-induced secretions were measured simultaneously by applying another preweighed collection disc to the analogous area of the contralateral nostril. The weight of nasal secretions generated by each challenge was calculated by subtracting the weights of the disc/microtube combination before challenge from that after it. A sham challenge with saline preceded the histamine challenges, and each challenge was separated by 5 min.

A single, 60-mg dose of terfenadine given 4 h prior to challenge completely inhibited ipsilateral and contralateral histamine-induced secretions for all doses of histamine (35). Although three histamine receptors have been defined, the completeness of the suppression by terfenadine suggested that the direct nasal secretory response as well as the initiation of the central nasonasal reflex are mainly mediated by H_1-receptors.

We investigated whether terfenadine given at the same dose had any anticholinergic properties. Using a similar technique, we challenged the nasal mucosa with methacholine (0.06–5 mg), a cholinoceptor agonist (35). In contrast to histamine, challenging with increasing doses of methacholine produced only an ipsilateral increase in secretion weights. This ipsilateral secretory response was not reduced after terfenadine premedication, supporting both the lack of anticholinergic activity and the observation that the inhibition of the response to histamine provocation was not related to anticholinergic effects of terfenadine.

Using the localized challenge technique, we studied the effect of terfenadine on the response of the nasal mucosa to antigen. Upon challenge with antigen, physiological changes such as secretion weights, nasal airway resistance, sneezes, and symptoms can be monitored and inflammatory mediators simultaneously measured.

Eight volunteers allergic to ragweed or grass were given a single dose of either terfenadine, 60 mg, or placebo 2 h prior to challenge out of season (36). Using paper discs, a unilateral antigen challenge was performed with a single dose of 333 PNU of the relevant antigen. The response was monitored by counting the number of sneezes, measuring nasal airway resistance, and determining the amount of secretions collected in a fixed time interval. Histamine levels were obtained from secretions eluted from the collection discs.

Terfenadine significantly reduced the number of sneezes but had no effect on nasal airway resistance. Partial inhibition of both the ipsilateral and contralateral secretory responses occurred compared to placebo, but it was less than the reduction after histamine challenge. Histamine release, as expected, increased on the ipsilateral, but not on the contralateral, side. Terfenadine pretreatment lowered the levels of histamine in the challenged nostril, again suggesting an inhibition of mast-cell histamine release (Fig. 4).

These results concur with the clinical effects of H₁-antihistamines: excellent efficacy for sneezing, some effect on rhinorrhea, and no effect on nasal congestion. The partial reduction in the secretory response to antigen with the dramatic reduction after histamine provocation suggests that

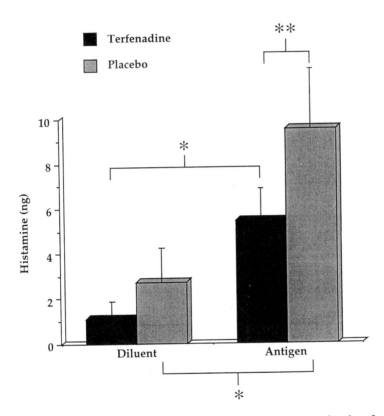

Figure 4 The effect of terfenadine on histamine release at the site of provocation using the localized antigen challenge technique in 12 allergic subjects. A dose of 333 PNU was used for this challenge. The values depicted represent the total amount of histamine recovered from eluted secretions obtained at two time points (0.5 and 2 min) after provocation with allergen or diluent. There was a significant increase in histamine levels after antigen challenge compared to diluent* ($p < 0.05$) when the patients were premedicated with both placebo and terfenadine. After allergen challenge, however, levels of histamine were significantly lower in the terfenadine-treated group** ($p < 0.05$). (Reprinted with permission from Ref. 36.)

other mediators or mechanisms contribute significantly to the secretory response generated after antigen provocation. The marked suppression of histamine and antigen-induced sneezing by terfenadine suggests that the sneezing response probably results from direct stimulation of nerve endings via H_1-receptors. Furthermore, the inhibition of sneezing, as well as the inhibition of the histamine-induced nasonasal reflex secretory response, suggests that these neural responses are generated by the action of histamine on H_1-receptors.

The lack of efficacy of H_1-antihistamines to treat nasal congestion also emphasizes the role of other mediators or receptors in allergic rhinitis. Our observation that treatment with terfenadine had no effect on nasal airway resistance measured by anterior rhinomanometry is consistent with this concept.

Albumin, an indicator of vascular permeability, increases after histamine challenge, consistent with the presence of H_1-receptors on vessels and the role of transudation in the secretory response. This H_1-mediated stimulation of blood vessels is completely suppressed by terfenadine. TAME-esterase activity, which largely represents plasma-derived enzymes in our model, was also reduced (37). Kinin generation is dependent, in part, upon the transudation of its substrate kininogen from plasma. Thus, the observed reduction in kinin levels could result from a reduction in vascular permeability due to antagonism of the histamine released from mast cells. Since mast-cell tryptase contributes to kinin generation, inhibition of its release from mast cells could also affect the amount of kinins recovered after antigen challenge. The reduction in the amount of kinins generated after antigen exposure illustrates another mechanism through which H_1-antihistamines may affect the pathophysiology of allergic rhinitis. Since kinins induce rhinorrhea and nasal congestion when instilled onto the nasal mucosa (38), a reduction in this secondarily generated inflammatory mediator may further decrease the symptoms associated with acute mast-cell degranulation and contribute favorably to the effectiveness of H_1-antihistamines.

Conceptually, the reduction of histamine levels in recovered lavage fluids by terfenadine could result from mechanisms other than inhibition of mast-cell degranulation, such as suppression of neuronal reflexes. The results from the disc experiments support the inability of the nasonasal reflex to induce histamine release. When discs were used to collect secretions from the challenged and nonchallenged sides, secretions increased on both sides, while histamine increased only on the challenged side. This suggests that histamine is released by mast cells due to direct contact with antigen at the challenge site. The mechanism by which terfenadine suppressed histamine release is not apparent from our studies. The duration of pretreatment or the activity of a metabolite of terfenadine (terfenadine is rapidly metabolized) could be important variables. In vitro studies showed inhibition of histamine release from mast cells treated with terfenadine, supporting our in vivo observations. The above findings illustrate that terfenadine, which is classified as an H_1-antihistamine, also has antiallergic properties.

V. CETIRIZINE

Cetirizine, the oxidative metabolite of hydroxyzine, is a cyclizine class H_1-receptor antagonist. Receptor-binding studies showed that cetirizine, compared to terfenadine and hydroxyzine, was more selective for H_1-sites and had less binding to serotonin, dopamine, alpha adrenergic, and calcium channel receptors (39). Studies evaluating the allergic response in the skin have shown that it reduces eosinophilic infiltration at the site of antigen challenge (40). This surprising finding has been reproduced, but is not a general property of H_1-antihistamines. Why cetirizine would prevent the selective recruitment of eosinophils to the skin is unknown. We describe the results of several clinical trials aimed at understanding the mechanism of action of cetirizine during the nasal allergic reaction.

We performed a double-blind, placebo-controlled, cross-over study in 10 asymptomatic, seasonal allergic subjects, comparing pretreatment with 20 mg of cetirizine daily for three doses, or placebo, on the response to intranasal challenge with antigen (41). The 2 study days were separated by approximately 2 weeks. The protocol for the nasal challenge involved nasal lavage to focus on the pattern of mediator release during the immediate allergic reaction. Serial challenges with 10, 100, and 1000 PNU of dialyzed ragweed or mixed grass pollen extract were performed. The number of sneezes after each antigen dose was counted. The resultant lavage fluid/secretion mixture was assayed for histamine, TAME-esterase activity, albumin, LTC_4, and prostaglandin D_2 (PGD_2).

After placebo pretreatment, the subjects sneezed in response to antigen stimulation. Concordant with the sneezing was a dose-dependent increase in the levels of histamine, PGD_2, LTC_4, albumin, and TAME-esterase activity recovered in the lavages postchallenge. After cetirizine pretreatment, there was a dramatic reduction in the number of sneezes. The levels of histamine and PGD_2 recovered in postchallenge lavages were not reduced, whereas the levels of albumin, TAME-esterase activity, and LTC_4 were (Fig. 5). These findings support the notion that cetirizine, an H_1-antihistamine, competes with histamine for H_1-receptors on nerves (reduced sneezing) and blood vessels (decreased vascular permeability).

The absence of an effect on histamine and PGD_2 production argues

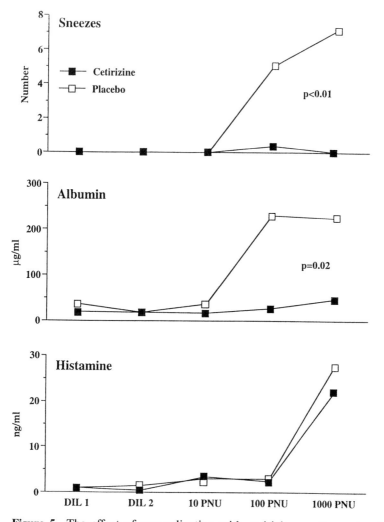

Figure 5 The effect of premedication with cetirizine on the early allergic response. Ten subjects with allergic rhinitis were premedicated with 20 mg daily of active drug or placebo for 2 days prior to challenge. The challenge protocol is identical to that in Figure 3. The median response for the group is plotted. The parameters evaluated were the number of sneezes and the levels of albumin and histamine in recovered nasal lavages. Statistical comparison between treatments is performed for the sum of net changes from diluent. Cetirizine pretreatment significantly inhibited sneezing and the levels of albumin in nasal lavages but had no inhibitory effect on histamine release after allergen challenge. (Reprinted with permission from Ref. 41.)

that mast-cell activation by antigen provocation was equivalent on the 2 study days. This observation, in addition to inhibition of sneezing, further suggests that histamine release from mast cells resulted directly from mast-cell activation and argues against neuronal reflexes stimulating mast-cell release.

The surprising result from this study was the inhibition of sulfidopeptide leukotriene production by an antihistamine. Mast-cell activation in vitro causes the generation of LTC_4. As seen in this study, LTC_4 is generated during the acute allergic response, the obvious implication being that mast cells are the source of leukotriene production. Since cetirizine in vitro did not inhibit leukotriene, histamine, or PGD_2 production during IgE-mediated mast-cell stimulation, other cells within the nasal mucosa might contribute to leukotriene generation. Furthermore, since leukotriene production is inhibited by cetirizine, we can hypothesize that it results from histamine stimulation of this other cell(s) source. However, in separate studies, histamine stimulation of the nasal mucosa did not lead to leukotriene generation. Another potential explanation for the results is that cetirizine may have some 5-lipoxygenase inhibitory activity that leads to a reduction of leukotriene generation from mast cells without affecting histamine or prostaglandin release or that this mediator is produced by other cells. In support of that notion is that activated macrophages, eosinophils, and endothelial cells generate leukotrienes. The clinical significance of the inhibition of leukotriene production by antihistamines requires further study. Our results emphasize the advantage of correlating in vivo and in vitro results and the greater complexity of in vivo systems.

Klementsson and colleagues evaluated the effect of cetirizine on eosinophil infiltration into nasal secretions hours after the immediate allergic reaction (42). They compared the effects of terfenadine, cetirizine, and placebo. Both drugs inhibited the symptoms of the immediate reaction, while neither drug had an effect on the number of recovered eosinophils. The authors also investigated allergen-induced hyperresponsiveness to methacholine. Since neither drug affects the nasal response to methacholine, this was a valid test of hyperresponsiveness. Surprisingly, both drugs blocked this phenomenon. A related observation was made by Majchel, who showed that histamine does not induce hyperresponsiveness to methacholine (43). Thus the mechanism by which these two antihistamines reduce hyperresponsiveness is probably not related to their H_1-antagonistic properties and remains unclear.

The effects of cetirizine (20 mg daily for 10 days) on the immediate and late nasal response to antigen and on pulmonary function in allergic asthmatics were also investigated (44). Cetirizine significantly reduced sneezing, and the levels of albumin and LTC_4 during the early reaction, confirming our observations reported above. It had no effect on late nasal symptoms, vascular permeability, or eosinophil influx into the nose. This study supports the work of Klementsson by showing no effect on eosinophil influx into the nasal mucosa, in contrast to its effect on the response in the skin. While on placebo, the asthmatic subjects showed a decrease in airflow obstruction 6 h after nasal challenge with antigen, and this returned to normal by 24 h. Cetirizine prevented this decrease in pulmonary function. The placebo arm suggests a possible extension of an allergic response of the upper airway to the lower airway and cetirizine seems to inhibit this deterioration of lower airway function after upper airway challenge with allergen. The mechanism for this effect needs to be established.

VI. AZELASTINE

Azelastine 4-[(p-chlorobenzyl)methyl]-2-(hexahydro-1-methyl-1H-azepin-4-yl)-1(2H)-phthalazinone hydrochloride] is an orally effective, long-acting antiallergic drug with multiple actions. In addition to histamine H_1-receptor blockade (45), it has been shown to exert blocking activities on leukotriene (46, 47) and serotonin receptors (48). Azelastine also inhibits histamine release from rat peritoneal mast cells (49, 50) and rabbit leukocytes (51), and it has a similar action on human chopped lung mast cells (52) and peripheral blood basophils (53) in response to a variety of stimuli such as anti-IgE, antigen, calcium ionophore, and compound 48/80. It has been shown to block guinea pig tracheal smooth muscle contraction to a variety of spasmogenic mediators (54) and to possess anti-inflammatory properties (55, 56). In vivo human clinical trials have shown oral azelastine to relax human airway smooth muscle (57), to inhibit skin test reactivity to histamine, codeine, and antigen, and to reduce antigen-induced histamine release in a skin chamber (58). Furthermore, in clinical trials designed to study seasonal grass and ragweed allergic rhinitis, azelastine, in doses ranging from 0.5 to 2 mg twice daily proved more effective than placebo and chlorpheniramine in relieving symptoms (59).

Because of this wide range of actions of azelastine, we initiated a clinical trial to investigate its effects on the immediate nasal allergic reaction in vivo. We performed a double-blind, placebo-controlled cross-over study comparing treatment with azelastine or placebo on the immediate response to intranasal challenge with antigen in 13 subjects (60). A challenge was performed 4 h after a single 2-mg dose of oral azelastine or placebo. A washout period of at least 2 weeks was observed between the two arms of the study. The protocol for the nasal challenge was similar to that described previously. The number of sneezes was recorded. Lavages were quantified for histamine, TAME-esterase activity, LTC_4, and PGD_2.

There was a dose-dependent increase in the number of sneezes in response to antigen challenge with patients on placebo. This increase was accompanied by a rise in the levels of histamine, TAME-esterase activity, PGD_2, LTC_4, and kinins in recovered nasal lavages. Pretreatment with a single dose of azelastine resulted in a dramatic reduction in the number of sneezes as well as in the levels of TAME-esterase activity, LTC_4, and kinins.

In contrast, azelastine had minimal effect on the levels of histamine and PGD_2 recovered in postchallenge lavages. These data suggest that mast cells were probably activated to similar extents on the two treatment days and that azelastine did not significantly affect this event. As mentioned previously, azelastine has been shown to inhibit the release of histamine from animal and human lung mast cells in vitro and we might have been able to demonstrate significant inhibition with a greater number of subjects or a larger dose of azelastine. However, the in vitro and in vivo findings need not necessarily be contradictory, since the amount of azelastine used to block anti-IgE-induced histamine release from human chopped lung mast cells (5–100 μM) exceeds by several orders of magnitude the concentration that the drug is expected to achieve in the nasal mucosa in vivo after oral administration. After treatment with a single 2-mg dose of oral azelastine, the significant reduction in sneezing and in the levels of TAME-esterase and kinins after antigen challenge suggests that azelastine acts as a competitive inhibitor of histamine leading to reduction of its effects on nerves and blood vessels.

An interesting finding in this study was that the production of leukotrienes was inhibited by azelastine as it was by cetirizine (41). Both drugs inhibited leukotriene production after nasal antigen challenge without having a measurable effect on the generation of histamine and PGD_2. This observation might result because the in vivo concentration of the drug was sufficient to inhibit LTC_4 but not histamine release. In vitro experiments with human lung mast cells previously showed that azatadine and azelastine inhibited LTC_4 release at a dose at least a log lower than necessary to inhibit histamine release (8, 61). Alternatively, this differential inhibition might suggest that cells other than the mast cell are responsible for the production of LTC_4 after antigen challenge. For instance, histamine, produced by mast cells, might stimulate other cells in the nasal mucosa, such as eosinophils, macrophages, and/or endothelial cells to produce LTC_4. Thus, effective H_1-antihistamines would reduce the levels of this mediator. Against this hypothesis is the observation that challenging the nasal mucosa with exogenous histamine does not lead to the production of LTC_4 (43). The reduction of LTC_4 by azelastine may alternatively result from direct inhibition by the drug of the production of leukotrienes from other cells in the nasal mucosa, such as eosinophils. Support for this hypothesis comes from in vitro studies that have shown that azelastine inhibits antigen-induced production of LTC_4 and LTD_4 by human and guinea pig lung fragments (54, 62). These studies suggest that azelastine achieves its effect by inhibiting the 5-lipoxygenase pathway of arachidonic acid metabolism in these cells (63). If azelastine truly has 5-lipoxygenase inhibitory properties, then it might also be directly inhibiting that pathway of arachidonic acid metabolism in nasal mast cells without affecting the cyclooxygenase pathway or the process of degranulation and histamine release. This would lead to preferential inhibition of leukotriene, but not histamine or PGD_2 generation from nasal mast cells and this has been shown to occur in vitro in human lung mast cells by pretreatment with L651-392, a specific inhibitor of the 5-lipoxygenase pathway (64).

In antigen challenges of allergic asthmatics, the increase in substance P levels during the early reaction was inhibited by treatment with azelastine (65). Since substance P is associated with sensory nerve fibers, this reduction may be the direct or indirect result of the drug's effect on neuronal activation. Substance P is thought to be involved in the flare response to skin testing. What the equivalent response in the nasal mucosa might be has not been determined yet.

Azelastine has also undergone investigation as a topical intranasal spray. By switching the route of administration from an oral to a topical preparation, the local concentration of the drug in the nasal mucosa may be increased. In one nasal antigen challenge study, a single dose of azelastine (0.28 mg) had no effect on nasal airway resistance after challenge, while it significantly reduced sneezing (66). It will be interesting to observe whether increasing the concentration of the drug in the nasal mucosa will increase the in vivo expression of some of the in vitro properties.

VII. KETOTIFEN

Ketotifen is an orally active antihistamine of the benzocycloheptathiophene class (67). It has several additional properties that support its use in the treatment of allergic rhinitis. In addition to being a competitive (68) and noncompetitive antagonist of the H_1-receptors (69), it inhibits the release of histamine from rat peritoneal mast cells (70). It also inhibits histamine release during antigen-induced skin reactions in guinea pigs (71), during passive cutaneous anaphylaxis reactions in rats (72), and following cold-induced urticaria in humans (73). In addition, it inhibits the release of sulfidopeptide leukotrienes from human leukocytes stimulated with calcium ionophore A23187 or rabbit anti-IgE (74). Some studies suggest that the drug antagonizes the effects of platelet-activating factor (PAF) on bronchospasm and on cellular infiltration, especially eosinophils, into the lungs (75), while other studies do not (67).

We examined the effect of pretreatment with ketotifen on the release of mediators and symptoms during the early reaction to nasal challenge with antigen, a mast-cell dominated reaction (76). One- and 2-mg doses, with pretreatment ranging from 1 h to 4 weeks, were evaluated. We recruited 10 volunteers between the ages of 23 and 40 years with a history of allergic rhinitis to grass or ragweed and a positive intradermal skin test to 10 PNU/ml of the appropriate extract. A two-treatment, cross-over study was performed. During each treatment period, 6 weekly nasal challenge protocols were performed. The first challenge, administered before the drug was given, was considered baseline. The subjects were started on 1 or 2 mg of ketotifen 1 h before the second challenge. They were maintained on the dose of the drug b.i.d. for the remainder of the treatment period. Thus, the sixth challenge occurred after 4 weeks of continuous treatment. A 2-week washout period separated the two treatment periods. The total duration of the study was 12 weeks. During each nasal challenge, sneezes were counted, symptom scores recorded, and levels of histamine and TAME-esterase activity in recovered lavages measured.

Challenge with the diluent for the antigen extract induced no response. In contrast, the challenges with antigen extract induced a dose-dependent increase in sneezing. Concordant with the physiological response was a dose-dependent increase in the levels of histamine and TAME-esterase activity. No significant differences between the two baselines were observed for any of the parameters.

One hour after drug ingestion, the number of sneezes decreased significantly. No difference between the two doses of drug was seen. Increasing the duration of the drug administration did not further increase the suppression of sneezing. After 4 weeks of treatment with 2 mg of ketotifen, the suppression of sneezing was no longer significant. Symptoms other than sneezing (runny nose, congestion, itchy nose or throat) increased with challenge but did not show a significant decrease with treatment. The levels of histamine and TAME-esterase activity during the baseline nasal challenges were compared with those after the challenges occurring 1 h and 1, 2, 3, and 4 weeks after drug administration. With both dosages of drug (1 and 2 mg), there was no significant difference in the amount of histamine and TAME-esterase activity recovered, either between doses or between durations of drug administration.

Despite the efficacy of ketotifen in inhibiting the sneezing response, it did not block the release of histamine or TAME-esterase activity during the early phase response to antigen challenge. We believe that this reflects the failure of the drug to inhibit mast-cell activation. Since other mast-cell mediators were not measured, we cannot exclude the possibility that ketotifen selectively inhibits other mast-cell products.

It has been reported in the literature that ketotifen is maximally effective in asthma after 6 to 12 weeks of continuous use (77). This led us to evaluate the drug's effect after prolonged usage. We found no increased benefit in the prolonged administration of ketotifen. Asthma, unlike nasal challenge of patients with seasonal allergic rhinitis out of season, is marked by underlying inflammation. Underlying inflammation causes increased mediator release in allergic rhinitis and, possibly, in asthma (78). Thus, reduction of inflammation, as opposed to a reduction in mast-cell activation, could explain the result reported in the study of asthma. However, it is not the entire explanation, since a shorter duration of drug exposure inhibited histamine release in physical urticarias with doses of 2 mg b.i.d. (73). This may reflect mast-cell heterogeneity. In this model, the effect of ketotifen on the early reaction was studied. The drug's effect on other aspects of the inflammatory response that follows the early reaction or its effects on subjects with preexisting inflammatory changes were not investigated.

Surprisingly, TAME-esterase activity was not reduced. The majority of this activity relates to plasma kallikrein and is thus a reflection of changes in vascular permeability. In the studies described above, TAME-esterase activity was routinely reduced by the other H_1-antihistamines. This routine inhibition was presumed to reflect antagonism of the effects of histamine on vascular permeability. A potential explanation is that the potency of ketotifen's H_1-antihistaminic activity in reducing changes in vascular permeability differs from its ability to suppress sneezing.

Two criticisms of this study can be raised: the limited number of subjects and the absence of a placebo group. This study was designed as a pilot to determine the dose and duration of ketotifen treatment needed to inhibit mast-cell activation for a subsequent study. The results convinced us not to pursue a placebo-controlled, double-blind study. The number of subjects chosen was based on prior experience with this model. Negative studies can always benefit from a larger number of subjects, but our 10 subjects were observed on 12 different occasions. These multiple observations, in addition to the knowledge that the model used in this study has successfully demonstrated reduction of the parameters assessed in prior studies with similar numbers of subjects, strengthened our confidence in the conclusions. Placebo controls are most useful for evaluating positive subjective effects. The absence of an effect on the recovery of histamine or TAME-esterase activity, two objective parameters, is the important observation.

VIII. LORATADINE

Loratadine is an 8-chloro-N-ethoxycarbonyl derivative of azatadine. Bousquet and colleagues performed a three-way cross-over experiment comparing the effects of 1 week of treatment with loratadine, terfenadine, and placebo on mediator release during the immediate response to nasal challenge with antigen (79). Both active drugs suppressed symptoms and the release of histamine and PGD_2, suggesting that they affected mast-cell activation. Using a similar model, Andersson showed a significant reduction in histamine release during the early allergic reaction after loratadine treatment (80).

We performed a double-blind three-way cross-over study comparing the effects of 1 week of pretreatment with loratadine 10 mg once daily, terfenadine 60 mg b.i.d., or placebo on the acute nasal response to antigen challenge, the cellular influx after challenge, and the induction of increased responsiveness to methacholine 24 h later (unpublished). Fourteen patients completed the protocol. We evaluated the number of sneezes and the levels of histamine, tryptase, PGD_2, LTC_4, kinins, albumin, and TAME-esterase activity in recovered nasal lavage fluids during the acute allergic reaction. Total cell count and the numbers of eosinophils, polymorphonuclear cells, and mononuclear cells in nasal lavage fluids were counted before and 24 h after antigen challenge, and the weight of secretions generated in response to methacholine challenge 24 h after antigen challenge was recorded.

The number of sneezes and the levels of mediators recovered in nasal lavage fluids were compared for all treatment limbs of the study. Pretreatment with loratadine resulted in significant reductions in the number of sneezes after challenge. Furthermore, when looking at the maximal increase above the diluent response, pretreatment with loratadine resulted in a significant reduction in the levels of histamine, kinins, and albumin in recovered lavages. There was a tendency for loratadine to decrease the levels of tryptase, PGD_2, LTC_4, and TAME-esterase activity but this did not reach statistical significance. Terfenadine significantly reduced sneezes and the levels of histamine, kinins, albumin, and TAME-esterase in lavage fluids. There was no statistical difference in the response parameters after pretreatment with loratadine or terfenadine.

The total number of cells and of eosinophils, polymorphonuclear, and mononuclear cells were counted before antigen challenge and 24 h later. When the patients were on placebo, there was a significant increase in the total number of cells 24 h after antigen challenge. In contrast, no significant increase occurred when the patients were on loratadine or terfenadine. Total eosinophils increased significantly 24 h after antigen challenge with the patients on placebo, and neither terfenadine nor loratadine had any inhibitory effect on this event. Total neutrophils and total mononuclear cells did not increase significantly after either placebo or terfenadine but decreased significantly after loratadine. Thus, both of the antihistamines studied had an inhibitory effect on total cellular influx 24 h after nasal antigen challenge, but neither had an inhibitory effect on eosinophil influx. Furthermore, loratadine but not terfenadine decreased the influx of neutrophils and mononuclear cells 24 h after antigen challenge.

Reactivity to methacholine was assessed by quantifying the weight of secretions generated after the application of increasing doses of methacholine to the nasal septum using filter paper discs. Comparing the baseline methacholine challenge performed during the screening visit to that performed 24 h after antigen provocation with the patients on placebo, there was a significant increase in reactivity to methacholine 24 h after antigen challenge, with the responses after baseline and the lowest dose of methacholine (0.41 mg/ml) being significantly increased. Comparing the responses after screening to those after antigen with the patients on loratadine and terfenadine, there was no significant difference between the curves, suggesting that both antihistamines prevented the increased reactivity to methacholine observed 24 h after antigen challenge.

As the reactivity seemed to consist of both an increase in the response to the sham saline challenge and an increase in the response to the lowest dose of methacholine, we compared the secretion weights obtained after the saline challenge for all treatment limbs. When the patients were on placebo, loratadine, or terfenadine, the response after sham saline challenge (24 h after antigen challenge) was significantly increased over the response obtained on the baseline screening visit suggesting that the antihistamines did not inhibit the reactivity component that resulted from the sham saline challenge.

As the weight of generated secretions tended to plateau with the highest doses of methacholine, the curves were compared after excluding the response to the two highest doses of the secretagogue. Comparing the responses after the screening visit to those 24 h after antigen challenge with the patients on loratadine and terfenadine, there was a significant increase in secretions 24 h after antigen challenge with the patients on terfenadine but not loratadine, which suggests that loratadine, but not terfenadine, inhibited reactivity to methacholine after antigen challenge.

Finally, since the secretions after the sham saline challenge were markedly elevated 24 h after antigen challenge in all treatment limbs as compared to the same response during the screening visit, we compared the response after the lowest dose of methacholine after subtracting the secretion weight obtained after the saline challenge in all treatment groups. There was a reduction in this difference after pretreatment with both terfenadine and loratadine as compared to placebo. There was a greater decrease with loratadine, approaching significance ($p = 0.06$), than with terfenadine ($p = 0.1$).

These results suggest that loratadine might not only antagonize the effects of histamine following its release from mast cells after acute antigen provocation, but may also inhibit subsequent cellular influx and hyperresponsiveness to methacholine. In a conjunctival provocation model, loratadine had a significant effect on the inflammatory cell infiltrate after antigen challenge (81).

Raptopoulou and colleagues treated allergic rhinitis patients with either loratadine 10 mg daily or placebo for 1 month (82). Loratadine-treated patients had a significantly lower symptom score. The number of IL-2R and HLA-DR positive cells was significantly decreased in the loratadine-treated group. How loratadine exerted its effect on T-cells remains to be determined.

IX. LEVOCABASTINE

The H₁-receptor antagonist levocabastine was developed for topical treatment of allergic rhinitis. Holmberg and colleagues studied the effect of levocabastine alone and in combination with the H₂-receptor antagonist ranitidine (83). The H₁-receptor antagonist alone reduced sneezing and rhinorrhea, while the H₂-receptor antagonist reduced rhinorrhea. Alone or in combination, neither treatment affected nasal congestion. The study also examined the effect of these treatments on nasal blood flow as assessed by the Xenon washout technique. Combined, the treatments inhibited the reduction in blood flow that accompanied exposure to allergen while on placebo. The reduction in blood flow after antigen challenge is surprising since increased blood flow is a hallmark of inflammation. In contrast, using the laser Doppler technique to assess blood flow, Druce found that oral terfenadine had no effect on blood flow, even though it reduced vascular permeability induced by antigen challenge (84).

Topical preparations could affect ciliary function and hence mucociliary transport in the nose. Such an effect would be expected to have a negative impact on nasal immune function. However, Merkus and colleagues noted no effect of levocabastine on ciliary beat frequency on ex vivo preparations or on saccharin dye clearance in healthy volunteers (85).

X. SUMMARY

Based on in vitro and animal experiments, drugs classified as H_1-receptor antagonists have long been recognized to have additional pharmacological properties. It has been only within the last decade that we have had the ability to test whether these properties occur in vivo in humans. Most of the human studies have focused on the newly developed agents. It is clear that a number of H_1-antihistamines have effects on multiple aspects of the inflammatory response. There are many aspects of the allergic response such as IgE synthesis, neuropeptides, and water and electrolyte transport, that have not been studied.

It is equally clear that these antiallergic effects are not uniformly shared among the classes of drugs. We need to establish the clinical significance of these results. Simply stated, is an H_1-antihistamine that blocks histamine release a better drug clinically compared to an H_1-antihistamine without this property? Most clinical trials comparing different H_1-antihistamines have not consistently found advantages of one drug over another. While this lack of difference may relate to the current limits of our testing systems to distinguish subtle differences, the clinical observation may be emphasizing the greater relevance of the effects of released histamine to the allergic response. If the structure of the H_1-antihistamines that relates to the other antiallergic properties can be determined, then novel agents for the treatment of allergic rhinitis may be developed.

REFERENCES

1. Dale HH, Laidlaw PP. The physiological action of β-imidozolylethylamine. J Physiol (Lond) 1910;41:318–44.
2. Best CH, Dale HH, Dudley HW, Thorpe WV. The nature of the vasodilator constituents of certain tissue extracts. J Physiol (Lond) 1927;62:397–417.
3. Lewis T. The Blood Vessels of the Human Skin. London: Shaw & Sons Ltd, 1927.
4. Bovet D, Staub A. Action protectrice des ethers phenoliques au cours de l'intoxication histaminique. CRS Soc Biol (Paris) 1937;124:547–549.
5. Bovet D, Horclois R, Walthert F. Propriétés antihistaminiques de la N-p-méthoxybenzyl N-diméthylaminoéthyl α amino-pyridine. CR Soc Biol (Paris) 1944;138:99–100.
6. Douglas WW. Histamine and 5-hydroxytryptamine (serotonin) and their antagonists. In: Gilman A, Goodman L, Rall T, Murad F, eds. The Pharmacologic Basis of Therapeutics. 7th ed. New York. Macmillan and Co, 1987; 605–638.
7. Togias A, Naclerio RM, Warner J, Proud D, Kagey-Sobotka A, Nimmagadda I, Norman PS, Lichtenstein LM. Demonstration of inhibition of mediator

release from human mast cells by azatadine base: in vivo and in vitro evaluation. JAMA 1986;255:225–229.

8. Lichtenstein LM, Gillespie E. The effects of the H₁ and H₂ antihistamines on "allergic" histamine release and its inhibition by histamine. J Pharmacol Exp Ther 1975;192:441–450.

9. Stedman's Medical Dictionary. 22d ed. Baltimore. Williams & Wilkins Company, 1972.

10. Tada T, Ishizaka K. Distribution of gamma E-forming cells in lymphoid tissues of the human and monkey. J Immunol 1970;104:377–387.

11. Grangette C, Gruart V, Quassi MA, Rizvi F et al. IgE receptor on human eosinophils (fcₑRII): comparison with B cell CD23 and association with an adhesion molecule. J Immunol 1989;143:3580–3588.

12. Melewicz FM, Spiegelberg HL. Fc receptors for IgE on a subpopulation of human peripheral blood monocytes. J Immunol 1980;125:1026–1029.

13. Cines DB, van der Keyl H, Levinson AI. In vitro binding of an IgE protein to human platelets. J Immunol 1986;136:3433–3440.

14. Gomez E, Corrado OJ, Galdwin DL, Swanston AR, Davies RJ. Direct in vivo evidence for mast cell degranulation during allergen-induced reactions in man. J Allery Clin Immunol 1986;637–645.

15. Iliopoulos O, Proud D, Adkinson NF Jr, Norman PS, Kagey-Sobotka A, Lichtenstein LM, Naclerio RM. Relationship between the early, late and rechallenge reaction to nasal challenge with antigen. Observations on the role of inflammatory mediators and cells. J Allergy Clin Immunol 1990;86: 851–861.

16. McLean J, Matthews K, Solomon WR, et al. Effect of histamine and methacholine on nasal airway resistance in atopic and nonatopic subjects. J Allergy Clin Immunol 1977;59:165–170.

17. Blackley CH. Experimental researches on the cause and nature of catarrhus aestivus. London: Bailliere, Tindal, Cox Ltd, 1873.

18. Gleich GJ. The late phase of the immunoglobulin E-mediated reaction: a link between anaphylaxis and common allergic disease? J Allergy Clin Immunol 1982;70:160–169.

19. Naclerio RM, Proud D, Togias AG, Adkinson NF Jr, Meyers DA, Kagey-Sobotka A, Plaut M, Norman PS, Lichtenstein LM. Inflammatory mediators in late antigen-induced rhinitis. N Engl J Med 1985;313:65–70.

20. Pipkorn U, Proud D, Lichtenstein LM, Schleimer RP, Peters SP, Adkinson NF Jr, Kagey-Sobotka A, Norman PS, Naclerio RM. Effect of short-term systemic glucocorticoid treatment on human nasal mediator release after antigen challenge. J Clin Invest 1987;80:957–961.

21. Bascom R, Wachs M, Naclerio RM, Pipkorn U, Galli SJ, Lichtenstein LM. Basophil influx occurs after nasal antigen challenge: Effects of topical corticosteroid pretreatment. J Allergy Clin Immunol 1988;81:580–589.

22. Bascom R, Pipkorn U, Lichtenstein LM, Naclerio RM. The influx of inflammatory cells into nasal washings during the late response to antigen challenge: Effect of systemic steroid pretreatment. Am Rev Respir Dis 1988;138: 406–412.

23. Linder A, Venge P, Deuschl H. Eosinophil cationic protein and myeloperoxidase in nasal secretions as markers of inflammation in allergic rhinitis. Allergy 1987;42:583–590.

24. Svensson C, Andersson M, Persson CGA, Venge P, Alkner U, Pipkorn U. Albumin, bradykinins, and eosinophil cationic protein on the nasal mucosa surface in patients with hay fever during natural allergen exposure. J Allergy Clin Immunol 1990;85:828–833.

25. Denburg JA, Dolovich J, Harnish D. Basophil mast cell and eosinophil growth and differentiation factors in human allergic disease. Clin Exp Allergy 1989;19:249–254.

26. Connell JT. Quantitative intranasal pollen challenge. III. The priming effect in allergic rhinitis. J Allergy 1969;43:33–44.

27. Hastie R, Chir B, Heroy JH, Levy DA. Basophil leukocytes and mast cells in human nasal secretions and scrapings studied by light microscopy. Lab Invest 1979;40:554–561.

28. Enerback L, Pipkorn U, Olofsson A. Intraepithelial migration of mucosal mast cells in hay fever: ultrastructural observations. Int Arch Allergy Appl Immunol 1986;19:17–43.

29. Walden SM, Proud D, Lichtenstein LM, Kagey-Sobotka A, Naclerio RM. Antigen-provoked increase in histamine reactivity: observations on mechanisms. Am Rev Respir Dis 1991;144:642–8.

30. Konno A, Togawa K, Nishihira S. Seasonal variation of sensitivity of nasal mucosa in pollinosis. Arch Otorhinolaryngol 1981;232:253–261.

31. Majchel AM, Proud D, Freidhoff L, Creticos PS, Norman PS, Naclerio RM. The nasal response to histamine challenge: effect of the pollen season and immunotherapy. J Allergy Clin Immunol. 1992;90:85–91.

32. Naclerio RM, Meier HL, Kagey-Sobotka A, Adkinson NF Jr, Meyers DA, Norman PS, Lichtenstein LM: Mediator release after nasal airway challenge with allergen. Am Rev Respir Dis 1983;128:597–602.

33. Naclerio RM, Kagey-Sobotka A, Lichtenstein LM, Freidhoff L, Proud D. Terfenadine, an H_1-antihistamine, inhibits histamine release in vivo in the human. Am Rev Respir Dis 1990;142:167–171.

34. Mygind N. Nasal Allergy. Oxford: Blackwell Scientific, 1978.

35. Baroody FM, Wagenmann M, Naclerio RM. A comparison of the secretory response of the nasal mucosa to methacholine and histamine. J Appl Physiol 1993;74:2661–2671.

36. Wagenmann M, Baroody FM, Kagey-Sobotka A, Lichtenstein LM, Naclerio RM. The effect of terfenadine on unilateral nasal challenge with allergen. J Allergy Clin Immunol 1994;93:594–605.

37. Proud D, Henley JO, Gwaltney JM, Naclerio RM. Recent studies on the role of kinins in inflammatory diseases of human airways. Adv Exp Med Biol 1989;247A:117–123.

38. Proud D, Reynolds CJ, LaCapra S, Kagey-Sobotka A, Lichtenstein LM, Naclerio RM. Nasal provocation with bradykinin induces symptoms of rhinitis and a sore throat. Am Rev Respir Dis 1988;37:613–616.

39. Spector S, Altman R: Cetirizine. A Novel Antihistamine. Am J Rhinol 1987; 1:147–149.

40. Michel L, De Vos C, Rihoux J-P, Burtin C, Benveniste J, Dubertret L. Inhibitory effect of oral cetirizine on in vivo antigen-induced histamine and PAF-acether release and eosinophil recruitment in human skin. J Allergy Clin Immunol 1988;82:101–109.

41. Naclerio RM, Proud D, Kagey-Sobotka A, Freidhoff L, Norman PS, Lichtenstein LM. The effect of cetirizine on early allergic response. Laryngoscope 1989;99:596–599.

42. Klementsson H, Andersson M, Pipkorn U. Allergen-induced increase in non-specific nasal reactivity is blocked by antihistamines without a clear-cut relationship to eosinophil influx. J Allergy Clin Immunol 1990;86:466–472.

43. Majchel AM, Proud D, Hubbard WC, Naclerio RM. Histamine stimulation of the nasal mucosa does not induce prostaglandin or leukotriene generation or induce methacholine hyperresponsiveness. Int Arch Allergy Appl Immunol 1991;95:145–55.

44. Noureddine G, Thompson M, Brennan F, Proud D, Kagey-Sobotka A, Lichtenstein LM, Naclerio RM, Togias A. Nasal antigen challenge in asthmatics: Effects of cetirizine on nasal and pulmonary responses. J Allergy Clin Immunol 1994;93:177 (abstract).

45. Diamantis W, Harrison JE, Melton J, Perhach JL, Sofia RD. In vivo and in vitro H₁ antagonist properties of azelastine. Pharmacologist 1981;23:149–154.

46. Chand N, Diamantis W, Sofia RD. Antagonism of leukotrienes, calcium and other spasmogens by azelastine. Pharmacologist 1984;26:152–157.

47. Diamantis W, Chand N, Harrison JE, et al. Inhibition of release of SRS-A and its antagonism by azelastine, an H₁-antagonist-antiallergic agent. Pharmacologist 1982;24:200–9.

48. Zechel HJ, Brock N, Lenke D, Achterrath-Tuckermann U. Pharmacological and toxicological properties of azelastine: a novel antiallergic agent. Arzneim Forsch 1981;31:1184–1193.

49. Fields DAS, Pillar J, Diamantis W, et al. Inhibition by azelastine of nonallergic histamine release from rat peritoneal mast cells. J Allergy Clin Immunol 1984;73:400–403.

50. Chand N, Pillar J, Diamantis W, Perhach JL, Sofia RD. Inhibition of calcium ionophore (A23187)-stimulated histamine release from rat peritoneal mast cells by azelastine: implications for its mode of action. Eur J Pharmacol 1983;96:227–233.

51. Chand N, Pillar J, Diamantis W, Sofia RD. Inhibition of allergic histamine release by azelastine and selected antiallergic drugs from rabbit leukocytes. Int Arch Allergy Appl Immunol 1985;77:451–455.

52. Little MM, Wood DR, Casale TB. Azelastine inhibits stimulated histamine release from human lung tissue in vitro but does not alter cyclic nucleotide content. Agents Actions 1989;28:16–21.

53. Little MM, Casale TB. Azelastine inhibits IgE-mediated human basophil histamine release. J Allergy Clin Immunol 1989;83:862–865.

54. Chand N, Diamantis W, Sofia RD. Modulation of in vitro anaphylaxis of guinea pig isolated tracheal segments by azelastine, inhibitors of arachidonic acid metabolism, and selected antiallergic drugs. Br J Pharmacol 1986;87: 443–448.

55. Taniguchi K, Takanaka K. Inhibitory effects of various drugs on phorbol myristate acetate- and N-formyl-methionyl-leucyl-phenylalanine-induced O_2 production in polymorphonuclear leukocytes. Biochem Pharmacol 1984;33: 3165–3169.

56. Busse W, Randlev B, Sedgwick J. The effect of azelastine on neutrophil and eosinophil generation of superoxide. J Allergy Clin Immunol 1989;83: 400–405.

57. Kemp JP, Meltzer EO, Orgel HA et al. A dose-response study of the broncho-dilator action of azelastine in asthma. J Allergy Clin Immunol 1987;79: 893–899.

58. Atkins P, Merton H, Karpink P, Weliky I, Zweiman B. Azelastine inhibition of skin test reactivity in humans [Abstract]. J Allergy Clin Immunol 1985; 75:167.

59. Weiler JM, Donnelly A, Campbell BH, Connell JT, Diamond L, Hamilton LH, Rosenthal RR, Hemsworth GR, Perhach JL Jr. Multicenter, double-blind, multiple-dose, parallel-groups efficacy and safety trial of azelastine, chlorpheniramine, and placebo in the treatment of spring allergic rhinitis. J Allergy Clin Immunol 1988;82:801–811.

60. Shin M-H, Baroody F, Proud D, Kagey-Sobotka A, Lichtenstein LM, Naclerio RM. The effect of azelastine on the early allergic response. Clin Exp Allergy 1992;22:289–295.

61. Daniels C, Temple DM. The inhibition by azatadine of the immunological release of leukotrienes and histamine from human lung fragments. Eur J Pharmacol 1986;123:463–465.

62. Achterrath-Tuckermann U, Simmet T, Luck W, Szelenyi I, Peskar BA. Inhi-bition of cysteinyl-leukotriene production by azelastine and its biologic sig-nificance. Agents Actions 1988;24:217–223.

63. Katayama S, Tsunoda H, Sakuma Y, et al. Effect of azelastine on the release and action of leukotriene C_4 and D_4. Int Arch Allergy Appl Immunol 1987; 83:284–289.

64. Warner JA, Lichtenstein LM, MacGlashan DW Jr. Effects of a specific inhib-itor of the 5-lipoxygenase pathway on mediator release from human basophils and mast cells. J Pharmacol Exp Ther 1988;247:218–222.

65. Nieber K, Baumgarten C, Rathsack R, Furkert J, Laake E, Muller S, Kunkel G. Effect of azelastine on substance P content in bronchoalveolar and nasal lavage fluids of patients with allergic asthma. Clin Exp Allergy 1993;23: 69–71.

66. Thomas KE, Ollier S, Ferguson H, Davies RJ. The effect of intranasal azelas-tine, Rhinolast, on nasal airways obstruction and sneezing following provoca-tion testing with histamine and allergen. Clin Exp Allergy 1992;22:642–647.

67. Morley J. Ketotifen: New Pharmacological Aspects. Sandoz Pharmaceutical Division. April 1985.
68. Trzeciakowski JP, Mendelsohn N, Levi R. Antihistamines. In: Middleton E, Reed CE, Ellis EF, Adkinson NF Jr, Yunginger JW, eds. Allergy: Principles and Practice. St Louis: CV Mosby, 1988:715–738.
69. Rafferty P, Holgate ST. Histamine and its antagonists in asthma. J Allergy Clin Immunol 1989;84:144–151.
70. Martin M, Romer D. Antianaphylactic properties of ketotifen in animal experiments. Triangle 1978;17:141–147.
71. Paul W, Beets JL, Bray MA, Morley J. Anti-inflammatory drug action on allergic responses in guinea-pig skin. Agents Actions 1978;8:509–514.
72. Martin U, Baggiolini M. Dissociation between the anti-anaphylactic and the anti-histamine actions of ketotifen. Arch Pharmacol 1981;316:186–189.
73. Huston DP, Bressler RB, Kaliner M, Sowell LK, Baylor MW. Prevention of mast-cell degranulation by ketotifen in patients with physical urticarias. Ann Intern Med 1986;104:507–510.
74. Tomioka H, Yoshida S, Tanaka M, et al. Inhibition of chemical mediator release from human leukocytes by a new antiasthma drug HC 20-511 (ketotifen). Monogr Allergy 1979;14:313–317.
75. Morley J, Page CP, Mazzoni L, Sanjar S. Effects of ketotifen upon responses to platelet activating factor: a basis for asthma prophylaxis. Ann Allergy 1986;56:335–340.
76. Majchel AM, Proud D, Kagey-Sobotka A, Lichtenstein LM, Naclerio RM. Ketotifen reduces sneezing but not histamine release following nasal challenge with antigen. Clin Exp Allergy 1990;20:701–705.
77. Rackham A, Brown CA, Chandra RK, et al. A Canadian multicenter study with Zaditen (ketotifen) in the treatment of bronchial asthma in children aged 5 to 17 years. J Allergy Clin Immunol 1989;84:286–296.
78. Wachs M, Proud D, Lichtenstein LM, Kagey-Sobotka A, Norman PS, Naclerio RM. Observations on the pathogenesis of nasal priming. J Allergy Clin Immunol 1989;84:492–501.
79. Bousquet J, Lebel B, Chanal I, Morel A, Michel FB. Antiallergic activity of H₁-receptor antagonists assessed by nasal challenge. J Allergy Clin Immunol 1988;82:881–887.
80. Andersson M, Nolte H, Baumgarten C, Pipkorn U. Suppressive effect of loratadine on allergen-induced histamine release in the nose. Allergy 1991;46:540–546.
81. Ciprandi G, Buscaglia S, Pesce GP, Marchesi E, Canonica GW. Protective effect of loratadine on specific conjunctival provocation test. Int Arch Allergy Appl Immunol 1991;96:344–347.
82. Raptopoulou G, Ilonidis G, Orphanou H, Preponis C, Sidiropoulos J, Lazaridis T, Goulis G. The effects of loratadine on activated cells of the nasal mucosa in patients with allergic rhinitis. J Invest Allergy Clin Immunol 1993;3:192–197.
83. Holmberg K, Pipkorn U, Bake B, Blychert LO. Effects of topical treatment

with H_1- and H_2-antagonists on clinical symptoms and nasal vascular reactions in patients with allergic rhinitis. Allergy 1989;44:281–287.

84. Druce HM. The effects of an H_1-receptor antagonist, terfenadine, on histamine-induced microcirculatory changes and vasopermeability in nasal mucosa. J Allergy Clin Immunol 1990;86:344–352.

85. Merkus FWHM, Schüsler-van-Hess MTIW. Influence of levocabastine suspension on ciliary beat frequency and mucociliary clearance. Allergy 1992; 47:230–233.

7

H_1-Receptor Antagonists: Pharmacokinetics and Clinical Pharmacology

Keith J. Simons and F. Estelle R. Simons
The University of Manitoba, Winnipeg, Manitoba, Canada

I. INTRODUCTION

The pharmacokinetics and clinical pharmacology of some of the first-generation H_1-receptor antagonists and most of the second-generation H_1-receptor antagonists have been well studied (Table 1). H_1-receptor antagonists are generally present in low concentrations in serum and other body fluids and are not measured routinely. Sensitive and specific assays are now available for plasma measurement of some of the older medications and for all of the newer ones. We will provide an overview of the pharmacokinetic studies and drug interaction studies that have been performed using these assays.

We will also review representative clinical pharmacology studies that have greatly increased our understanding of the onset, duration, and offset of action, and of the relative effectiveness of H_1-receptor antagonists. In most of these studies, the unique ability of H_1-receptor antagonists to suppress the wheal and flare response produced by histamine or allergen in the skin has been utilized as an end point. In some studies, their ability to suppress the histamine or allergen-induced response in the nose, the eye, or the lower airways has been evaluated.

II. FIRST-GENERATION H₁-RECEPTOR ANTAGONISTS

Some of the first-generation H_1-receptor antagonists have been in use for four or five decades. For many of these medications, there is little pharmacokinetic or clinical pharmacology information available, as they were introduced in an era when regulatory agencies did not require these studies before approving a new drug (Table 1) (1–50).

Some first-generation H_1-receptor antagonists such as triprolidine and tripelennamine are not used as much as they were years ago, because of

Table 1 Overview of Studies of Pharmacokinetics and Clinical Pharmacology of H_1-Receptor Antagonists

H₁-receptor antagonist	Assay	Pharmacokinetics		Clinical pharmacology				Refs.
		Healthy adults	Other subjects	Skin	Lung	Nose		
First-generation								1–50
Azatadine	HPLC	+				+		
Brompheniramine	HPLC	+		+		+		
Chlorpheniramine	GLC, HPLC	+ + +	+ +	+ +	+			22–35
Cyproheptadine	GLC-MS	+						
Diphenhydramine	GLC-NP, HPLC	+ +	+	+		+		44–50
Hydroxyzine	HPLC	+ +	+ + +	+ +				36–43
Mequitazine	GLC-MS							
Promethazine	HPLC	+ +						
Tripelennamine	GLC-NP	+						
Triprolidine	HPLC, RIA	+ +	+					
Second-generation								51–202
Terfenadine (TAM)ᵃ	RIA, HPLCᵃ, GLC-MS	+ + +	+ + +	+ + +	+ +	+ +		51–84
Astemizole (DMA)ᵃ	RIA, HPLCᵃ	+ + +	+	+ +	+			85–110
Loratadine (DCL)ᵃ	RIA, HPLCᵃ	+ + +	+ + +	+ + +	+	+		111–132
Cetirizine	GLC, HPLC	+ + +	+ + +	+ + +	+	+		133–148
Acrivastine	HPLC, GLC-MS	+		+				149–158
Ketotifen	RIA, GLC-MS	+	+	+	+	+		159–171
Ebastine (carebastine)ᵃ	GLC-MS, HPLCᵃ	+	+ + +	+	+			172–180
Azelastine (DMAz)ᵃ	HPLC, GLC-MS	+		+	+ +	+		181–198
Levocabastine	RIA, HPLC	+	+	+		+		199–202

ᵃ Active metabolite.
Number and quality of studies: + = minimal; + + = adequate; + + + = good.

TAM = terfenadine acid metabolite; DMA = desmethylastemizole; DCL = descarbo-ethoxyloratadine; DMAz = desmethylazelastine; HPLC = high performance liquid chromatography; GLC = gas liquid chromatography; GLC-MS = gas liquid chromatography-mass spectroscopy; RIA = radioimmunoassay.

their short plasma elimination half-life values and requirement for three or four times daily administration, combined with relative lack of efficacy and a high incidence of sedation and other adverse effects. Others, such as chlorpheniramine, hydroxyzine, and diphenhydramine, are still widely used and are among the best studied of the older H_1-receptor antagonists. They will be reviewed in further detail.

A. Chlorpheniramine

1. Pharmacokinetics

Chlorpheniramine (21–35) is well absorbed after oral administration. Its absolute bioavailability (25–59%) is low, due to a considerable first-pass effect following oral administration. Chlorpheniramine terminal elimination half-life values are surprisingly long, with mean values of 21–28 h reported in adults and 9–13 h reported in children.

2. Clinical Pharmacology

The duration of action of a single dose of chlorpheniramine as assessed by suppression of the histamine-induced wheal and flare is at least 24 h. There is a direct correlation between mean serum chlorpheniramine concentrations and allergic rhinitis symptom relief ($r = 0.75$), and an inverse correlation between mean serum chlorpheniramine concentrations and mean wheal and flare diameters ($r = -0.68$ and $r = -0.92$, respectively). The rationale for using time-release or repeat-action formulations of chlorpheniramine is to reduce peak serum and CNS concentrations, and thus reduce CNS adverse effects (31–35).

Tachyphylaxis to chlorpheniramine has been reported, and seems to be related to decreased compliance, rather than to an increased rate of metabolism (35).

B. Hydroxyzine

1. Pharmacokinetics

Hydroxyzine (36–43) is well absorbed after oral administration. The pharmacokinetics of hydroxyzine and of the active metabolite cetirizine arising from hydroxyzine in vivo have been extensively studied. After a single dose of hydroxyzine in healthy young adults, mean terminal half-life values of 14–20 h have been reported. In children with atopic dermatitis, the mean half-life of this medication is 7 h. In elderly adults, the mean half-life is 29 h, and in adults with hepatic dysfunction secondary to primary biliary cirrhosis, a mean half-life of 37 h has been reported. In an animal model, the pharmacokinetics of hydroxyzine administered daily for 150 days did not change over time and there was no evidence for autoinduction of hepatic enzyme systems, more rapid clearance of the drug, or lower serum concentrations of the drug over time.

Hydroxyzine enters the skin rapidly (Fig. 1) and sustained high concentrations in skin after single or multiple doses may contribute to its well-known efficacy in the treatment of skin disorders in which histamine plays a role.

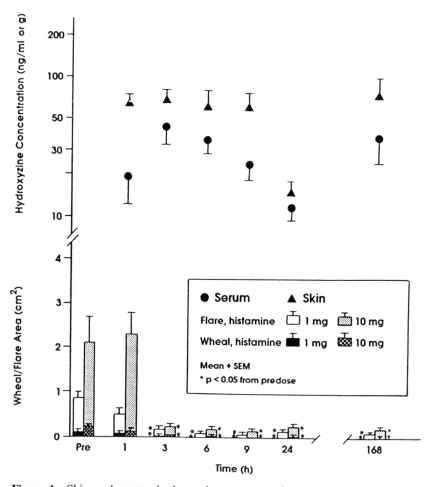

Figure 1 Skin and serum hydroxyzine concentrations versus time plots, and wheal and flare areas after epicutaneous tests with histamine phosphate 1 mg/ml and 10 mg/ml. Tests were performed at baseline and 1, 3, 6, 9, and 24 h after the initial dose of hydroxyzine 50 mg. Subjects then took hydroxyzine 50 mg at 2100 h for 6 consecutive days, and the tests were repeated at 168 h (steady state), exactly 12 h after the seventh and last dose. (From Ref. 43.)

2. Clinical Pharmacology

Significant suppression of the histamine-induced wheal and flare following a single 50-mg hydroxyzine dose lasts from 1–36 h in healthy young adults, from 1–144 h in healthy elderly adults, and from 1–120 h (wheal) and 1–144 h (flare) in subjects with hepatic dysfunction (36–43).

C. Diphenhydramine

1. Pharmacokinetics

Diphenhydramine (44–50) is well absorbed after oral administration. Its absolute bioavailability is 69%. The mean terminal elimination half-life value of diphenhydramine in young adults is 9 h, longer than the half-life of 5 h obtained in children, but shorter than the half-life of about 14 h obtained in elderly adults (Fig. 2). The half-life of diphenhydramine is prolonged in subjects with chronic liver disease.

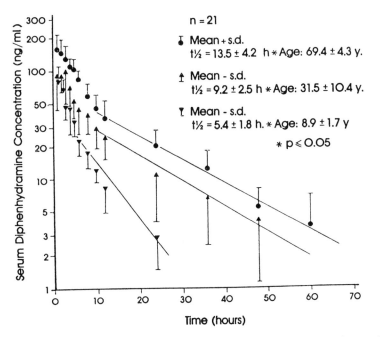

Figure 2 Serum diphenhydramine concentration versus time plots and terminal elimination half-life values (t½) after a single dose of diphenhydramine 1.25 mg/kg in elderly adults (mean age 69.4 ± 4.3 years), young adults (mean age 31.5 ± 10.4 years), and children (mean age 8.9 ± 1.7 years). There were seven subjects in each group. (From Ref. 48.)

2. Clinical Pharmacology

Diphenhydramine appears to have a shorter duration of action than chlorpheniramine or hydroxyzine. After a single dose, histamine-induced wheals and flares are significantly suppressed for 12 h. Significant protection against symptoms produced by nasal allergen challenge lasts at least 10 h.

III. SECOND-GENERATION H₁-RECEPTOR ANTAGONISTS

A. Terfenadine

1. Pharmacokinetics

Terfenadine (51–69) is rapidly absorbed after oral administration as a tablet or liquid suspension. Mean peak plasma terfenadine concentrations of 1.5 ng/mL and 4.5 ng/mL have been determined by radioimmunoassay 1 h after a 60-mg and a 180-mg dose, respectively (Table 2). Absorption is not affected by food.

Terfenadine is 97% bound to plasma protein and terfenadine acid metabolite (TAM) is 70% protein-bound.

In studies in which ^{14}C-terfenadine has been administered, the terminal elimination half-life of terfenadine is reported to be 16–23 h. Terfenadine is rapidly and extensively (99%) metabolized by the CYP3A4 isoenzyme in the cytochrome P_{450} system on the first pass through the liver. Unmetabolized terfenadine is not usually found in the plasma after a 60-mg dose. The two major metabolites of terfenadine are: TAM, also known as the carboxylic acid derivative or metabolite I, which has antihistaminic activity and a piperidine-carbinol compound, metabolite II, which has no apparent antihistaminic activity. Peak plasma concentration of TAM is reached approximately 3 h after dosing with terfenadine and is not significantly affected by food ingestion. Following administration of single oral 60-, 120-, and 180-mg terfenadine doses, peak plasma TAM concentrations increase linearly with dose. In young adults, the terminal elimination half-life of TAM is approximately 17 h, but it may be longer in adults with hepatic dysfunction and in elderly adults and shorter in children. The pharmacokinetics of TAM do not change during regular terfenadine administration over 56 days.

Potentially serious drug interactions may occur when terfenadine is coadministered with a macrolide antibiotic such as erythromycin or clarithromycin, an imidazole antifungal such as ketoconazole or itraconazole, or any other medication or other substance, e.g., the naringenin in grapefruit juice that inhibits the hepatic mixed-function oxygenase cytochrome

Table 2 Pharmacokinetics of Representative H_1-Receptor Antagonists

H_1-receptor antagonist[a] (metabolite)	t_{max}[b] (h)	$t_{1/2}$[c] (h)	Cl[d] (ml/min/kg)
First-generation			
Chlorpheniramine	2.8 ± 0.8	27.9 ± 8.7	1.8 ± 0.1
Hydroxyzine	2.1 ± 0.4	20.0 ± 4.1	9.8 ± 3.2
Diphenhydramine	1.7 ± 1.0	9.2 ± 2.5	23.3 ± 9.4
Second-generation			
Terfenadine (terfenadine acid metabolite)	0.78–1.1 (3)	16–23 (17)	n/a[+] (598–697 mL/min)
Astemizole (desmethyl-astemizole)	0.5 ± 0.2 to 0.7 ± 0.3	1.1 days (9.5 days)	1500 mL/min
Loratadine (descarboethoxy-loratadine)	1.0 ± 0.3 (1.5 ± 0.7)	11.0 ± 9.4 (17.3 ± 6.9)	142–202 n/a[+]
Cetirizine	1.0 ± 0.5	7.4 ± 1.6	1.0 ± 0.2
Acrivastine	1.4 ± 0.4 0.85–1.4	1.7 ± 0.2 1.4–2.1	4.4 ± 0.6 4.56
Ketotifen[++]	2.3 ± 1.2–3.9 ± 2.2	12.2 ± 4.5–18.3 ± 6.7	n/a[+]
Azelastine (desmethyl-azelastine)	5.3 ± 1.6 (20.5)	22 ± 4 (54 ± 15)	8.5 ± 3.2 n/a[+]
Ebastine[++] (carebastine)	(3.6 ± 1.0)	(13–15)	n/a[+]

Results are mean $\pm$ standard deviation.
[a] In normal young adults.
[b] Time from oral intake to peak plasma concentration.
[c] Terminal elimination half-life.
[d] Clearance.
[+] Data not available.
[++] Not available in the U.S. at time of publication.

P_{450} CYP3A4 system (Table 3). When coadministered with a cytochrome P_{450} inhibitor, terfenadine parent compound concentrations accumulate in serum and in cardiac tissue, and may lead to prolongation of the QTc interval (Fig. 3) and ventricular dysrhythmias such as torsade de pointes as discussed extensively in Chapter 14 of this book. There is no evidence for a pharmacokinetic interaction between terfenadine and theophylline or between terfenadine and phenytoin. Coadministration of terfenadine and pseudoephedrine does not affect the pharmacokinetics of terfenadine.

2. Clinical Pharmacology

Current recommended dosage regimens for terfenadine (21, 32, 70–84) are 60 mg every 12 h, or 120 mg every 24 h (Table 4). Ingestion of terfena-

Table 3 Interaction of Terfenadine 60 mg q 12 h. × 7 Days* with Other
Pharmacological Agents in Healthy Subjects

Drug coadministered for 7 days	Dose (mg)	n	↑ Terf. parent compound	↑ Terf. acid metabolite (c_{max})(%)	↑ Terf. acid metabolite (t_{max})(%)	↑ AUC terf. acid metabolite (%)	EKG changes
Macrolide Antibiotics							
Erythromycin	500 q8h	9	3/9	100	82	170	yes
Clarithromycin	500 q12h	6	4/6	109	21	156	yes
Azithromycin	500 load 250 qd	6	0/6	no	28	no	no
Imidazole Antifungals							
Ketoconazole	200 q12h	6	6/6	no	no	57	yes
Itraconazole	200 qd	6	6/6	100	148	30 (ns)	yes
Fluconazole	200 qd	6	0/6	26 (ns)	74	34	no
H₂-Receptor Antagonists							
Cimetidine	600 q12h	6	0/6	no	no	12 (ns)	no
Ranitidine	150 q12h	6	0/6	−27 (ns)	no	−21 (ns)	no

Abbreviations: qd = daily; C_{max} = maximum plasma concentration; t_{max} = time at which C_{max} occurs; AUC = area under the curve; ns = not significant.
From Refs. 63–68.
*Itraconazole: single dose study of terfenadine 120 ng.

dine 60 mg or 120 mg produces prompt suppression of the histamine-induced wheal and flare. After terfenadine dosing, peak plasma TAM concentrations precede maximum antihistaminic activity by 1–3 h. Significant suppression of the histamine-induced wheals and flares has been shown to persist for 24 h following the administration of a single 60- or 120-mg oral dose in young adults and also in healthy elderly subjects given the 60-mg dose. A single dose of 60 or 120 mg is significantly more effective than a single dose of astemizole 4 mg, loratadine 10 mg, or chlorpheniramine 4 mg but less effective than cetirizine 10 mg (Fig. 4).

Terfenadine pretreatment protects against histamine or allergen-induced nasal symptoms. Inhibition of histamine-induced bronchoconstriction after pretreatment with terfenadine is dose-related. Mean histamine doses required to reduce the FEV_1 by 20% (PC_{20}) are 3.8 mg/mL after terfenadine 60 mg, 8.1 mg/mL after 120 mg, and 8.7 mg/mL after 180 mg, compared with 0.6 mg/mL after placebo.

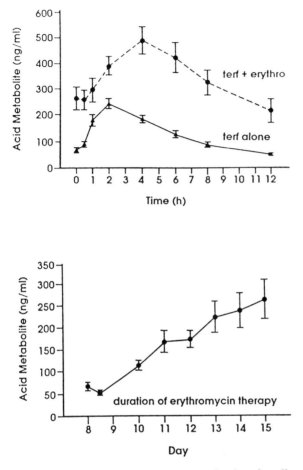

Figure 3 Nine healthy subjects received terfenadine 60 mg q 12 h for 7 days before initiation of concurrent treatment with erythromycin 500 mg q 8 h. Terfenadine acid metabolite (terfenadine carboxylate) concentrations (ng/mL), mean ± SEM, are shown in the top graph, after dosing with terfenadine alone and after terfenadine/erythromycin coadministration from days 9 through 15. While subjects were receiving erythromycin concomitantly, the average C_{max} for the terfenadine acid metabolite was 507 ng/mL, an increase of 107%; the t_{max} was delayed to 4.0 h from 2.2 h, and the mean AUC was increased 170%, from 1577 ng h/ml to 4255 ng h/ml. Trough terfenadine acid metabolite concentrations (mean ± SEM), shown in the bottom graph, increased progressively after erythromycin was added to the treatment regimen during days 9 to 15. Three subjects accumulated the unmetabolized terfenadine and developed QTc prolongation (not shown). (From Ref. 63.)

A 7-day course of terfenadine produces histamine blockade that persists for another 7 days after the medication is stopped; therefore, terfenadine should be discontinued at least 1 week before skin tests or inhalation challenge tests with histamine or antigen are performed. Healthy subjects receiving terfenadine 60 mg orally every 12 h for 56 days do not develop tachyphylaxis or subsensitivity to the suppressive effect of terfenadine on the histamine-induced wheals and flares.

B. Astemizole

1. Pharmacokinetics

Astemizole (85–98) is absorbed well. Intake with food does not affect the extent of absorption, although peak plasma levels are obtained at 2.2 h, about 1 h later than after intake on an empty stomach. Prompt administration of 25 g of activated charcoal effectively prevents the absorption of astemizole from the gastrointestinal tract; this information is relevant in the management of astemizole overdose. Astemizole is about 97% bound to plasma proteins.

Astemizole undergoes extensive first-pass metabolism. One of its major metabolites, desmethylastemizole (DMA), has clinically significant antihistaminic activity. Peak plasma concentrations of astemizole and DMA are reached in 1–4 h. Steady-state plasma concentrations of unchanged astemizole are reached within 1 week, and those of astemizole plus metabolites within 4 weeks. Further accumulation does not occur after steady state is reached.

After a single 30-mg dose of astemizole, the terminal elimination half-life value of unchanged astemizole is 1.1 days, and the terminal elimination half-life value of DMA is 9.5 days. In a multiple-dose study, the terminal elimination half-life of astemizole and DMA has been reported as 13 days.

There are few studies of the pharmacokinetics of astemizole in the elderly, or in subjects with impaired hepatic or renal function. The pharmacokinetics of astemizole in children are similar to those reported in adults. Astemizole is not eliminated by hemodialysis.

The metabolism of astemizole has been investigated in human liver microsome preparations in vitro. In this system, the metabolism of astemizole to DMA is less sensitive to the effects of ketoconazole than the metabolism of terfenadine to hydroxyterfenadine is, and itraconazole does not inhibit astemizole metabolism significantly, although it does inhibit terfenadine metabolism. Few prospective clinical studies of potential drug interactions between astemizole and macrolide antibiotics, imidazole antifungals, or other cytochrome P_{450} inhibitors have been performed to date. Astemizole does not affect salivary antipyrine clearance, 6-β-hydrocorti-

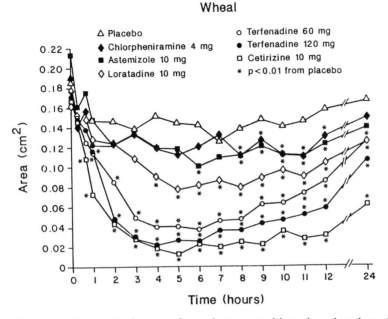

Figure 4 Mean wheal areas after epicutaneous histamine phosphate (1 mg/ml) before and up to 24 h after a single oral dose of placebo or H₁-receptor antagonist, in a double-blind, cross-over study in 20 healthy men. The rank order of suppression was, from most effective to least effective: cetirizine 10 mg > terfenadine 120 mg > terfenadine 60 mg > loratadine 10 mg > astemizole 10 mg > chlorpheniramine 4 mg > placebo. (Ref. 32.)

sone excretion, ethanol elimination, or indocyanine green clearance. The pharmacokinetics of astemizole administered concomitantly with pseudoephedrine do not differ from the pharmacokinetics of astemizole administered alone.

2. Clinical Pharmacology

The recommended dose of astemizole (21, 32, 99–110) is 10 mg daily (Table 4). The mean time to peak effect is dose-related, as are onset and duration of significant inhibitory effects, and overall effect on suppression of the histamine- or antigen-induced wheal and flare. The maximal inhibition of the initial wheal response coincides with peak astemizole concentrations in the peripheral compartment, but later inhibition of wheal and flare correlates with peak DMA in the peripheral compartment (97). Loading doses of astemizole are no longer recommended because of the possibility of producing high levels in cardiac tissue and associated dys-rhythmias.

In single-dose studies, astemizole produces less wheal and flare suppression than cetirizine, terfenadine, or loratadine, but in multiple-dose studies, astemizole produces significantly greater wheal and flare suppression than terfenadine does. Suppression of histamine- or antigen-induced wheals and flares correlates with suppression of nasal symptoms. Astemizole reduces antigen- and histamine-induced bronchoconstriction.

Following a 7-day course of astemizole, inhibition of the wheal and, to a lesser extent, the flare response (Fig. 5) and suppression of histamine-induced bronchial hyperreactivity is rather prolonged. Astemizole should therefore be discontinued at least 4–6 weeks before diagnostic skin tests or challenge tests with histamine or allergen are performed. Tachyphylaxis to astemizole has not been reported.

C. Loratadine

1. Pharmacokinetics

Loratadine (111–122) is well absorbed following oral administration, with peak plasma concentrations of 24.3–30.5 μg/L occurring 1–1.5 h after ingestion of a 40-mg dose. Absorption is not affected by concurrent food ingestion.

Loratadine has a large volume of distribution, 119 L/kg. Its distribution half-life values range from 0.9–1.0 h in single-dose studies of 20 or 40 mg and in multiple-dose studies in which 40 mg has been administered once daily. It exhibits linear dose-proportional pharmacokinetics after single and multiple doses. It is about 98% plasma protein-bound.

Loratadine is extensively metabolized. After administration of single doses of 20–40 mg, terminal elimination half-life values range from 7.8–11 h. After administration of 40 mg daily for 10 days, a terminal elimination half-life value of 14.4 h is found at steady state, and little accumulation of loratadine occurs. The major metabolite, descarbo-ethoxyloratadine (DCL), has an elimination half-life of 17.3–24 h. It is 73–76% protein-bound. It is converted to an inactive metabolite that is excreted primarily in the urine. Like its parent compound, DCL has dose-proportional phar-macokinetics and does not accumulate to any extent during multiple dosing.

The elimination of loratadine is not significantly decreased in subjects with renal impairment, although a doubling of the AUC (area under the curve) for DCL occurs in those whose creatinine clearance is <1.8 L/h (30 ml/min). Dose adjustment is not necessarily required in these subjects.

Terminal elimination half-life values of loratadine and DCL tend to be

Table 4 Formulations and Dosages of Representative H₁-Receptor Antagonists

	Formulation	Recommended dose
First-generation		
Chlorpheniramine maleate (Chlortrimeton®)	Tablets 4 mg, 8 mg[+], 12 mg[+] Syrup 2.5 mg/5 ml Parenteral solution 10 mg/ml	Adult: 8–12 mg b.i.d.[+] Pediatric[a]: 0.35 mg/kg/24 h
Hydroxyzine hydrochloride (Atarax®)	Capsules 10, 25, 50 mg Syrup 10 mg/5 ml Parenteral solution 50 mg/ml	Adult: 25–50 mg qd (h) b.i.d. Pediatric: 2 mg/kg/24 h
Diphenhydramine hydrochloride (Benadryl®)	Capsules 25 or 50 mg Elixir 12.5 mg/5 ml Syrup 6.25 mg/5 ml Parenteral solution 50 mg/ml	Adult: 25–50 mg t.i.d. Pediatric: 5 mg/kg/24 h
Second-generation		
Terfenadine (Seldane®)	Tablets 60 mg, 120 mg[b] Suspension 30 mg/5 ml[b]	Adult: 60 mg b.i.d. *or* 120 mg q.d.[b] Pediatric: (3–6 yrs): 15 mg b.i.d. (7–12 yrs): 30 mg b.i.d.[b]
Astemizole (Hismanal®)	Tablets 10 mg Suspension 10 mg/5 ml[b]	Adult: 10 mg q.d. Pediatric: 0.2 mg/kg/24 h[b]
Loratadine (Claritin®)	Tablets 10 mg[b] Syrup 1 mg/ml[b]	Adult: 10 mg q.d. Pediatric: (2–12 yrs): 5 mg/day (>12 yrs & > 30 kg): 10 mg/day[b]
Cetirizine (Reactine®)	Tablets 10 mg	Adult: 5–10 mg q.d.
Acrivastine (Semprex®)	Tablets 8 mg[c]	Adult: 8 mg t.i.d.
Ketotifen[b] (Zaditen®)	Tablets 1 mg; 2 mg[+] Syrup 1 mg/5 ml[b]	Subjects >3 yrs: 1 mg b.i.d. or 2 mg q.d.[+] 4 mg q.d.[+] is used in urticaria
Ebastine[b] (Ebastel®)	Tablets 10 mg	Adult: 10 mg q.d.
Azelastine (Astelin®)	Nasal solution 0.1% (0.137 mg/spray) Tablets 2 mg	Topical: 2 sprays in each nostril q.d. or b.i.d.; Adult: 2–4 mg b.i.d.
Levocabastine (Livostin®)	Microsuspension 0.5 mg/ml	Topical: 2 sprays in each nostril b.i.d.-q.i.d.; 1 drop in each eye b.i.d-q.i.d

Abbreviations: q.d., once daily; b.i.d, twice daily; t.i.d, three times daily; q.i.d., four times daily.
[a] For subjects ≤40 kg.
[b] Not approved in the U.S. at time of publication.
[c] Available as Semprex D in combination with pseudoephedrine 60 mg.
[+] Sustained release.

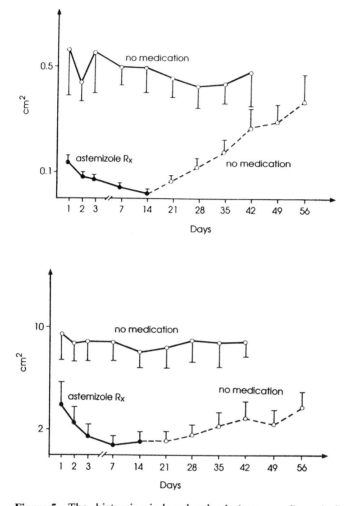

Figure 5 The histamine-induced wheal (top panel) and flare or erythema (bottom panel) monitored over 8 weeks in control subjects (top line in each graph) and in subjects treated with astemizole 30 mg on day 1 followed by 10 mg on days 2–14, inclusive (bottom line in each graph). The astemizole treatment was discontinued on day 14. Astemizole produced inhibition of skin tests starting 24 h after the loading dose on day 1, peaking at the end of the treatment on day 14, and disappearing very slowly during the following 6 weeks when no treatment was given. Flares were inhibited for a longer period of time than wheals. Values are expressed as mean ± 2 SEM. (From Ref. 109.)

prolonged in healthy elderly subjects and in subjects with alcoholic liver disease, although in both these populations values may still fall within the wide range seen in normal healthy subjects. The pharmacokinetics of loratadine and DCL have been studied in children, in whom the terminal elimination half-life of DCL is 13.8 h. Hemodialysis has a negligible effect on the clearance of loratadine and DCL.

Like terfenadine, loratadine is metabolized by the cytochrome P$_{450}$ CYP3A4, but in the presence of cytochrome P$_{450}$ CYP3A4 inhibitors, it may also be metabolized by cytochrome P$_{450}$ CYP2D6. Ketoconazole 200 mg twice daily for 5 days in healthy volunteers inhibits the metabolism of a single oral 20-mg dose of loratadine, as does coadministration of erythromycin 500 mg every 8 hours with loratadine 10 mg daily for 10 days. No adverse ECG effects have been associated with elevated plasma concentrations of loratadine and DCL. The pharmacokinetics of loratadine do not change when loratadine is coadministered with pseudoephedrine.

2. Clinical Pharmacology

The recommended dose of loratadine (32, 121–132) is 10 mg once daily (Table 4). As with all H$_1$-receptor antagonists, the magnitude and duration

Table 5 Pharmacokinetics of Oral Cetirizine in Healthy Adults, Elderly Adults, Children, and in Subjects with Renal Dysfunction or Hepatic Dysfunction

Study participants	Dose (mg)	t$_{max}$ (h)	t$_{1/2}\beta$ (h)	Total body clearance (L/h/kg)	Vd (L or L/kg)	Urinary excretion in 24 h (%)
Healthy adults	10	0.6	6.5–10	0.04–0.05	33.2	58
	20	1.0	7–11	0.04–0.05	40.8	—
Healthy elderly	10	0.9	11.8	1.68	0.38	—
Children[a]	5,10	0.8	6.9–7.1	0.062–0.066	0.7	39–40
[b]	5	1.4	4.9	0.09	0.7	38
Renal dysfunction	10	2.2	20.9		0.54	
Hepatic dysfunction	10	1.0[†]	13.8[†]	0.03[†]	0.44[†]	32[†]
	20	2.5[††]–3.1[†]	13.9[†]–14.3[††]	0.41[†]–0.48[††] ml/min/kg	0.47[†]–0.67[††]	69[††]

Abbreviations: t$_{max}$ = time from oral intake to peak serum/plasma concentration; t$_{1/2}\beta$ = terminal elimination half-life; Vd = volume of distribution
[a] Mean age 8 yrs.
[b] Age 2–4 years (infants [Ref. 143] not shown).
[†] Primary biliary cirrhosis.
[††] Hepatocellular liver disease.
From Refs. 133–142.

of suppression of the histamine-induced wheal and flare by loratadine is dose-related. In single-dose studies, histamine-induced wheals and flares are suppressed for 12–24 h following the ingestion of 10 mg, and suppression for more than 24 h has been reported following doses of 20 to 80 mg. Onset of wheal suppression is observed within 1–2 h of loratadine administration. Loratadine 10 mg is more effective in suppressing the wheal and flare than astemizole 10 mg, chlorpheniramine 4 mg, or placebo, but not as effective as cetirizine 10 mg, terfenadine 120 mg, or terfenadine 60 mg. In other studies, cetirizine 2.5–10 mg has been more effective than loratadine 10–40 mg in suppressing wheals and flares in healthy and atopic subjects. The onset of significant inhibitory effects may occur later (4 h) after loratadine than after the other agents (1–2 h). In children, loratadine and terfenadine have similar effects on wheal and flare suppression.

Relief of nasal symptoms in subjects with allergic rhinitis occurs more rapidly and to a greater extent after loratadine administration than would be anticipated from data obtained in wheal and flare studies. Loratadine prevents the bronchoconstrictor response to histamine or allergen.

The offset of action of a short course of loratadine takes about 1 week (i.e., if the subject has been taking loratadine 10 mg daily for 1 week or more, the medication should be discontinued for 7 days before skin tests or bronchial challenge tests with histamine or allergen are performed). Regular treatment with loratadine 10 mg daily for 3 months does not result in tachyphylaxis (Fig. 6).

D. Cetirizine

1. Pharmacokinetics

The pharmacokinetics of cetirizine (133–144) have been thoroughly studied in normal and allergic subjects of different ages and in subjects with renal or hepatic dysfunction (Table 5). After oral administration of a 10- or 20-mg dose, cetirizine is rapidly absorbed from the gastrointestinal tract. After a single 10-mg cetirizine dose in adults, mean peak plasma concentrations of 257 μg/L to 460 μg/L are achieved within 1 h. Food ingestion does not interfere with the extent of absorption, but may decrease the rate of absorption.

The volume of distribution of cetirizine, 0.56 L/kg, is low compared with that of other H_1-receptor antagonists. Twenty-four hours after a single dose and at steady state, skin cetirizine concentrations are similar to, or exceed, serum concentrations (Fig. 7). Plasma protein binding is reported to be 93% at plasma cetirizine concentrations of 25–1000 μg/L.

Only a small percentage of cetirizine is metabolized. After administration of ^{14}C-cetirizine, over 90% of plasma radioactivity is attributed to unchanged cetirizine at 2 h, 80% at 10 h, and 70% at 24 h. A metabolite produced by oxidative O-dealkylation appears in the plasma at 10 h and in the feces from 24–48 h after dosing. About 50% of administered radioactivity is excreted in the urine during the first 24 h after dosing, and a further 10% is excreted in the next four days. The major source of urinary radioactivity is unchanged cetirizine, accounting for 93% of the radioactivity during the first 2 h after dosing and 83% from 12–24 h. The remaining radioactivity is attributable to small amounts of unidentified metabolites.

Mean terminal elimination half-life values for cetirizine range from 7–11 h after a single 10-mg dose in healthy adults. Approximately 60% of the administered dose is recovered in the urine within 24 h. Steady-state concentrations are achieved within 3 days, and no accumulation occurs during long-term administration. Cetirizine is eliminated no more rapidly after 5–8 weeks of regular administration than at the onset of treatment.

In elderly subjects, the mean cetirizine terminal elimination half-life is 11.8 h; this slight prolongation is associated with reduction in total body clearance and is dependent on renal function rather than age. In subjects with renal dysfunction the terminal elimination half-life is 20–20.9 h and

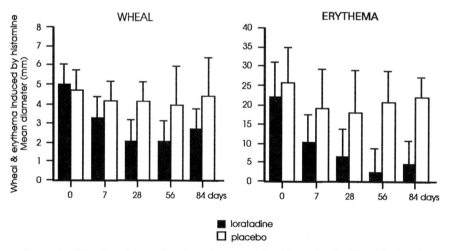

Figure 6 Wheal and flare (erythema) induced by histamine in 20 subjects who received either placebo or loratadine 10 mg daily for 12 weeks. The subjects treated with loratadine had significantly smaller wheal and flare reactions after 7 days. This effect was maximal at 28 days and lasted throughout the study. Subsensitivity to loratadine did not develop during the 12-week period. (From Ref. 132.)

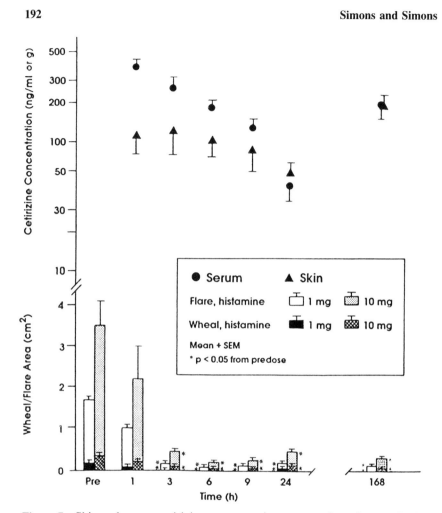

Figure 7 Skin and serum cetirizine concentrations versus time plots, and wheal and flare areas after epicutaneous tests with histamine phosphate 1 mg/ml and 10 mg/ml. Tests were performed at baseline and 1, 3, 6, 9, and 24 h after the initial dose of cetirizine 10 mg. Subjects then took cetirizine 10 mg at 2100 h for 6 consecutive days and the tests were repeated at 168 h (steady state), exactly 12 h after the seventh and last dose. (From Ref. 43.)

in subjects with hepatic dysfunction it is 13.8–14.3 h. The pharmacokinetics of cetirizine have been well studied in children. In 19 allergic children with a mean age of 8 years, given a 5- or 10-mg dose, the terminal elimination half-life is approximately 7 h. In younger children with a mean age of 2–4 years, given a 5-mg dose (0.3 mg/kg), the terminal elimination half-life is 4.9 h. In infants with a mean age of 12 months, it is 2.8 h, significantly shorter than in adults.

Cetirizine is not eliminated to any extent by hemodialysis. Cetirizine elimination is not inhibited by other drugs such as cimetidine.

2. Clinical Pharmacology

The recommended dose of cetirizine is 5–10 mg once daily (Table 4). Cetirizine 10 mg suppresses the histamine- or allergen-induced wheal and flare response rapidly with a significant effect in less than 1 h, and a peak effect 4–8 h after administration. Significant suppression lasts for 24 h (21, 32, 145–148). It is more effective in suppressing the wheal and flare than: diphenhydramine 50 mg, hydroxyzine 25 mg, chlorpheniramine 6 mg, mequitazine 5 mg, terfenadine 60 or 120 mg, loratadine 10 mg, astemizole 10 mg, ebastine 10 mg, ketotifen 1 mg, brompheniramine 4 mg, cyproheptadine 4 mg or clemastine 1 mg. In contrast to other H_1-receptor antagonists, cetirizine significantly decreases the wheal and flare even if concentrations of histamine phosphate as high as 500 mg/ml are used in the epicutaneous tests.

Cetirizine 10 mg twice daily for 3 days attenuates the increase in nasal airway resistance produced by histamine challenge in subjects with allergic rhinitis. Cetirizine 5–20 mg provides a dose-dependent protective effect against histamine-induced bronchospasm in subjects with mild asthma. It is superior in this regard to hydroxyzine 25 mg and placebo.

Subjects taking cetirizine regularly should discontinue it 3–4 days before having skin tests with histamine or allergen. Tachyphylaxis to cetirizine has not been reported.

E. Acrivastine

1. Pharmacokinetics

Absorption of acrivastine (149–154) is rapid after oral dosing. In healthy subjects, the mean peak plasma concentration is 73 μg/L, 1.4 h after administration of a single capsule containing acrivastine 4 mg. Absorption is not significantly decreased by concomitant food ingestion. During administration of acrivastine 2, 4, 8, 12, 16, 24, or 32 mg three times daily for 7 days, the t_{max} and the AUC increase linearly with dose, accumulation does not occur, and t_{max} is 1.9 h.

The mean apparent volume of distribution is 0.64 and 0.75 L/kg after single and multiple doses, respectively. Acrivastine is approximately 50% protein-bound.

The principal acrivastine metabolite is a propionic acid analogue formed by reduction of the acrylic acid side chain. This metabolite accounts for about 10% of total plasma drug concentration and is two to three times as active as acrivastine in inhibiting histamine-induced bronchospasm in the guinea pig model. Unchanged acrivastine accounts for about 60% of the administered dose recovered in the urine. After a single 8-mg dose of ^{14}C-acrivastine, 88% of the ingested radioactivity is recovered in the urine within a 48-h period; the remainder is excreted in the feces within 5 days.

The apparent total body clearance is 4.3 ml/min/kg after a single oral acrivastine dose, and 4.5 ml/min/kg after repeated doses. The terminal elimination half-life ranges between 1.4 and 3.1 h after single or repeated doses, and the mean elimination half-life of the metabolite is 2.3 h.

In 36 elderly volunteers receiving acrivastine 8 mg or 16 mg three times daily for 22 doses, acrivastine clearance is about 25% lower than reported for healthy young volunteers, and the elimination half-life is about 35% longer, probably related to the decreased creatinine clearance found in the elderly. In this population, compared to younger subjects receiving acrivastine 8 mg three times daily for 5 days, AUC doubles, t_{max} increases by about 34%, and the elimination half-life after 5 days is 2.17 h versus 1.71 h.

There are few studies of the pharmacokinetics of acrivastine in subjects with renal or hepatic failure, or in children. Coadministered pseudoephedrine does not appear to influence the pharmacokinetics of acrivastine.

2. Clinical Pharmacology

The recommended dose of acrivastine (155–158) is 8 mg three times daily (Table 4), usually combined with pseudoephedrine 60 mg. Acrivastine suppresses the histamine-induced wheal within 15 min and the flare within 25 min. Peak inhibition of flare occurs at 90 min and peak inhibition of wheal at 120 min. The effect is dose-related over the range of 1–8 mg. The suppressive effects are comparable to those of terfenadine or triprolidine.

Acrivastine 4 mg, but not 2 mg, reduces the response to histamine in nasal challenge studies. Acrivastine 8 mg reduces the conjunctival response to histamine. In subjects with asthma, acrivastine 8 mg protects against histamine-induced bronchoconstriction.

No studies of potential development of tachyphylaxis to acrivastine have been published.

F. Ketotifen

1. Pharmacokinetics

Ketotifen (159–163) is well absorbed after oral administration of a 2-mg dose. For capsule and syrup formulations, mean peak serum concentrations range from 270–421 pg/mL, respectively, and the mean time at which peak concentrations are achieved ranges from 2.3–3.9 h. Bioavailability is only about 50%, due to a considerable first-pass effect. Peak plasma concentrations after multiple oral doses of 1 mg twice daily are 1.92 μg/L in adults and 3.25 μg/L in children, achieved at 1.33 and 2.39 h, respectively.

The volume of distribution of ketotifen is unknown. The drug is 75% protein-bound.

Ketotifen is metabolized to ketotifen-N-glucuronide, which is inactive, and nor-ketotifen (N-demethyl ketotifen), which is pharmacologically active. Oxidation and reduction of ketotifen also occurs. Recovery of ketotifen-N-glucuronide and nor-ketotifen in urine accounts for 50% and 10% of the administered ketotifen dose. Unchanged parent compound ketotifen accounts for approximately 1%. Disposition of ketotifen in plasma after oral administration is biphasic, with a distribution half-life of 3 h and mean terminal elimination half-life values ranging from 13.1–18.3, depending on the formulation studied.

There are no published studies of ketotifen pharmacokinetics in the elderly or in subjects with hepatic or renal dysfunction. The pharmacokinetics of ketotifen have been incompletely studied in children; half-life values and clearance rates have not been reported in this population. Mean AUC_{0-12} values are lower in children than in adults (16.98 μg/L/h versus 20.72 μg/L/h); therefore, relatively higher doses of ketotifen on a mg/kg basis are recommended for children.

Few drug interaction studies have been published for ketotifen; ketotifen coadministered with theophylline does not influence serum theophylline concentrations.

2. Clinical Pharmacology

The recommended ketotifen dose is 1 mg twice daily in adults and children (164–171) (Table 4), although ketotifen has a reasonably long terminal elimination half-life, and a single 1-mg oral dose significantly suppresses the wheal and flare response to histamine or allergen for up to 48 h. Oral ketotifen also reduces the wheal and flare produced by codeine, dextromethorphan, and weighted stroke-induced dermographism. Ketotifen 1 mg twice daily is more effective than clemastine 1 mg twice daily in wheal and flare suppression.

Ketotifen pretreatment protects against bronchial challenge with histamine or allergen in asthmatics. Tachyphylaxis to ketotifen has not been optimally studied. Its efficacy in chronic asthma is believed to increase gradually during the first 8–12 weeks of treatment.

G. Ebastine

1. Pharmacokinetics

After administration of ebastine (172–179), the parent compound is present in extremely low concentrations in plasma; therefore, in pharmacokinetic studies, the carboxylic acid metabolite, carebastine, is measured. After ingestion of a single 10-mg ebastine tablet, peak plasma carebastine concentrations of 120 μg/L are obtained at 2.6 h. Extrapolating from plasma carebastine concentrations, ebastine seems to be highly bioavailable and food ingestion does not impair its absorption from the gastrointestinal tract.

Ebastine undergoes extensive hepatic metabolism, principally to carebastine, with 75–95% of a dose being recovered in the feces, and 4–9% in the urine. The terminal carebastine elimination half-life is 13–15 h. The pharmacokinetics of carebastine appear to be linear following ebastine doses of 10–15 mg. During multiple dosing, the C_{max} of 86 μg/L increases to 162 μg/L at steady state.

The terminal carebastine elimination half-life is similar in young and elderly adults. The effect of hepatic cirrhosis on the pharmacokinetics of carebastine is minimal, as is the effect of impaired renal function, and dose adjustments may not be necessary under these circumstances. In children, the terminal elimination half-life of carebastine is 10–11 h. No interactions of ebastine with alcohol or diazepam have been found.

2. Clinical Pharmacology

Ebastine (179, 180), over the range 3–30 mg, has a dose-related suppressive effect on the histamine-induced wheal and flare. After a single dose, the maximum suppressive effect of approximately 60% occurs at 6–8 h, and about 30% inhibition is still present at 24 h. The maximum effect occurs hours later than the peak plasma carebastine concentrations, as found with other H_1-receptor antagonists. The effect of ebastine is more prolonged than that of terfenadine. The rate of onset of action is slower than that of cetirizine.

A single oral dose of ebastine 10 mg or 30 mg administered before bronchial challenge with histamine shifts the dose-response curve 3- to 27-fold. Correlation of wheal and flare suppression, and of suppression of histamine-induced bronchoconstriction, with plasma concentrations appears to be excellent.

H. Azelastine

1. Pharmacokinetics

A single oral dose of azelastine (181–184) is rapidly and almost completely absorbed. The absolute bioavailability is about 80% from oral tablets.

Following doses of 1, 2, 3, or 4 mg of azelastine hydrochloride in healthy adults, peak plasma concentrations of 0.6, 1.1, 2.0, and 2.7 μg/L, respectively, are attained. Maximum concentrations occur 4 h after doses of 1–3 mg and 6 h or more after a dose of 4 mg. In subjects with asthma who receive single oral azelastine doses ranging from 2–16 mg, peak plasma azelastine concentrations are linearly related to dose. After administration of azelastine nasal spray, plasma concentrations are about 100-fold less than after oral administration.

The volume of distribution of azelastine at steady state is 14.5 ± 4.0 L/kg. Azelastine is 78–88% bound to plasma proteins.

Azelastine is almost completely metabolized by hepatic oxidation. The major metabolite, desmethylazelastine, is pharmacologically active. Additional metabolites include the 2- and 7- acid derivatives formed by oxidation and subsequent azepinyl ring opening. Following administration of a single oral dose of ^{14}C azelastine, 75% of the radioactivity is recovered within 120 h; 50% in the feces, and 25% in the urine.

The pharmacokinetics of azelastine have been determined after an oral dose of 9.9 mg and after an intravenous dose of 4.5 mg, using an HPLC assay with fluorometric detection and a limit of sensitivity of 0.3 μg/L and 0.5 μg/L for azelastine and desmethylazelastine, respectively. The pharmacokinetics of azelastine appear to be linear over the therapeutic dose range. The terminal elimination half-life is 22 and 54 h for azelastine and desmethylazelastine, respectively, and does not differ following oral or intravenous administration.

At steady state, after multiple oral azelastine doses of 4 mg twice daily, a mean peak plasma concentration of 10 μg/L is achieved, and the terminal elimination half-life is 35.5 h, possibly due to accumulation of demethylazelastine.

There is no published information about the pharmacokinetics of azelastine in the elderly, in subjects with hepatic or renal dysfunction, or in children.

Azelastine does not affect antipyrine clearance, suggesting that it does not induce hepatic metabolism in therapeutic doses.

2. Clinical Pharmacology

The recommended dose of azelastine is 2–4 mg by mouth twice daily or two sprays of a topical nasal solution daily or twice daily (185–198) (Table 4). Wheal and flare suppression by orally administered azelastine is dose-related, peaking at about 4 h after dosing and persisting for at least 1 week after the last dose of a short course of treatment. Few comparative studies of the effects of azelastine and other H_1-receptor antagonists have been performed.

In subjects with perennial allergic rhinitis, azelastine significantly inhibits the response to intranasal histamine in a dose-related manner. A single oral dose of 1 or 2 mg significantly decreases the effect of histamine on inspiratory and expiratory power, as measured by computerized posterior rhinomanometry.

Azelastine produces bronchodilation over the dose range 2–16 mg by mouth; 4 mg appears to be optimal for this purpose. Azelastine 2–8 mg orally in single- and multiple-dose studies protects against bronchial challenge with histamine in subjects with mild asthma. The PD_{20} histamine shifts in a dose-dependent manner, up to a maximum of 28- to 45-fold compared to the protection provided by placebo.

Tachyphylaxis to azelastine does not develop.

I. Levocabastine

1. Pharmacokinetics

The pharmacokinetics of levocabastine (199–202) have been studied after administration of oral and intravenous formulations, which are not used clinically, as well as after application of the nasal and ocular formulations. Single- and multiple-dose studies have been performed in healthy subjects and in subjects with allergic rhinitis and allergic conjunctivitis.

After levocabastine given orally in doses of 0.5, 1.0, or 2.0 mg, peak plasma concentrations are attained within 2 h and systemic bioavailability is 100–120%, indicative of a negligible first-pass effect.

At steady state, intravenous levocabastine has a mean volume of distribution of 82 L (1.14 L/kg) and a total plasma clearance of 30 ml/min. The plasma protein binding of levocabastine is approximately 55%, with most of the drug being bound to albumin.

After a bolus intravenous dose of 0.2 mg in healthy volunteers, the pharmacokinetic profile fits the description of a two-compartment open model. The distribution phase is approximately 0.6 h while the terminal elimination half-life is approximately 33 h. The metabolism of levocabastine is of minor significance. About 65–70% of absorbed levocabastine is excreted in the urine as unchanged drug; 10–20% appears unchanged in the feces, probably due to biliary excretion. The remainder is recovered in the urine as the acylglucuronide metabolite. The plasma elimination half-life is 35–40 h in multiple-dose studies, indicating that no accumulation occurs.

Absorption of topical levocabastine occurs within 1–2 h and is incomplete. Bioavailability after intranasal or ocular application is 60–80% and 30–60%, respectively, in healthy subjects. Nasal absorption is lower in subjects with allergic rhinitis. Swallowing of the nasal formulation may contribute to the overall systemic availability of the nasal spray. Because of the minute amount of drug applied when it is administered topically, the AUC or systemic availability of levocabastine is very low. Steady-state plasma concentrations are achieved in 7–10 days and are approximately 10.4 μg/L after two sprays per nostril, three times daily, and 1.6 μg/L after application of 1 drop per eye three times daily. Steady-state plasma concentrations are lower in allergic subjects (i.e., 3.5–5.2 μg/L after the nasal spray). After single or repeated doses, the elimination half-life of levocabastine is between 35 and 40 h, regardless of the route of administration.

In subjects with renal impairment, compared to healthy subjects, the terminal elimination half-life is prolonged to 95 h, urinary excretion of unchanged levocabastine is reduced, and the AUC is increased by 56%. There are few studies of levocabastine pharmacokinetics in the elderly or in subjects with hepatic impairment, but no reductions in topical levocabastine doses or dose frequency are needed in these populations. Steady-state plasma levocabastine concentrations do not differ in adults and infants. Hemodialysis removes 10% of a levocabastine dose, but does not alter the effect of impaired renal clearance on levocabastine pharmacokinetics. Minute amounts of levocabastine applied to the nasal mucosa or the eyes have been detected in the saliva and breast milk of nursing mothers.

Few levocabastine drug interaction studies have been performed.

2. Clinical Pharmacology

In allergic rhinoconjunctivitis, the recommended dose of levocabastine (201) is 0.5 mg/ml nasal spray is two sprays in each nostril two to four times daily (Table 4). The minimum effective dose of levocabastine 0.5 mg/ml eye drops is one drop in each eye twice daily; up to three to four applications daily may be required.

Levocabastine eye drops do not significantly inhibit the wheal and flare reaction to intradermal histamine in healthy subjects, suggesting that little systemic absorption occurs after topical administration. Levocabastine nasal spray administered regularly has a similar minimal suppressive effect on histamine skin tests. Topically applied levocabastine prevents histamine- or allergen challenge-induced nasal or ocular symptoms.

IV. SUMMARY

The pharmacokinetics and clinical pharmacology of some of the first-generation and most of the second-generation H₁-receptor antagonists have now been investigated. While many studies of H₁-receptor antagonists have been performed in unique groups of subjects (e.g., those with hepatic and renal dysfunction, the elderly, and children), more studies in these special populations are needed. While terfenadine has been well investigated with regard to potential interactions with other medications eliminated through the cytochrome P_{450} system, other H_1-receptor antagonists eliminated primarily through this system have not been as optimally investigated.

There have been many clinical pharmacology studies of the suppressive effects of H_1-receptor antagonists on the histamine- or allergen-induced wheal and flare in the skin, but there are still relatively few objective studies of their suppressive effects in the nose, eyes, and lower airways. There is also a paucity of true pharmacodynamic studies, in which objectively documented H_1-receptor antagonist effects are correlated with serum or plasma H_1-receptor antagonist concentrations or, where relevant, H_1-receptor antagonist active metabolite concentrations (203).

Available pharmacokinetic and clinical pharmacology information about H_1-receptor antagonists has greatly influenced the clinical usage of this class of medications. Their diversity in pharmacokinetics and clinical pharmacology is an advantage rather than a disadvantage, as it enables physicians to make a rational choice of medications for use in various clinical circumstances. Fortunately, despite their unique pharmacokinetic and pharmacodynamic profiles, once-daily dosing is possible for many H_1-receptor antagonists.

REFERENCES

1. Paton DM, Webster DR. Clinical pharmacokinetics of H_1-receptor antagonists (the antihistamines). Clin Pharmacokinet 1985;10:477–97.
2. Tozzi S, Roth FE, Tabachnick IIA. The pharmacology of azatadine, a potential antiallergy drug. Agents Actions 1974;4:264–70.
3. Simons FER, Frith EM, Simons KJ. The pharmacokinetics and antihistaminic effects of brompheniramine. J Allergy Clin Immunol 1982;70:458–64.
4. Georgitis JW, Shen D. Nasal pharmacodynamics of brompheniramine in perennial rhinitis. Arch Otolaryngol Head Neck Surg 1988;114:63–7.
5. Tham R, Norlander B, Hagermark O, Fransson L. Gas chromatography of clemastine. A study of plasma kinetics and biological effect. Arzneim-Forsch/Drug Res 1978;28:1017–20.
6. Porter CC, Arison BH, Gruber VF, Titus DC, Vandenheuvel WJA. Human metabolism of cyproheptadine. Drug Metab Dispos 1975;3:189–97.
7. Kennedy KA, Halmi KA, Fischer LJ. Urinary excretion of a quaternary-ammonium glucuronide metabolite of cyproheptadine in humans undergoing chronic drug therapy. Life Sci 1977;21:1813–19.
8. Hintze KL, Wold JS, Fischer LJ. Disposition of cyproheptadine in rats,

mice and humans and identification of a stable epoxide-metabolite. Drug Metab Dispos 1975;3:1–9.

9. Richards DM, Brogden RN, Heel RC, Speight TM, Avery GS. Oxatomide. A review of its pharmacodynamic properties and therapeutic efficacy. Drugs 1984;27:210–31.

10. DiGregorio GJ, Ruch E. Human whole blood and parotid saliva concentrations of oral and intramuscular promethazine. J Pharm Sci 1980;69:1457–9.

11. Taylor G, Houston JB, Shaffer J, Mawer G. Pharmacokinetics of promethazine and its sulphoxide metabolite after intravenous and oral administration to man. Br J Clin Pharmacol 1983;15:287–93.

12. Moolenaar F, Ensing JG, Bolhuis BG, Visser J. Absorption rate and bioavailability of promethazine from rectal and oral dosage forms. Int J Pharm 1981;9:353–7.

13. Quinn J, Calvert R. The disposition of promethazine in man. J Pharm Pharmacol 1976;28:59P.

14. Schwinghammer TL, Juhl RP, Dittert LW, Melethil SK, Kroboth FJ, Chungi VS. Comparison of the bioavailability of oral, rectal and intramuscular promethazine. Biopharm Drug Disp 1984;5:185–94.

15. Chaudhuri NK, Servando OA, Manniello MJ, Luders RC, Chao DK, Bartlett MF. Metabolism of tripelennamine in man. Drug Metab Dispos 1976; 4:372–8.

16. Findlay JWA, Butz RF, Coker GG, DeAngelis RL, Welch RM. Triprolidine radioimmunoassay: disposition in animals and humans. J Pharm Sci 1984; 73:1339–44.

17. Findlay JWA, Butz RF, Sailstad JM, Warren JT, Welch RM. Pseudoephedrine and triprolidine in plasma and breast milk of nursing mothers. Br J Clin Pharmacol 1984;18:901–6.

18. Gibson JR, Medder KT, McDonnell KA, Bye CE, Hughes DTD. The effect of orally administered triprolidine and pseudoephedrine singly and combined on histamine-induced skin reactions. Eur J Clin Pharmacol 1982;22: 411–2.

19. Peck AW, Fowle ASE, Bye C. A comparison of triprolidine and clemastine on histamine antagonism and performance tests in man: implications for the mechanism of drug induced drowsiness. Eur J Clin Pharmacol 1975;8: 455–63.

20. Simons KJ, Singh M, Gillespie CA, Simons FER. An investigation of the H₁-receptor antagonist triprolidine: Pharmacokinetics and antihistaminic effects. J Allergy Clin Immunol 1986;77:326–30.

21. Wood-Baker R, Holgate ST. The comparative actions and adverse effect profile of single doses of H₁-receptor antihistamines in the airways and skin of subjects with asthma. J Allergy Clin Immunol 1993;91:1005–14.

22. Rumore MM. Clinical pharmacokinetics of chlorpheniramine. DICP 1984; 18:701–7.

23. Yacobi A, Stoll RG, Chao GC, et al. Evaluation of sustained-action chlorpheniramine-pseudoephedrine dosage form in humans. J Pharm Sci 1980; 69:1077–81.

24. Vallner JJ, Needham TE, Chan W, Viswanathan CT. Intravenous administration of chlorpheniramine to seven subjects. Curr Ther Res 1979;26: 449–53.

25. Huang SM, Athanikar NK, Sridhar K, Huang YC, Chiou WL. Pharmacokinetics of chlorpheniramine after intravenous and oral administration in normal adults. Eur J Clin Pharmacol 1982;22:359–65.

26. Dube LM, Bloch R, Warner RN, Hyslop RM, Popovich NG, Gonzalez MA. Pharmacokinetics of chlorpheniramine in chronic renal failure: effect of hemodialysis and peritoneal dialysis. Am Pharm Assoc 1980;10:84 (abstract).

27. Simons KJ, Martin TJ, Watson WTA, Simons FER. Pharmacokinetics and pharmacodynamics of terfenadine and chlorpheniramine in the elderly. J Allergy Clin Immunol 1990;85:540–7.

28. Simons FER, Luciuk GH, Simons KJ. Pharmacokinetics and efficacy of chlorpheniramine in children. J Allergy Clin Immunol 1982;69:376–81.

29. Thompson JA, Bloedow DC, Leffert FH. Pharmacokinetics of intravenous chlorpheniramine in children. J Pharm Sci 1981;70:1284–6.

30. Simons KJ, Simons FER, Luciuk GH, Frith EM. Urinary excretion of chlorpheniramine and its metabolites in children. J Pharm Sci 1984;73: 595–9.

31. Müller FO, deK Botha JJ, van Dyk M, Luus HG, Groenewoud G. Attenuation of cutaneous reactivity to histamine by cetirizine and dexchlorpheniramine. Eur J Clin Pharmacol 1988;35:319–21.

32. Simons FER, McMillan JL, Simons KJ. A double-blind, single-dose, crossover comparison of cetirizine, terfenadine, loratadine, astemizole, and chlorpheniramine versus placebo: Suppressive effects on histamine-induced wheals and flares during 24 hours in normal subjects. J Allergy Clin Immunol 1990;86:540–7.

33. Long WF, Taylor RJ, Wagner CJ, Leavengood DC, Nelson HS. Skin test suppression by antihistamines and the development of subsensitivity. J Allergy Clin Immunol 1985;76:113–7.

34. Taylor RJ, Long WF, Nelson HS. The development of subsensitivity to chlorpheniramine. J Allergy Clin Immunol 1985;76:103–7.

35. Bantz EW, Dolen WK, Chadwick EW, Nelson HS. Chronic chlorpheniramine therapy: subsensitivity, drug metabolism, and compliance. Ann Allergy 1987;59:341–6.

36. Simons FER, Simons KJ, Frith EM. The pharmacokinetics and antihistaminic effects of the H_1-receptor antagonist hydroxyzine. J Allergy Clin Immunol 1984;73:69–75.

37. Gengo FM, Dabronzo J, Yurchak A, Love S, Miller JK. The relative antihistaminic and psychomotor effects of hydroxyzine and cetirizine. Clin Pharmacol Ther 1987;42:265–72.

38. Simons KJ, Watson WTA, Chen XY, Simons FER. Pharmacokinetic and pharmacodynamic studies of the H_1-receptor antagonist hydroxyzine in the elderly. Clin Pharmacol Ther 1989;45:9–14.

39. Simons FER, Watson WTA, Chen XY, Minuk GY, Simons KJ. The pharmacokinetics and pharmacodynamics of hydroxyzine in patients with primary biliary cirrhosis. J Clin Pharmacol 1989;29:809–15.

40. Simons FER, Simons KJ, Becker AB, Haydey RP. Pharmacokinetics and antipruritic effects of hydroxyzine in children with atopic dermatitis. J Pediatr 1984;104:123–7.

41. Simons FER, Sussman GL, Simons KJ. Effect of the H₂-antagonist cimetidine on the pharmacokinetics and pharmacodynamics of the H₁-antagonists hydroxyzine and cetirizine in patients with chronic urticaria. J Allergy Clin Immunol 1995;95:685–693.

42. Simons KJ, Simons FER. The effect of chronic administration of hydroxyzine on hydroxyzine pharmacokinetics in dogs. J Allergy Clin Immunol 1987;79:928–32.

43. Simons FER, Murray HE, Simons KJ. Quantitation of H₁-receptor antagonists in skin and serum. J Allergy Clin Immunol 1995;95:759–64.

44. Abernethy DR, Greenblatt DJ. Diphenhydramine determination in human plasma by gas-liquid chromatography using nitrogen-phosphorus detection: application to single low-dose pharmacokinetic studies. J Pharm Sci 1983; 72:941–3.

45. Carruthers SG, Shoeman DW, Hignite CE, Azarnoff DL. Correlation between plasma diphenhydramine level and sedative and antihistamine effects. Clin Pharmacol Ther 1978;23:375–82.

46. Spector R, Choudhury AK, Chiang CK, Goldberg MJ, Ghoneim MM. Diphenhydramine in Orientals and Caucasians. Clin Pharmacol Ther 1980;28: 229–34.

47. Berlinger WG, Goldberg MJ, Spector R, Chiang CK, Ghoneim MM. Diphenhydramine: kinetics and psychomotor effects in elderly women. Clin Pharmacol Ther 1982;32:387–91.

48. Simons KJ, Watson WTA, Martin TJ, Chen XY, Simons FER. Diphenhydramine: pharmacokinetics and pharmacodynamics in elderly adults, young adults, and children. J Clin Pharmacol 1990;30:665–71.

49. Meredith CG, Christian Jr CD, Johnson RF, Madhavan SV, Schenker S. Diphenhydramine disposition in chronic liver disease. Clin Pharmacol Ther 1984;35;474–9.

50. Majchel AM, Proud D, Kagey-Sobotka A, Witek TJ Jr, Lichtenstein LM, Naclerio RM. Evaluation of a bedtime dose of a combination antihistamine/analgesic/decongestant product on antigen challenge the next morning. Laryngoscope 1992;102:330–4.

51. Cook CE, Williams DL, Myers M, et al. Radioimmunoassay for terfenadine in human plasma. J Pharm Sci 1980;69:1419–23.

52. Van Landeghem VH, Burke JT, Thebault J. The use of a human bioassay in determining the bioequivalence of two formulations of the antihistamine terfenadine. Clin Pharmacol Ther 1980;27:290–1.

53. Okerholm RA, Weiner DL, Hook RH, et al. Bioavailability of terfenadine in man. Biopharm Drug Disp 1981;2:185–90.

54. Garteiz DA, Hook RH, Walker BJ, Okerholm RA. Pharmacokinetics and biotransformation studies of terfenadine in man. Arzneim-Forsch/Drug Res 1982;32:1185–90.

55. Coutant JE, Westmark PA, Nardella PA, Walter SM, Okerholm RA. Determination of terfenadine and terfenadine acid metabolite in plasma using solid-phase extraction and high-performance liquid chromatography with fluorescence detection. J Chromatogr 1991;570:139–48.

56. Chan KY, George RC, Chen T-M, Okerholm RA. Direct enantiomeric separation of terfenadine and its major acid metabolite by high-performance liquid chromatography, and the lack of stereoselective terfenadine enantiomer biotransformation in man. J Chromatogr 1991;571:291–7.

57. Surapaneni S, Khalil SKW. A preliminary pharmacokinetic study of the enantiomers of the terfenadine acid metabolite in humans. Chirality 1994; 6:479–83.

58. Eller MG, Walker BJ, Yuh L, Antony KK, McNutt BE, Okerholm RA. Absence of food effects on the pharmacokinetics of terfenadine. Biopharm Drug Disp 1992;13:171–7.

59. Simons FER, Watson WTA, Simons KJ. The pharmacokinetics and pharmacodynamics of terfenadine in children. J Allergy Clin Immunol 1987;80: 884–90.

60. Simons FER, Watson WTA, Simons KJ. Lack of subsensitivity to terfenadine during long-term terfenadine treatment. J Allergy Clin Immunol 1988; 82:1068–75.

61. Yun C-H, Okerholm RA, Guengerich FP. Oxidation of the antihistaminic drug terfenadine in human liver microsomes: Role of cytochrome P-450 3A(4) in N-dealkylation and C-hydroxylation. Drug Metab Dispos 1993;21: 403–9.

62. Eller M, Stoltz M, Okerholm R, McNutt B. Effect of hepatic disease on terfenadine and terfenadine metabolite pharmacokinetics. Clin Pharmacol Ther 1993;53:162.

63. Honig PK, Woosley RL, Zamani K, Conner DP, Cantilena LR Jr. Changes in the pharmacokinetics and electrocardiographic pharmacodynamics of terfenadine with concomitant administration of erythromycin. Clin Pharmacol Ther 1992;52:231–8.

64. Honig PK, Wortham DC, Zamani K, Conner DP, Mullin JC, Cantilena LR. Terfenadine-ketoconazole interaction. Pharmacokinetic and electrocardiographic consequences. JAMA 1993;269:1513–8.

65. Honig PK, Wortham DC, Zamani K, Mullin JC, Conner DP, Cantilena LR. The effect of fluconazole on the steady-state pharmacokinetics and electrocardiographic pharmacodynamics of terfenadine in humans. Clin Pharmacol Ther 1993;53:630–6.

66. Honig PK, Wortham DC, Zamani K, Cantilena LR. Comparison of the effect of the macrolide antibiotics erythromycin, clarithromycin and azithromycin on terfenadine steady-state pharmacokinetics and electrocardiographic parameters. Drug Invest 1994;7:148–56.

67. Honig PK, Wortham DC, Zamani K, Conner DP, Mullin JC, Cantilena LR.

Effect of concomitant administration of cimetidine and ranitidine on the pharmacokinetics and electrocardiographic effects of terfenadine. Eur J Clin Pharmacol 1993;45:41–6.

68. Honig PK, Wortham DC, Hull R, Zamani K, Smith JE, Cantilena LR. Itraconazole affects single-dose terfenadine pharmacokinetics and cardiac repolarization pharmacodynamics. J Clin Pharmacol 1993;33:1201–1206.

69. Luskin SS, Fitzsimmons WE, MacLeod CM, Luskin AT. Pharmacokinetic evaluation of the terfenadine-theophylline interaction. J Allergy Clin Immunol 1989;83:406–11.

70. Janssens MM-L, Howarth PH. The antihistamines of the nineties. Clin Rev Allergy 1993;11:111–53.

71. Hüther KJ, Renftle G, Barraud N, Burke JT, Koch-Weser J. Inhibitory activity of terfenadine on histamine-induced skin wheals in man. Eur J Clin Pharmacol 1977;12:195–9.

72. Juniper EF, White J, Dolovich J. Efficacy of continuous treatment with astemizole (Hismanal) and terfenadine (Seldane) in ragweed pollen-induced rhinoconjunctivitis. J Allergy Clin Immunol 1988;82:670–5.

73. Työlahti H, Lahti A. Start and end of the effects of terfenadine and astemizole on histamine-induced wheals in human skin. Acta Derm Venereol (Stockh) 1989;69:269–71.

74. Van Rooy P, Janssens M. Comparison of trough effects of five non-sedating antihistamines on histamine-induced wheal and flare reactions. J Allergy Clin Immunol 1991;87:151.

75. Shall L, Newcombe RG, Marks R. Assessment of the duration of action of terfenadine on histamine-induced weals. Br J Dermatol 1988;119:525–31.

76. Small P. Suppression of epicutaneous reactivity by terfenadine and loratadine. Ann Allergy 1992;68:30–4.

77. Di Lorenzo G, Ingrassia A, Crescimanno G, Cutrò S, Purello D'Ambrosio F, Barbagallo Sangiorgi G. A single-dose, double-blind, crossover comparison of cetirizine, loratadine and terfenadine versus placebo: suppressive effect on both histamine- and allergen-induced wheals and flares in atopic subjects. Allergy 1992;47 (Suppl.):60.

78. Rihoux J-P, Ghys L, Mühlethaler K, Wüthrich B. The skin as a target organ for the investigation of antiallergic drugs: comparison between cetirizine and terfenadine. Dermatology 1992;184:111–5.

79. Leynadier F, Murrieta M, Dry J, Colin JN, Gillotin C, Steru D. Effects of acrivastine and terfenadine on skin reactivity to histamine. Ann Allergy 1994;72:520–4.

80. Simons FER, Lukowski JL, Becker AB, Simons KJ. Comparison of the effects of single doses of the new H$_1$-receptor antagonists loratadine and terfenadine versus placebo in children. J Pediatr 1991;118:298–300.

81. Rafferty P, Holgate ST. Terfenadine (Seldane) is a potent and selective histamine H$_1$ receptor antagonist in asthmatic airways. Am Rev Respir Dis 1987;135:181–4.

82. Chan TB, Shelton DM, Eiser NM. Effect of an oral H$_1$-receptor antagonist, terfenadine, on antigen-induced asthma. Br J Dis Chest 1986;80:375–84.

83. Badier M, Beaumont D, Orehek J. Attenuation of hyperventilation-induced bronchospasm by terfenadine: a new antihistamine. J Allergy Clin Immunol 1988;81:437–40.

84. Labrecque M, Ghezzo H, L'Archevêque J, Trudeau C, Cartier A, Malo J-L. Duration of effect of loratadine and terfenadine administered once a day for one week on cutaneous and inhaled reactivity to histamine. Chest 1993; 103:777–81.

85. Meuldermans W, Hendrickx J, Lauwers W, et al. Excretion and biotransformation of astemizole in rats, guinea-pigs, dogs and man. Drug Develop Res 1986;8:37–51.

86. Heykants J, Van Peer A, Woestenborghs R, Jageneau A, Vanden Bussche G. Dose-proportionality, bioavailability, and steady-state kinetics of astemizole in man. Drug Develop Res 1986;8:71–8.

87. Al-Deeb OA, Abdel-Moety EM, Bayomi SM, Khattab NA. Spectrophotometric quantification of astemizole and its demethylated metabolite in urine after TLC separation. Eur J Drug Metab Pharmacokinet 1992;17:251–5.

88. Mangalan S, Patel RB, Gandhi TP, Chakravarthy BK. Detection and determination of free and plasma protein-bound astemizole by thin-layer chromatography: a useful technique for bioavailability studies. J Chromatogr 1991; 567:498–503.

89. Laine K, Kivistö KT, Neuvonen PJ. The effect of activated charcoal on the absorption and elimination of astemizole. Human Exp Toxicol 1994;13: 502–5.

90. Möller C, Andlin-Sobocki P, Blychert L-O. Pharmacokinetics of astemizole in children. Rhinology 1992;13:21–5.

91. Zazgornik J, Scholz N, Heykants J, Vanden Bussche G. Plasma concentrations of astemizole in patients with terminal renal insufficiency, before, during and after hemodialysis. Int J Clin Pharmacol Ther Toxicol 1986;24: 246–8.

92. Lavrijsen K, Van Houdt J, Meuldermans W, et al. The interaction of ketoconazole, itraconazole and erythromycin with the in vitro metabolism of antihistamines in human liver microsomes. Allergy 1993;48:34.

93. Bateman DN, Chapman PH, Rawlins MD. The acute and chronic effects of H$_1$-receptor blockade with astemizole on indocyanine green clearance. Br J Clin Pharmacol 1983;16:241–4.

94. Bateman DN, Chapman PH, Rawlins MD. Lack of effect of astemizole on ethanol dynamics or kinetics. Eur J Clin Pharmacol 1983;25:567–8.

95. Bateman DN, Rawlins MD. The effects of astemizole on antipyrine clearance. Br J Clin Pharmacol 1983;16:759–60.

96. Holmes GB, Adams MA, Hunt TL, Shand DG, Eisen GF, Kravec WG. Eighteen-week, steady-state astemizole/pseudoephedrine bioequivalency study. Pharmacotherapy 1991;11:109.

97. Janssens MM-L. Astemizole. A nonsedating antihistamine with fast and sustained activity. Clin Rev Allergy 1993;11:35–63.

98. Rombaut N, Heykants J, Vanden Bussche G. Potential of interaction between the H$_1$-antagonist astemizole and other drugs. Ann Allergy 1986;57: 321–4.

99. Bateman DN, Chapman PH, Rawlins MD. The effects of astemizole on histamine-induced weal and flare. Eur J Clin Pharmacol 1983;25:547–51.
100. Gendreau-Reid L, Simons KJ, Simons FER. Comparison of the suppressive effect of astemizole, terfenadine, and hydroxyzine on histamine-induced wheals and flares in humans. J Allergy Clin Immunol 1986;77:335–40.
101. Howarth PH, Emanuel MB, Holgate ST. Astemizole, a potent histamine H_1-receptor antagonist: effect in allergic rhinoconjunctivitis, on antigen and histamine-induced skin weal responses and relationship to serum levels. Br J Clin Pharmacol 1984;18:1–8.
102. Howarth PH, Holgate ST. Comparative trial of two non-sedative H_1-antihistamines, terfenadine and astemizole, for hay fever. Thorax 1984;39:668–72.
103. Rihoux J-P, Dupont P. Pharmacological modulation by astemizole and cetirizine 2 HCl of the skin reactivity to histamine. Acta Ther 1989;15:265–70.
104. Humphreys F, Hunter JAA. The effects of astemizole, cetirizine and loratadine on the time course of weal and flare reactions to histamine, codeine and antigen. Br J Dermatol 1991;125:364–7.
105. Almind M, Dirksen A, Nielsen NH, Svendsen UG. Duration of the inhibitory activity on histamine-induced skin weals of sedative and non-sedative antihistamines. Allergy 1988;43:593–6.
106. Lobaton P, Moreno F, Coulie P. Comparison of cetirizine with astemizole in the treatment of perennial allergic rhinitis and study of the concomitant effect on histamine and allergen-induced wheal responses. Ann Allergy 1990;65:401–5.
107. Simons FER, Watson WTA, Becker AB, Simons KJ. Histamine blockade after astemizole in children: a single-dose, placebo-controlled study. Ped Allergy Immunol 1994;5:214–7.
108. Benoît C, Malo J-L, Ghezzo H, Cartier A. Effect of a single dose of astemizole (10 mg) on the bronchoconstriction induced by histamine in asthmatic subjects. J Allergy Clin Immunol 1991;87:337.
109. Lantin JP, Huguenot CH, Pécoud A. Effect of the H_1-antagonist astemizole on the skin reactions induced by histamine, codeine, and allergens. Curr Ther Res 1990;47:683–92.
110. Malo J-L, Fu CL, L'Archevêque J, Ghezzo H, Cartier A. Duration of the effect of astemizole on histamine-inhalation tests. J Allergy Clin Immunol 1990;85:729–36.
111. Hilbert J, Radwanski E, Weglein R, Perentesis G, Symchowicz S, Zampaglione N. Pharmacokinetics and dose proportionality of loratadine. J Clin Pharmacol 1987;27:694–8.
112. Radwanski E, Hilbert J, Symchowicz S, Zampaglione N. Loratadine: multiple-dose pharmacokinetics. J Clin Pharmacol 1987;27:530–3.
113. Hilbert J, Moritzen V, Parks A, et al. The pharmacokinetics of loratadine in normal geriatric volunteers. J Int Med Res 1988;16:50–60.
114. Matzke GR, Halstenson CE, Opsahl JA, et al. Pharmacokinetics of loratadine in patients with renal insufficiency. J Clin Pharmacol 1990;30:364–71.
115. Lin CC, Radwanski E, Affrime MB, Cayen MN. Pharmacokinetics of loratidine in pediatric subjects. Am J Therapeutics 1995;2:504–508.

116. Hilbert J, Radwanski E, Affrime MB, Perentesis G, Symchowicz S, Zampaglione N. Excretion of loratadine in human breast milk. J Clin Pharmacol 1988;28:234–9.
117. Katchen B, Cramer J, Chung M, et al. Disposition of ^{14}C-SCH 29851 in humans. Ann Allergy 1985;55:393.
118. Parkinson A, Clement RP, Casciano CN, Cayen MN. Evaluation of loratadine as an inducer of liver microsomal cytochrome P450 in rats and mice. Biochem Pharmacol 1992;43:2169–80.
119. Yumibe N, Huie K, Chen KJ, Clement RP, Cayen MN. Identification of human liver cytochrome P450s involved in the microsomal metabolism of the antihistaminic drug loratadine. J Allergy Clin Immunol 1994;93:234.
120. Van Peer A, Crabbé R, Woestenborghs R, Heykants J, Janssens M. Ketoconazole inhibits loratadine metabolism in man. Allergy 1993;48:34.
121. Brannan MD, Affrime MB, Reidenberg P, et al. Evaluation of the pharmacokinetics and electrocardiographic pharmacokinetics of loratadine with concomitant administration of cimetidine. Pharmacotherapy 1994;14:347.
122. Brannan MD, Reidenberg P, Radwanski E, et al. Loratadine administered concomitantly with erythromycin: pharmacokinetic and electrocardiographic evaluations. Clin Pharmacol Ther 1995;58:269–78.
123. Kassem N, Roman I, Gural R, Dyer JG, Robillard N. Effects of loratadine (SCH 29851) in suppression of histamine-induced skin wheals. Ann Allergy 1988;60:505–7.
124. Roman IJ, Kassem N, Gural RP, Herron J. Suppression of histamine-induced wheal response by loratadine (SCH 29851) over 28 days in man. Ann Allergy 1986;57:253–6.
125. Skassa-Brociek W, O'Quigley J, Blizard R, Cougnard J, Michel FB, Bousquet J. Differentiation between the anti-allergic and anti-histaminic effects of loratadine and astemizole by skin tests. J Allergy Clin Immunol 1988;81:175.
126. Rihoux JP, Ghys L, Coulie P. Compared peripheral H_1 inhibiting effects of cetirizine 2 HCl and loratadine. Ann Allergy 1990;65:139–42.
127. Kontou-Fili K, Paleologos G, Herakleous M. Suppression of histamine-induced skin reactions by loratadine and cetirizine diHCl. Eur J Clin Pharmacol 1989;36:617–9.
128. Pechadre JC, Beudin P, Eschalier A, Trolese JF, Rihoux J-P. A comparison of central and peripheral effects of cetirizine and loratadine. J Int Med Res 1991;19:289–95.
129. Fadel R, Herpin-Richard N, Dufresne F, Rihoux J-P. Pharmacological modulation by cetirizine and loratadine of antigen and histamine-induced skin wheals and flares, and late accumulation of eosinophils. J Int Med Res 1990;18:366–71.
130. Town GI, Holgate ST. Comparison of the effect of loratadine on the airway and skin responses to histamine, methacholine, and allergen in subjects with asthma. J Allergy Clin Immunol 1990;86:886–93.
131. Debelic M. Protection against histamine-induced bronchoconstriction by loratadine. Allergol Immunopathol Madr 1992;20:97–100.
132. Bousquet J, Chanal I, Skassa-Brociek W, Lemonier C, Michel FB. Lack

of subsensitivity to loratadine during long-term dosing during 12 weeks. J Allergy Clin Immunol 1990;86:248–53.

133. Wood SG, John BA, Chasseaud LF, Yeh J, Chung ML. The metabolism and pharmacokinetics of ¹⁴C-cetirizine in humans. Ann Allergy 1987;59: 31–4.

134. Lefebvre RA, Rosseel MT, Bernheim J. Single dose pharmacokinetics of cetirizine in young and elderly volunteers. Int J Clin Pharm Res 1988;8: 463–70.

135. Baltes E, Coupez R, Brouwers L, Gobert J. Gas chromatographic method for the determination of cetirizine in plasma. J Chromatogr 1988;430: 149–55.

136. Matzke GR, Yeh J, Awni WM, Halstenson CE, Chung M. Pharmacokinetics of cetirizine in the elderly and patients with renal insufficiency. Ann Allergy 1987;59:25–30.

137. Pellegrin PL, Jacqmin PE, Van Bel P, Van Wymersch I. Mechanism of the renal clearance of cetirizine. Eur J Drug Metab Pharmacokinet 1990;15:31.

138. Coulie PJ, Ghys L, Rihoux J-P. Inhibitory effects of orally or sublingually administered cetirizine on histamine-induced weals and flares and their correlation with cetirizine plasma concentrations. J Int Med Res 1991;19:174–9.

139. Simons FER, Watson WTA, Minuk GY, Simons KJ. Cetirizine pharmacokinetics and pharmacodynamics in primary biliary cirrhosis. J Clin Pharmacol 1993;33:949–54.

140. Horsmans Y, Desager JP, Hulhoven R, Harvengt C. Single-dose pharmacokinetics of cetirizine in patients with chronic liver disease. J Clin Pharmacol 1993;33:929–32.

141. Watson WTA, Simons KJ, Chen XY, Simons FER. Cetirizine: a pharmacokinetic and pharmacodynamic evaluation in children with seasonal allergic rhinitis. J Allergy Clin Immunol 1989;84:457–64.

142. Desager JP, Dab I, Horsmans Y, Harvengt C. A pharmacokinetic evaluation of the second-generation H₁-receptor antagonist cetirizine in very young children. Clin Pharmacol Ther 1993;53:431–5.

143. Spicak V. The pharmacokinetics of cetirizine in infants. J Allergy Clin Immunol 1995;95:197.

144. Awni WM, Yeh J, Halstenson CE, Opsahl JA, Chung M, Matzke GR. Effect of haemodialysis on the pharmacokinetics of cetirizine. Eur J Clin Pharmacol 1990;38:67–9.

145. Juhlin L, de Vos C, Rihoux J-P. Inhibiting effect of cetirizine on histamine-induced and 48/80-induced wheals and flares, experimental dermographism, and cold-induced urticaria. J Allergy Clin Immunol 1987;80:599–602.

146. Rivest J, Despontin K, Ghys L, Rihoux JP, Lachapelle JM. Pharmacological modulation by cetirizine and ebastine of the cutaneous reactivity to histamine. Dermatologica 1991;183:208–11.

147. Tashkin DP, Brik A, Gong H Jr. Cetirizine inhibition of histamine-induced bronchospasm. Ann Allergy 1987;59:49–52.

148. Brik A, Tashkin DP, Gong H Jr, Dauphinee B, Lee E. Effect of cetirizine, a new histamine H₁ antagonist, on airway dynamics and responsiveness to inhaled histamine in mild asthma. J Allergy Clin Immunol 1987;80:51–6.

149. Cohen AF, Hamilton M, Burke C, Findlay J, Peck AW. A new H_1-receptor antagonist, BW825C: effects on skin histamine response, vigilance and autonomic variables. Br J Clin Pharmacol 1984;17:647P–8P.

150. Cohen AF, Hamilton MJ, Liao SHT, Findlay JWA, Peck AW. Pharmacodynamics and pharmacokinetics of BW 825C: a new antihistamine. Eur J Clin Pharmacol 1985;28:197–204.

151. Jallad NS, Garg DC, Fleck RJ, et al. Pharmacokinetics of acrivastine, a new H_1-antagonist, following ascending doses. J Clin Pharmacol 1985;25: 629–37.

152. Balasubramanian R, Klein KB, Pittman AW, Liao SHT, Findlay JWA, Frosolono MF. Pharmacokinetics of acrivastine after oral and colonic administration. J Clin Pharmacol 1989;29:444–7.

153. Chang SY, Nelson FR, Findlay JWA, Taylor LCE. Quantitative gas chromatographic-mass spectrometric analysis of acrivastine and a metabolite in human plasma. J Chromatogr 1989;497:288–95.

154. Liao SHT, Fleck RJ, Blum MR, Frosolono MF, Findlay JWA. Clinical pharmacokinetics and safety of acrivastine with pseudoephedrine in geriatric volunteers. J Pharm Sci 1987;76:S113.

155. Marks P, Manna VK, Gibson JR. Acrivastine—an evaluation of initial and peak activity in human skin. J Int Med Res 1989;17:3B–8B.

156. Hamilton MJ, Ashby L, Cohen AF, Letley E, Peck AW. A comparative study of acrivastine and terfenadine on histamine (H_1-receptor) antagonism in skin and central actions in man. Br J Clin Pharmacol 1985;20:283P–4P.

157. Petersen LJ, Bindslev-Jensen C, Poulsen LK, Malling H-J. Time of onset of action of acrivastine in the skin of pollen-allergic subjects. A double-blind, randomized, placebo-controlled comparative study. Allergy 1994;49: 27–30.

158. Rolan PE, Adams J, Posner J. Comparison of the onset of H_1-antagonism with acrivastine and terfenadine by histamine bronchial challenge in volunteers. J Int Med Res 1989;17:35B–9B.

159. Le Bigot JF, Cresteil T, Kiechel JR, Beaune P. Metabolism of ketotifen by human liver microsomes. Drug Metab Dispos 1983;11:585–9.

160. Leis HJ et al. Deuterium-labelling and quantitative measurement of ketotifen in human plasma by gas chromatography/negative ion chemical ionization mass spectrometry. Biol Mass Spectrom 1991;20:467–70.

161. Grahnén A, Lönnebo A, Beck O, Eckernas S-A, Dahlström B, Lindström B. Pharmacokinetics of ketotifen after oral administration to healthy male subjects. Biopharm Drug Disp 1992;13:255–62.

162. Kennedy GR. Metabolism and pharmacokinetics of ketotifen in children. Res Clin Forums 1982;4:17–20.

163. Schmidt-Redemann B, Brenneisen P, Schmidt-Redemann W, Gonda S. The determination of pharmacokinetic parameters of ketotifen in steady state in young children. Int J Clin Pharmacol Ther Toxicol 1986;24:496–8.

164. Phillips MJ, Meyrick Thomas RH, Moodley I, Davies RJ. A comparison of the in vivo effects of ketotifen, clemastine, chlorpheniramine and sodium cromoglycate on histamine- and allergen-induced weals in human skin. Br J Clin Pharmacol 1983;15:277–86.

165. Wang SSM, Wang SR, Chiang BN. Suppressive effects of oral ketotifen on skin responses to histamine, codeine, and allergen skin tests. Ann Allergy 1985;55:57–61.

166. Mansfield LE, Taistra P, Santamauro J, Ting S, Andriano K. Inhibition of dermographia, histamine, and dextromethorphan skin tests by ketotifen. A possible effect on cutaneous vascular response to mediators. Ann Allergy 1989;63:201–6.

167. Esau S, del Carpio J, Martin JG. A comparison of the effects of ketotifen and clemastine on cutaneous and airway reactivity to histamine and allergen in atopic asthmatic subjects. J Allergy Clin Immunol 1984;74:270–4.

168. Phillips MJ, Ollier S, Gould CAL, Davies RJ. Effect of antihistamines and antiallergic drugs on responses to allergen and histamine provocation tests in asthma. Thorax 1984;39:345–51.

169. Palmen FMLHG. Antigen bronchial challenge and efficacy after 4 weeks of treatment with ketotifen and disodium cromoglycate. Respiration 1983; 44:103–8.

170. Baronti A, Grieco A, Virgilli G, Lelli IM, Vibelli C. Effects of acutely administered ketotifen on specific bronchial provocation tests. Curr Ther Res 1983;33:936–40.

171. Tamura G, Mue S, Takishima T. Protective effects of ketotifen on allergen-induced bronchoconstriction and skin wheal. Clin Allergy 1986;16:535–41.

172. Yamaguchi T, Hashizume T, Matsuda M, et al. Pharmacokinetics of the H₁-receptor antagonist ebastine and its active metabolite carebastine in healthy subjects. Arzneim-Forsch/Drug Res 1994;44:59–64.

173. Mattila MJ, Aranko K, Kuitunen T. Diazepam effects on the performance of healthy subjects are not enhanced by treatment with the antihistamine ebastine. Br J Clin Pharmacol 1993;35:272–7.

174. Vincent J, Liminana R, Meredith PA, Reid JL. The pharmacokinetics, antihistamine and concentration-effect relationship of ebastine in healthy subjects. Br J Clin Pharmacol 1988;26:497–502.

175. Huang M-Y, Argenti D, Wilson J, Heald D, Ziemniak J. Single dose and steady-state pharmacokinetics of carebastine following administration of 10 mg ebastine tablets once daily in healthy elderly and young adults. Pharmaceut Res 1993;10:S-391.

176. Van Rooij J, Schoemaker HC, Bruno R, Reinhoudt JF, Breimer DD, Cohen AF. Cimetidine does not influence the metabolism of the H₁-receptor antagonist ebastine to its active metabolite carebastine. Br J Clin Pharmacol 1993; 35:661–3.

177. Wilson J, Huang M-Y, Argenti D, Pandit B, Heald D, Ziemniak J. Comparative pharmacokinetics of carebastine/ebastine in moderately renally impaired and healthy volunteers following a single 10 mg dose of ebastine. Pharmaceut Res 1993;10:S-391.

178. Huang M-Y, Wilson J, Argenti D, Heald D, Ziemniak J. Comparative pharmacokinetics of ebastine/carebastine in liver cirrhosis and healthy volunteers following administration of a 10 mg ebastine tablet. Pharmaceut Res 1993;10:S-390.

179. Simons FER, Watson WTA, Simons KJ. Pharmacokinetics and pharmaco-dynamics of ebastine in children. J Pediatr 1993;122:641–6.
180. Wood-Baker R, Holgate ST. Dose-response relationship of the H_1-hista-mine antagonist, ebastine, against histamine- and methacholine-induced bronchoconstriction in patients with asthma. Agents Actions 1990;30:284–6.
181. Tatsumi K, Ou T, Yamada H, Yoshimura H. Studies on metabolic fate of a new antiallergic agent, azelastine (4-(p-chlorobenzyl)-2-[N-methylperhy-droazepinyl-(4)]-1-(2H)-phthalazinone hydrochloride). Jpn J Pharmacol 1980;30:37–48.
182. Weliky I, Howard JR, Wichmann JK. Absolute bioavailability and pharma-cokinetics of azelastine. Pharmacol Res 1990;7:S247.
183. Johnston A, Warrington SJ, Turner P, Aurich R. The effect of repeated oral doses of azelastine hydrochloride on antipyrine half-life in normal vol-unteers. J Pharm Pharmacol 1988;40:225.
184. Pivonka J, Segelman FH, Hartman CA, Segl WE, Kucharczyk N, Sofia RD. Determination of azelastine and desmethylazelastine in human plasma by high-performance liquid chromatography. J Chromatogr 1987;420:89–98.
185. Rafferty P, Harrison PJ, Aurich R, Holgate ST. The in vivo potency and selectivity of azelastine as an H_1-histamine-receptor antagonist in human airways and skin. J Allergy Clin Immunol 1988;82:1113–8.
186. Atkins P, Merton H, Karpink P, Weliky I, Zweiman B. Azelastine inhibition of skin test reactivity in humans. J Allergy Clin Immunol 1985;75:167.
187. Spector SL, Perhach JL, Rohr AS, Rachelefsky GS, Katz RM, Siegel SC. Pharmacodynamic evaluation of azelastine in subjects with asthma. J Al-lergy Clin Immunol 1987;80:75–80.
188. Lurie A, Saudubray F, Eychenne JL, et al. Azelastine reduces allergen-induced nasal hyperresponsiveness in patients with allergic rhinitis: a rhino-manometric and clinical assessment. Am Rev Respir Dis 1990;141:A117.
189. Shapiro GG, Bierman CW, Pierson WE, Furukawa CT, Altman LC. Azelas-tine differential induced blockade of histamine nasal challenge test. J Allergy Clin Immunol 1987;79:255.
190. Meltzer EO, Storms WW, Pierson WE, et al. Efficacy of azelastine in peren-nial allergic rhinitis: Clinical and rhinomanometric evaluation. J Allergy Clin Immunol 1988;82:447–55.
191. Kemp JP, Meltzer EO, Orgel HA, et al. A dose-response study of the bron-chodilator action of azelastine in asthma. J Allergy Clin Immunol 1987;79:893–9.
192. Gould CAL, Ollier S, Aurich R, Davies RJ. A study of the clinical efficacy of azelastine in patients with extrinsic asthma, and its effect on airway responsiveness. Br J Clin Pharmacol 1988;26:515–25.
193. Magnussen H. The inhibitory effect of azelastine and ketotifen on histamine-induced bronchoconstriction in asthmatic patients. Chest 1987;91:855–8.
194. Bauer K, Aurich R, Kaik G. Dose related protective effect of azelastine on histamine-induced bronchoconstriction in extrinsic asthma. Int J Clin Pharmacol Ther Toxicol 1990;28:333–8.

195. Magnussen H, Reuss G, Jörres R, Aurich R. Duration of the effect of a single dose of azelastine on histamine-induced bronchoconstriction. J Allergy Clin Immunol 1989;83:467–71.

196. Rafferty P, Holgate ST. The inhibitory effect of azelastine hydrochloride on histamine- and allergen-induced bronchoconstriction in atopic asthma. Clin Exp Allergy 1989;19:315–20.

197. Albazzaz MK, Patel KR. Effect of azelastine on bronchoconstriction induced by histamine and leukotriene C4 in patients with extrinsic asthma. Thorax 1988;43:306–11.

198. Rafferty P, Ng WH, Phillips G, et al. The inhibitory actions of azelastine hydrochloride on the early and late bronchoconstrictor responses to inhaled allergen in atopic asthma. J Allergy Clin Immunol 1989;84:649–57.

199. Heykants J, Van Peer A, Woestenborghs R, Geuens I, Rombaut N, Vanden Bussche G. Pharmacokinetics and bioavailability of levocabastine (R 50547) in man. Arch Int Pharmacodyn Ther 1985;274:329–30.

200. Zazgornik J, Huang ML, Van Peer A, Woestenborghs R, Heykants J, Stephen A. Pharmacokinetics of orally administered levocabastine in patients with renal insufficiency. J Clin Pharmacol 1993;33:1214–8.

201. Dechant KL, Goa KL. Levocabastine. A review of its pharmacological properties and therapeutic potential as a topical antihistamine in allergic rhinitis and conjunctivitis. Drugs 1991;41:202–24.

202. Heykants J, Vanpeer A, Vandevelde V, Snoeck E, Meuldermans W, Woestenborghs R. The pharmacokinetic properties of topical levocabastine: a review. Clin Pharmacokinet 1995;29:221–30.

203. Yasuda SU, Wellstein A, Likhari P, Barbey JT, Woosley RL. Chlorpheniramine plasma concentration and histamine H₁-receptor occupancy. Clin Pharmacol Ther 1995;58:210–20.

8

H₁-Receptor Antagonists in Rhinoconjunctivitis

Peter Howarth
University of Southampton, Southampton, England

I. INTRODUCTION

H_1-receptor antagonists have been the mainstay of therapy for allergic rhinitis since their early clinical use, following the demonstration by Staub and Bovet in 1937 that this class of compounds protected against allergen-induced anaphylaxis (1). The experimental use of antihistamines in allergic disease was a natural sequel to the suggestion by Dale that histamine was central to immediate anaphylaxis (2) and the demonstrated release of histamine following allergen exposure in vitro (3, 4). The relationship between rhinitis and conjunctivitis symptoms and allergen (grass pollen) exposure in "hay fever" sufferers had been established at the end of the previous century (5).

Although observational studies reported benefit in allergic rhinoconjunctivitis with the newly developed antihistamines, the widespread acceptance of the older medications in this class was limited by their undesirable pharmacological effects such as sedation, dry mouth, and blurred vision. In addition, there was concern that asthma, a condition often present in association with rhinitis, could be worsened by antihistaminic therapy (6, 7). This concern was compounded by the reports from in vitro studies that antihistamines may potentiate mast-cell degranulation (8). Careful subsequent studies, however, did not confirm the bronchoconstrictor effects of H_1-antihistamines, indeed suggesting bronchodilatation (9–12), and the in vitro effects were identified only at high (suprapharmacological) concentrations and were unrelated to the H_1-receptor inhibitory effects (13, 14).

While an ethylamine group is common to most H_1-receptor antagonists, it was realized that many of the additional properties of this class of compounds, with the exception of sedation, were related to side-chain structure. Thus it was possible to synthesize H_1-antihistamines without the anticholinergic (15), antiserotoninergic (16), α-adrenergic receptor antagonistic (17) or local anesthetic (18) activity that was evident in early compounds. While all H_1-antihistamines were beneficial, the major breakthrough in their development for clinical use came with the fortuitous discovery of the H_1-antihistamine terfenadine, which, while retaining peripheral H_1-receptor antagonistic activity, was devoid of central nervous system (CNS) antihistaminic effects and thus did not induce sedation or impair psychomotor function (19). Paradoxically, terfenadine was manufactured during a program to identify new butyrophenones for antipsychotic use. As it had no CNS activity, it was unsuccessful in this regard. Fortunately, it was realized that terfenadine contained an ethylamine group and might be a useful H_1-receptor antagonist and this was proven in isolated tissue preparations. Its H_1-antihistaminic activity was shown to be specific, in that there was no H_2-receptor antagonism, no α- or β-adrenergic activity, no antiserotonin effect and no antimuscarinic effect (20). Thus terfenadine, in 1981, became the first oral, nonsedating antihistamine for the treatment of rhinoconjunctivitis. This represented a major advance in the use of H_1-antihistamines in the treatment of this condition. Subsequently, astemizole (1983), cetirizine (1988), and loratadine (1989) followed as alternative nonsedating H_1-antihistamines and several additional relatively nonsedating antihistamines have been launched or are under advanced development for the treatment of rhinoconjunctivitis, including topical H_1-antihistamines (21).

This review of antihistamines in rhinoconjunctivitis will discuss the current understanding of mucosal inflammation and histamine release in rhinoconjunctivitis, the role of histamine and its receptor specificity in this disease, the influence of H_1-receptor blockade on experimental and clinical rhinoconjunctivitis, the clinical use and comparative effects of differing H_1-receptor antagonists, the adverse effects of antihistamines, and finally the newer developments in this area.

II. MUCOSAL INFLAMMATION AND HISTAMINE RELEASE

A. Rhinitis

Allergic rhinitis is characterized by mucosal inflammation associated with an epithelial accumulation of mast cells, basophils, and eosinophils along with endothelial and epithelial cell activation, an expansion of dendritic antigen-presenting cells and, in chronic disease, T-lymphocyte accumulation and activation (22). The end organ effects of nasal itch, sneeze, rhinorrhea, and nasal obstruction arise due to mediator release from activated effector cells, primarily mast cells, basophils, and eosinophils (23).

Histamine is a major granule constituent of both tissue mast cells and circulating basophils (23). Immunological activation of mast cells and basophils induces histamine release amounting to 3–5 $\mu g/10^6$ cells and 1 $\mu g/10^6$ cells, respectively. This far exceeds the reported ability of both mast cells and basophils to generate leukotrienes such as LTC_4 on immunological activation (each 60 $ng/10^6$ cells) and the ability of mast cells to generate prostaglandins such as PGD_2 or platelet activating factor (PAF) (60 $ng/10^6$ cells and 2 $pmol/10^6$ cells, respectively). Thus histamine is quantitatively the major mediator generated on immunological activation of both mast cells and basophils.

In the nose, mast cells represent the primary source of histamine (24). Recent studies on nasal polyps and turbinate tissue reveal a tissue histamine content of 6.4 $\mu g/g$ wet weight and 3.9 $\mu g/g$ wet weight, respectively, with histamine release induced by anti-IgE antibody and calcium ionophore but not formyl-methionyl-leucyl-phenylalanine (FMLP). As FMLP is a potent leucyl-histamine secretagogue from human basophils, this suggests that basophils, unlike mast cells, are not a constitutive component of nasal tissue. Basophils are only identified in allergic rhinitis in nasal secretions (22) and thus represent a smaller contribution to the histamine release in this condition.

Nasal lavage has been used to recover tissue lining fluid from the nose and mediators and cells can be studied in this fluid. After nasal allergen challenge, increments in histamine concentrations in nasal lavage can be identified when repeated lavage is undertaken prior to challenge, to obtain a low prechallenge baseline (25). In naturally occurring disease single lavage measurements have surprisingly not identified a disease-related difference (26). This lack of change in allergic rhinitis may occur due to both histamine generation in the normal nose by alternative sources such as bacteria (27), or to the rapid degradation of histamine in active rhinitis by histaminase (diamine oxidase) in the nasal cavity, derived either from the circulation in association with plasma protein leakage (28, 29) or released in association with eosinophil activation (30).

Alternative methods have therefore been used to document mast-cell degranulation in allergic rhinitis. Ultrastructural changes of degranulation are reported following nasal allergen challenge by use of transmission electron microscopic (EM) examination of nasal biopsies (31). Using this technique to quantify all granules present in each mast cell as intact or degranulating and representing the number of degranulating granules as a percentage of the total granules present, we have evidence of mast-cell degranulation in naturally occurring seasonal allergic rhinitis. These results are confirmed by measurement of alternative markers of mast-cell degranulation in nasal lavage, such as tryptase, which is less susceptible to metabolism than histamine and allows seasonal increments to be identified in naturally occurring allergic rhinitis.

B. Conjunctivitis

The cellular changes in allergic conjunctivitis are similar to those identified in allergic rhinitis, involving mast cells, eosinophils, and T-lymphocytes. Eosinophils are rarely evident in the normal conjunctiva while mast cells are a normal constituent. The normal conjunctiva contains an estimated 5000 to 6000 mast cells/mm^3, usually located below the basement membrane in the substantia propria (32). Mast cells are not generally found in the normal epithelium of the bulbar or tarsal conjunctiva (33, 34). In allergic conjunctivitis mast cells increase in number and infiltrate into the epithelium (34, 35). Eosinophils are also recruited into the conjunctival tissue and can be identified both in the substantia propria and the epithelium (34). T-lymphocytes are found in the epithelium and substantia propria of normal conjunctiva (33), consistent with their "immunological policing role" and as in rhinitis, are only increased in severe chronic disease, such as keratoconjunctivitis (36). T-cells, cloned from conjunctival biopsies of patients with vernal keratoconjunctivitis, have been shown to have a cytokine profile compatible with a TH$_2$ subset (37).

Histamine is present in normal tears (5 ng/ml) and is elevated in vernal keratoconjunctivitis (16 ng/ml), in which there is severe conjunctival inflammation (38). As in rhinitis, it has been more difficult to detect histamine increments in milder forms of conjunctivitis or indeed after conjunctival allergen challenge. However, it has been demonstrated that histaminase is present in human tears and with inactivation of this enzyme it has been possible to demonstrate acute increments in tear levels of histamine following conjunctival allergen challenge in sensitized individuals (38).

III. ROLE OF HISTAMINE AND RECEPTOR SPECIFICITY IN RHINOCONJUNCTIVITIS

A. Rhinitis

Topical challenge with histamine has been used to investigate its influence in the nose. Nasal histamine insufflation mimics acute allergic rhinitis in inducing immediate nasal itching, sneezing, and rhinorrhea as well as causing nasal obstruction (39). Following single nostril application of histamine, the rhinorrhea is bilateral, indicative of reflex neuronal stimulation. The nasal itching and sneezing are also neuronally mediated, due to stimulation of histamine-sensitive afferent receptors, while nasal obstruction is vascular. Histamine exerts several differing effects on the nasal vasculature as it increases mucosal blood flow (40), enhances nasal vascular permeability (41), induces plasma protein exudation from fenestrated superficial capillaries, and produces nasal venous engorgement through its regulatory effects on the nasal capacitance vessels (42). The increase in turbinate volume reduces nasal airflow and increases nasal airway resistance.

The neuronal effects of histamine are H$_1$-receptor mediated in that specific H$_1$-receptor antagonism prevents histamine-induced nasal itch, sneeze, and rhinorrhea (43). The receptor regulation of the nasal vascular effects is less precise. H$_1$-receptor blockade can prevent histamine-induced nasal vascular plasma protein leakage suggestive of a specific H$_1$-receptor effect in this regard (41). Histamine H$_2$-receptor antagonism can, however, also partially reduce plasma protein exudation. This effect may be indirect due to a regulatory effect on mucosal blood flow; in the skin H$_2$-receptor stimulation causes vasodilation (44) and in the nose H$_2$-receptor antagonism reduces the increase in mucosal blood flow in response to allergen (45). An H$_2$-receptor antagonist, by reducing mucosal blood flow, would indirectly limit the potential plasma protein exudation from the superficial mucosal capillaries.

Both H$_1$- and H$_2$-receptor antagonists also modify histamine-induced nasal obstruction. When combined, their effects have been reported to be greater than after administration of either alone (46). While this limited effect in inhibiting histamine-induced nasal blockage could relate to the histamine-induced kinin generation which occurs within the nose (47), as kinins also induce nasal blockage (48) as well as promoting plasma protein leakage (49), this is unlikely since H$_1$-receptor blockade completely inhibits plasma protein extravasation. More probable is an effect of histamine on H$_3$-receptors. In rodents, H$_3$-receptors have been identified on presynaptic perivascular nerve terminals where they regulate sympathetic tone

(50). As the nasal vasculature is under sympathetic control, with stimulation inducing constriction, inhibition of this effect by H_3-receptor activation would lead to vasodilation and nasal venous engorgement. Thus the nasal obstruction that is not H_1- or H_2-receptor-mediated following nasal histamine insufflation is likely to be H_3-receptor-mediated.

Histamine may also contribute to the cellular tissue recruitment that characterizes allergic mucosal inflammation. In cultured human umbilical vein endothelial cells, histamine upregulates the expression of the cell adhesion molecule P-selectin (51). It has been proposed that P-selectin, which induces a rolling margination of leukocytes when up-regulated on the vascular endothelium, is critical to tissue eosinophil recruitment (52). In vivo studies in rodents have demonstrated that histamine-induced rolling margination is inhibited by a P-selectin antibody and that the mobilization of P-selectin onto the vascular endothelium by histamine is mediated by H_1- rather than H_2-receptors (53). Histamine H_1-receptors also appear to be relevant to the epithelial expression of intercellular adhesion molecule-1 (ICAM-1), as terfenadine has been shown to reduce epithelial ICAM-1 expression in pollen-sensitive patients during seasonal pollen exposure (54). In addition to promoting eosinophil-endothelial adherence, histamine contributes to eosinophil activation (55). This action is H_3-receptor-mediated as the H_3-receptor antagonist thioperamide, but not the H_1- or H_2-receptor antagonists mepyramine (pyrilamine) or cimetidine, respectively, prevents histamine-induced eosinophil activation (55).

B. Conjunctivitis

As for rhinitis, instillation of histamine into the conjunctival sac reproduces many of the features of allergic conjunctivitis (56). Histamine induces itching, lacrimation, hyperemia, and conjunctival edema. These effects all occur rapidly. Investigation of the effects of the topical H_1-receptor antagonist, levocabastine, identified that itching, tearing, redness, and conjunctival edema are all H_1-receptor-mediated (57). In guinea pigs, the inhibition of histamine-induced conjunctival edema by levocabastine has been shown to be dose-related (58). Within the conjunctiva there is also an H_2-receptor-mediated effect of histamine on blood flow (59). As such, the hyperemia and conjunctival edema is in part H_2-receptor dependent. No details exist concerning H_3-receptors and the conjunctiva. The relevance of histamine to tissue cell recruitment and activation in allergic conjunctivitis is as described for rhinitis.

IV. ANTIHISTAMINES AND EXPERIMENTAL ALLERGIC RHINOCONJUNCTIVITIS

Allergen challenge in sensitized individuals has been used to investigate the effects of oral and topical antihistamine preparations in both the nose and eye. Nasal allergen challenge, either with an aqueous extract or with pollen grains, induces an immediate response characterized by nasal itch, sneeze, rhinorrhea, and nasal blockage. The nasal blockage is often long lasting but the biphasic obstructive response that typifies the immediate and late responses in the lower airways is not readily discernible within the upper airways. A symptomatic late-phase response is not a characteristic finding following ocular allergen challenge. There are, however, biochemical and cellular measures indicative of events extending beyond the immediate response.

During the immediate nasal response to allergen, an increase in histamine, tryptase, PGD_2, and kinins is evident, consistent with mast-cell degranulation, while 6–10 h postchallenge further increases in histamine and kinins but not PGD_2 or tryptase, are apparent (25). The late change in histamine but not the other mast-cell markers is indicative of basophil activation. Consistent with this, an accumulation of basophils is evident within nasal lavage fluid during the late response (60), although the most important cellular change 24 h postallergen challenge relates to eosinophil influx (61, 62). At this time there is allergen-induced nasal hyperresponsiveness which is suppressed by antihistamines (Fig. 1).

Treatment with H_1-receptor antagonists modifies the response to allergen challenge in both the nose and eye (45, 63). Relief of symptoms with single-dose medication is more rapid with topical than oral medication. Topical levocabastine is reported to afford protection within 5 min of application (64), whereas in a pollen challenge chamber, relief of symptoms took 2–4 h with oral astemizole (65). In the latter model astemizole 10 mg and 30 mg afforded comparable relief. The H_1-receptor-mediated events (e.g., nasal itch, sneeze, and rhinorrhea) are reduced by H_1-antihistamine therapy (45). Reduction of allergen-induced nasal obstruction is variable, with some studies reporting some reduction in comparison to placebo (64) while others have failed to find any effect of either H_1-receptor or H_2-receptor blockade (45, 66). Using laboratory allergen challenge as a model, few studies have assessed duration of antihistaminic effect at the organ site, usually investigating the H_1-receptor inhibitory effect at a single time point, the choice of time being derived from time course studies of skin wheal and flare responses. Topical levocabastine therapy, 0.05%

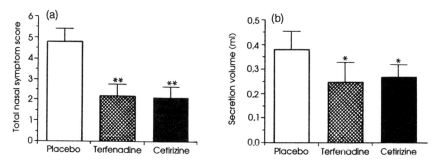

Figure 1 In a double-blind study, terfenadine 60 mg twice daily, cetirizine 10 mg daily, or placebo were given for 5 days, outside the pollen season. At steady state, a nasal methacholine challenge test was performed before and 24 h after nasal allergen challenge. (a) A composite nasal symptom score is shown (mean ± SEM). The differences between terfenadine or cetirizine and placebo are statistically significant ($p < 0.01$). (b) The volume of secretion after a nasal methacholine challenge, 24 h after an allergen challenge is shown (mean ± SEM). Compared to placebo, terfenadine and cetirizine significantly inhibited the increased response to methacholine ($p < 0.05$). (From Ref. 61.)

solution, has, however, been shown to reduce allergen-induced conjunctival itch, hyperemia, tearing, and chemosis 4 h after initial application (63), having also been effective 10 min after administration. In a separate study, a single, intranasal dose of levocabastine was protective for 10–12 h (58).

V. ANTIALLERGIC ACTIVITIES OF H_1-ANTIHISTAMINES IN RHINOCONJUNCTIVITIS

Nasal allergen challenge has also been used to investigate the antiallergic properties of H_1-antihistamines *in vivo* (67–70). These investigations are based on the *in vitro* findings that H_1-antihistamines inhibit anti-IgE-induced histamine release from human lung fragments at low concentrations even though at high concentrations they enhance histamine release (13). Comparative in vitro studies suggest that this effect does not relate to H_1-receptor antagonist potency but more to the lipophilicity of the compound. H_1-antihistamines with low lipophilicity have less effect on histamine release than those with greater lipophilicity. This antiallergic potential of antihistamines has been extended in other in vitro studies reporting inhibition of LTC$_4$ and LTD$_4$ release from dispersed human lung cells and from eosinophils, modification of macrophage function, inhibition of both neutrophil and basophil activation, and alteration in platelet cytotoxicity (71).

When administered orally at standard clinical doses, the piperidine group-containing antihistamines, azatadine, terfenadine, and loratadine have all been shown to reduce the allergen-induced increment in nasal lavage concentrations of histamine and kinins (67, 68, 70, 72). Where measured, this protective effect is also associated with a reduction in induced increments in albumin. The piperidine astemizole is, however, without effect on allergen-induced histamine release and the piperazine, cetirizine, is also without effect, although it does decrease allergen-induced changes in LTC$_4$ and LTD$_4$ (69). The interpretation of these findings is complicated in that kinins are likely to be generated in association with plasma protein extravasation (47). Thus the inhibitory action of antihistamines may purely reflect an H$_1$-mediated action on vascular permeability, with the inhibitory action on the histamine increase reflecting reduced tissue availability of histamine due to the reduced tissue fluid exudation. Against such an explanation, however, is the reported lack of inhibitory effect of cetirizine on the histamine rise in the presence of an inhibition of the allergen-induced albumin changes (69). This single study, which divorces the protein and histamine increments in nasal lavage following topical allergen exposure, is thus critical to the interpretation.

Should the findings in the nasal challenge model indicate inhibition of mast-cell degranulation with some but not all H$_1$-antihistamines, then those drugs with an additional action might be anticipated to achieve greater clinical benefit. Histamine and tryptase, mediators stored preformed in mast-cell granules (29), along with the preformed cytokines IL-4 and TNF-α (73, 74), would all be anticipated to participate in allergen-induced cell recruitment. "Mast-cell-stabilizing" antihistamines would thus inhibit allergen-induced cell recruitment. Paradoxically, in this respect, cetirizine, which does not inhibit allergen-induced histamine release, has been reported in a skin blister model to inhibit allergen-induced eosinophil and neutrophil recruitment (75, 76). When this has been investigated in the nose, however, cetirizine has not been found to influence either acute allergen-induced eosinophil recruitment or mucosal eosinophil accumulation in naturally occurring disease (61, 77). Furthermore, a comparative study of the antiallergic antihistamines terfenadine and azatadine with the antihistamine astemizole identified that after 1 week of treatment at standard clinical doses, all three antihistamines similarly inhibited the nasal provocation response to ragweed, as assessed by the clinical response, the amount of induced nasal secretions, and rhinomanometry (78). An inability to identify additional benefit with antiallergic antihistamines queries the true interpretation of the laboratory challenge findings, a difficulty that is also reflected in naturally occurring disease in which all newer H_1-antihistamines appear equally efficacious when used at standard clinical doses (vide infra).

VI. ORAL H_1-ANTIHISTAMINES AND NATURALLY OCCURRING RHINOCONJUNCTIVITIS

A. Seasonal Allergic Disease

While minor differences exist among clinical trials due to size of the trial, local environmental factors, patient selection criteria, and individual patient variability, in general, in seasonal allergic rhinitis H_1-antihistamines modify nasal itch, sneeze, and rhinorrhea with little or no effect on obstruction (79–82). The effect on the H_1-responsive symptoms is only partial, approximately a 50–60% reduction, as histamine is not the sole contributor to these symptoms (83). Oral H_1-antihistamines also modify conjunctival itch, watering, and redness (84, 85). The effect on conjunctival symptoms may be less marked than for rhinitis symptoms, possibly due to a higher local allergen load. Topical H_1-antihistamines have a similar profile (86) and while theoretically they may achieve a higher local concentration and create a greater effect (87), comparative trials of topical H_1-antihistamine therapy with oral H_1-antihistamine therapy report similar overall efficacy in reducing ocular symptoms (88–89).

1. Placebo-Controlled Studies

Terfenadine   Although early dose-ranging studies failed to reveal clinical benefit (90), subsequent randomized, double-blind, placebo-controlled studies of terfenadine 60 mg twice daily reported significant benefit over placebo when assessed as either global symptom reduction (91) or use of rescue medication (92). Treatment during a 7-day period revealed maximum symptom improvement by the end of 2 days of therapy, with the exception of the symptoms of itchy nose and nasal congestion (80). Subsequent comparisons of terfenadine 60 mg twice daily with either terfenadine 120 mg twice daily (93, 94) or terfenadine 120 mg daily (95) have shown that these treatment regimens do not significantly differ in their clinical efficacy. Thus a total dose of 120 mg terfenadine per day is sufficient whether given in single or divided doses.

Astemizole Randomized double-blind, placebo-controlled trials of astemizole 10 mg daily, have confirmed the benefit of this H_1-antihistamine over placebo in the treatment of seasonal allergic rhinoconjunctivitis, whether active drug is administered when patients are symptomatic (symptom resolution) (79) or administered prophylactically (symptom prevention) (83). Due to the long half-life of astemizole (96), the first placebo-controlled trial involving 155 patients investigated therapy given only once per week for hay fever. This compared 10 mg and 25 mg astemizole weekly versus placebo for 6 weeks with the use of a 50-mg loading dose in the first week (97). Both active doses were significantly better than placebo, but the clinical benefit declined toward the end of the week. Thus once-weekly therapy was considered inadequate and daily therapy was investigated; 10 mg daily was found to be optimal, with 5 mg daily providing poorer control (98). Due to the delay in reaching steady-state kinetics with astemizole therapy, a loading dose (30 mg) was initially considered necessary (99) but was subsequently not found to influence clinical benefit (65). The recommended dose of 10 mg daily for astemizole is supported by placebo-controlled studies (100). Astemizole is probably optimally used by initiating therapy prior to the onset of the pollen season (83, 101).

Cetirizine Placebo-controlled comparative studies of cetirizine 5 mg daily (102), 10 mg daily (102, 103), and 10 mg twice daily (104) have been conducted with placebo and all doses were found to be effective in seasonal allergic rhinitis. A subsequent study compared 5, 10, and 20 mg cetirizine doses in this disorder and found all active treatments to be effective with no significant dose-response relationship (81) (Fig. 2). Cetirizine is marketed at a dose of 5–10 mg daily for the treatment of seasonal allergic rhinoconjunctivitis. In the placebo-controlled trials, cetirizine, like other H_1-antihistamines, is effective for relief of nasal pruritus, runny nose, and sneezing but less conclusively effective for nasal obstruction or lacrimation.

Loratadine Placebo-controlled, double-blind, randomized trials of loratadine in seasonal allergic rhinitis have confirmed the efficacy of this H_1-antihistamine at a dose of 10 mg daily (82, 105, 106). Like other H_1-receptor antagonists, loratadine is less effective in relieving nasal blockage than itch, sneeze, and rhinorrhea (107). The choice of 10 mg daily is supported by the findings of a study that investigated the effect of 40 mg daily loratadine in seasonal allergic rhinitis and suggested that this dose was no more effective than lower doses and carried the potential risk of sedation (108).

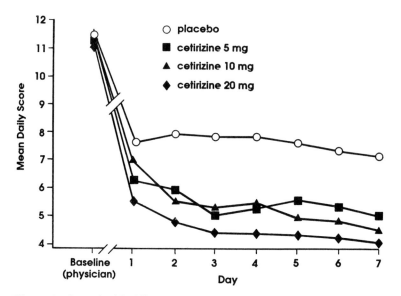

Figure 2 In a double-blind parallel-group study in 419 patients with seasonal allergic rhinitis, cetirizine 5, 10, or 20 mg or placebo was given once daily for 1 week. The patients' mean daily total symptom severity score ratings are shown. All cetirizine doses produced greater relief of symptoms than placebo did. For most symptoms, no significant dose-response relationship was found. (From Ref. 81.)

2. Comparative Studies

Comparisons have been made between nonsedating H_1-antihistamines and (1) sedating H_1-antihistamines such as chlorpheniramine and pheniramine; (2) alternative nonsedating H_1-receptor antagonists; and (3) nonantihistaminic therapies for allergic rhinitis. While sedating and nonsedating H_1-antihistamines often have similar clinical efficacy, the two generations of H_1-receptor antagonists are clearly distinguishable by their sedative adverse effect profiles (80, 107, 108–112); the nonsedating H_1-antihistamines have a much more favorable risk/benefit ratio. With minor differences among trials, the overall effectiveness of the nonsedating H_1-antihistamines is similar (106, 113–117). Astemizole has been found to have a slower onset of symptom relief than either terfenadine (118) or loratadine (119) although with continued regular therapy the comparator drugs were not significantly different.

In longer term studies, astemizole has been reported to be more effective than either loratadine (117) or terfenadine (83, 101) and loratadine to be more effective than terfenadine once treatment has continued for over 4 weeks (120). This difference may, however, relate to treatment compliance with a once-a-day medication as compared to the twice-a-day regimen, rather than reflect a fundamental difference in drug therapy. Similarly, factors other than pharmacological differences may account for the reported finding that cetirizine is more effective than astemizole (121), as the end point for comparison was the switch of therapy after 4 days medication, with greater numbers of patients opting to change from astemizole to cetirizine rather than vice versa. Those changing from astemizole to cetirizine, however, also failed to improve when followed subsequently and could be considered nonresponders to H$_1$-antihistamines. The trial results could be accounted for by an imbalance in the initial study groups.

Although one study found no difference between an oral H$_1$-antihistamine and nasal steroid therapy in seasonal allergic rhinitis (85), the consensus is that topical nasal steroid therapy is more effective than H$_1$-antihistamine therapy for the treatment of this disorder, especially for relief of nasal obstruction (84, 122) (Fig. 3). Oral H$_1$-antihistamines have the benefit of modifying conjunctival symptoms (84, 101). As might be anticipated, a decongestant has a greater effect on nasal obstruction than an H$_1$-antihistamine does (123), although not on other symptoms. Oral H$_1$-antihistamines and topical cromoglycate have comparable clinical benefit, although the frequency of treatment regimens favors H$_1$-antihistamines on the basis of improved compliance with regular medication administered once or twice daily rather than four times daily (124–125) (Fig. 4).

B. Perennial Allergic Rhinitis

H$_1$-antihistamines are effective in the treatment of this condition as they are for seasonal allergic rhinitis. With chronic persistent disease, however, nasal obstruction is often a more prominent symptom while nasal itch is less common, so the benefit from H$_1$-antihistamines may be less marked than that reported in seasonal allergic rhinitis and time-to-improvement may be more prolonged. Indeed an early study in perennial disease found no treatment benefit (126). Due to these differences from seasonal disease, patient selection for clinical trials is often more rigorous with either exclusion of patients in whom nasal obstruction is the most troublesome symptom or the exclusion of this symptom from an overall rhinitis evaluation score.

1. Placebo-Controlled Studies

Significant clinical benefit over placebo has been reported with terfenadine 60 mg twice daily (127), terfenadine 120 mg daily (128), astemizole

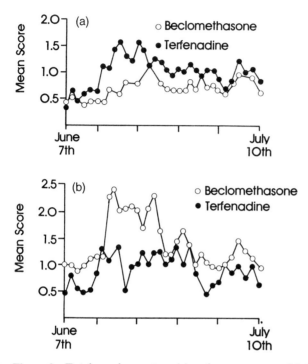

Figure 3 Total nasal symptom (a) and eye symptom (b) diary scores in a double-blind, parallel group comparison of beclomethasone dipropionate aqueous nasal spray and terfenadine tablets, 60 mg twice daily in seasonal allergic rhinitis. The terfenadine-treated group had lower eye symptom scores than the beclomethasone-treated group, and these differences were statistically significant during the first half of the study. The beclomethasone-treated group had lower nasal symptom scores than the terfenadine-treated group and these differences reached statistical significance on high pollen-count days. (From Ref. 84.)

10 mg daily (129), cetirizine 10 mg daily (128, 130), and loratadine 10 mg daily (127) (Fig. 5). These trials report most marked benefit for nasal pruritus, sneezing, and rhinorrhea with a lesser effect on nasal obstruction. The lack of benefit on nasal obstruction was highlighted in one study in which the nasal corticosteroid, beclomethasone dipropionate, was administered after the use of regular antihistamine treatment. This resulted in a further reduction in symptoms, in particular blocked nose (129). The improvement with H_1-antihistamine treatment alone can increase during 28 days of treatment (127).

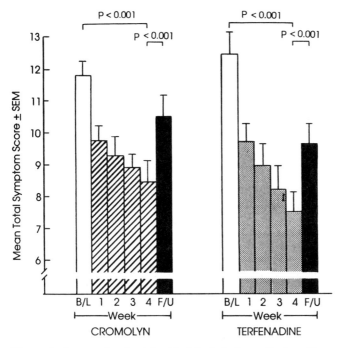

Figure 4 In a randomized double-blind study, patients with seasonal allergic rhinitis received either cromolyn 4% solution, one puff in each nostril four times daily or terfenadine 60 mg twice daily, along with the appropriate placebo spray or tablet for 4 weeks, following a 1-week baseline qualification period. Mean ± SEM total symptom scores at baseline, at the end of each week of therapy, and at follow-up visit after 1 week without therapy are shown. The P values are for differences between means within treatments. Cromolyn and terfenadine were comparably effective treatments for allergic rhinitis. (From Ref. 124.)

2. Comparative Studies with H₁-Antihistamines

As for seasonal allergic disease, comparative studies among the nonsedating H₁-antihistamines indicate that at standard doses they have similar clinical efficacy. Comparative trials of terfenadine 60 mg twice daily with cetirizine, 10 mg daily, over 2 weeks (131), and loratadine 10 mg daily, over 4 weeks (127), as well as between terfenadine 120 mg daily, and cetirizine 10 mg daily, over 8 weeks (132) reveal no difference between treatments. A shorter study (3 weeks) comparing terfenadine 120 mg daily, and cetirizine 10 mg daily, reported a greater effect of cetirizine on rhinorrhea, by clinician assessment (128), but this difference was not apparent

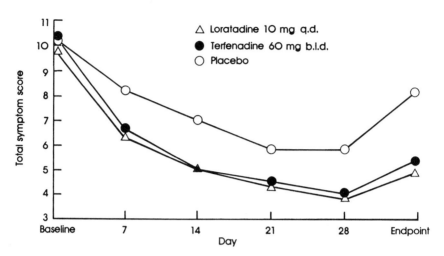

Figure 5 In a double-blind, parallel-group study, the efficacy of loratadine 10 mg q.d., terfenadine 60 mg b.i.d., and placebo was evaluated in 228 patients with perennial allergic rhinitis. Patients receiving loratadine or terfenadine had significantly improved mean combined symptom scores. Both active treatments were significantly different from placebo but not from each other ($p < 0.04$) at each evaluation after baseline. (From Ref. 127.)

on analysis of patient diary cards and no difference existed between the two treatments in their efficacy with respect to control of sneezing, itchy nose, itchy throat, itchy palate, nasal congestion, or ocular symptoms. Similar comparative studies between cetirizine 10 mg daily, and astemizole 10 mg daily, showed clinical improvement with both H_1-receptor antagonists (133).

With respect to comparisons with older H_1-antihistamines, investigator assessment suggested significantly greater benefit with astemizole 10 mg daily, than chlorpheniramine 4 mg three times daily, over a 4-week treatment period (134). A 6-month trial of loratadine 10 mg daily and clemastine 1 mg twice daily, reported comparable effects. Good or excellent responses were considered to have occurred in 63% of patients treated with loratadine and 64% of those receiving clemastine (135). The clemastine-treated patients did, however, experience more sedation (28%) compared to those on loratadine (4%). A similar difference in sedative profile between loratadine and clemastine was reported in a separate study (136).

The persistence of benefit with both loratadine and clemastine over the 6-month treatment period indicates that there is no tachyphylaxis to these H_1-antihistamines. Although tachyphylaxis to terfenadine was suggested in one small comparative study ($n = 27$) in which the benefit from terfenadine at 4 weeks (7% good/excellent response) was less than at 2 weeks and significantly less than for astemizole (54% good/excellent response) (137), this loss of effect with continued treatment has not been evident in other studies with terfenadine (127, 132), continued improvement over 28 days being evident in one (127).

3. Comparative Studies with Other Agents

Fewer studies have been undertaken in perennial disease than in seasonal disease. Topical nasal corticosteroid therapy is more effective than H₁-antihistaminic therapy (138–140), while no difference was evident in one comparative study between terfenadine alone and a terfenadine/decongestant preparation (terfenadine-D), which contained 60 mg terfenadine and pseudoephedrine hydrochloride 10 mg in an outer coat for immediate release and 110 mg pseudoephedrine in an extended release core (141). Both treatments were given twice daily. No good comparisons exist with respect to cromoglycate preparations.

VII. TOPICAL H₁-ANTIHISTAMINES AND NATURALLY OCCURRING RHINOCONJUNCTIVITIS

Two H₁-antihistamines have been developed for topical application, levocabastine for nasal and ocular administration and azelastine for nasal administration (58, 142). Both antihistamines are sedating when administered orally but are free from this adverse effect when administered topically because of the lower dose required to achieve clinical benefit with topical rather than systemic administration. The nasal dose for levocabastine and azelastine is 0.1 mg to each nostril twice daily and 0.14 mg to each nostril twice daily, respectively, while the conjunctival dose of levocabastine is 0.03 mg administered to each eye twice daily. The nasal efficacy of these topical H₁-antihistamines has been demonstrated under challenge conditions (21, 45, 64). Both have a rapid onset of effect (10–15 min) and a duration of action of up to 12 h, findings which have been confirmed with the ocular formulation of levocabastine using conjunctival challenge (57, 63).

A. Rhinitis

Both levocabastine and azelastine, when administered topically, are most effective against nasal itch, sneeze, and rhinorrhea, although some benefit is also reported for nasal obstruction with azelastine (143–145). There are few published placebo-controlled trials in seasonal allergic rhinitis (146), the majority of studies making comparison with alternative active medications, such as H_1-antihistamines (144, 145, 147, 150), cromoglycate (146, 151), or nasal corticosteroids (21, 148). Azelastine, administered as a nasal spray, is more effective than either itself or terfenadine administered orally in relieving nasal obstruction while producing comparable relief of other nasal symptoms. Levocabastine is reported to be more effective than an alternative topical antihistamine/decongestant (antazoline/naphazoline) preparation (148), oral terfenadine (149–150), or topical cromoglycate (146, 151).

Three double-blind, double-dummy, randomized parallel group studies have recently been published comparing the therapeutic efficacy of levocabastine eye drops (0.5 mg/ml, one drop to each eye twice daily) and nasal spray (0.5 mg/ml, two puffs each nostril twice daily) with oral terfenadine (60 mg twice daily) in the treatment of seasonal allergic rhinoconjunctivitis. These studies involved 267 adults and lasted for 8 weeks (88, 149, 150). Analysis of patient evaluations of global therapeutic efficacy with pooled data identified comparable efficacy (152) while analysis of investigator and patient assessments of individual symptom severity identified a number of statistically significant differences in favor of levocabastine (149, 150), particularly for the relief of nasal discharge and sneezing. The efficacy of levocabastine was maintained even when the pollen count was high, the levocabastine-treated patients experiencing more symptom-free days during these periods than those receiving terfenadine.

Comparative studies between levocabastine (0.5 mg/ml, two sprays each nostril four times daily) and cromoglycate (20 mg/ml, two sprays each nostril four times daily) involving 114 patients over 2 weeks found significantly fewer symptoms with levocabastine therapy (76% of patients improving vs 46% on sodium cromoglycate) (152) which was accompanied by more symptom-free days in the levocabastine-treated patients (146).

While both levocabastine (148) and azelastine (21) nasal sprays are reported to be as effective as topical nasal corticosteroids, further studies are required before valid comparisons can be made.

Studies in perennial rhinitis are also limited. One preliminary 2-week study reported improvement in sneezing and rhinorrhea with topical levocabastine, in comparison with placebo, which could not be improved further by the addition of topical nasal beclomethasone dipropionate, while nasal blockage responded to the additional therapy (153).

B. Conjunctivitis

Levocabastine (0.5 mg/ml, one drop into each eye twice daily) has been shown to be superior to placebo for the treatment of itchy eyes, runny eyes, and red eyes associated with seasonal allergic conjunctivitis (86). Relief of symptoms is rapid, occurring within 15 min of application (143).

With regular therapy, topical levocabastine and oral terfenadine have comparable overall response rates. Analysis of pooled data from trials involving 350 patients reported a good/excellent response of 80% with levocabastine and 72% with terfenadine (21). This is mirrored by the lack of difference between topical levocabastine (0.5 mg/ml one drop into each eye twice daily) and oral terfenadine (60 mg twice daily) when the individual symptom/signs of ocular itching, lacrimation, ocular redness, and eyelid edema are analyzed (150). In contrast, topical levocabastine (0.5 mg/ml) appears more effective than topical cromoglycate (20 mg/ml) when both preparations are administered 4 times a day for seasonal allergic conjunctivitis ($n = 158$) over a 4-week period (154–155). Patient diary cards identified fewer symptoms in the levocabastine-treated patients (152) and in one of the studies (155) the proportion of patients virtually symptom-free on high pollen count days was greater in the levocabastine- (33%) than in the cromoglycate- (6%) treated patients.

VIII. ADVERSE EFFECTS

The use of medication for a condition is based both on its beneficial effects and on its potential to cause adverse effects. The major advantages of the second-generation oral H$_1$-antihistamines, which considerably improved their risk/benefit profile, was their absence of CNS sedative effects when used at standard clinical doses. With recommended doses of terfenadine, astemizole, cetirizine, and loratadine, the reported side effects of tiredness and drowsiness are no different from placebo (156, 157). This clinical information is substantiated by laboratory assessment incorporating reaction time analysis, performance of complex sensorimotor tasks, such as simulated car driving, sleep latency studies, and electroencephalographic monitoring (158–162). Using such methods the topical H$_1$-antagonist, levocabastine, has also been shown to be devoid of adverse CNS effects (163).

Despite the lack of CNS sedative effects, astemizole has been reported to cause weight gain (79), an effect that is considered to be mediated centrally due to an action on the satiety center causing an increase in appetite. Although exceptionally rare, astemizole, along with terfenadine, has been reported to cause serious, occasionally fatal, cardiac arrhythmias (164). The tachyarrhythmia, classically torsade de pointes, is associated with a drug-induced prolongation of the QT$_c$ interval (165, 166) and has been reported in association with overdose (167, 168) and liver dysfunction. Both astemizole and terfenadine are hepatically metabolized, and when there is hepatic dysfunction or coadministration of drugs competing for the same cytochrome P_{450} isoenzyme, the metabolism of the antihistamine may be impaired (169–171). Such drugs include the imidazole antifungal agents, ketoconazole and itraconazole, and the macrolide antibiotics, erythromycin and clarithromycin. The coadministration of drugs that intrinsically lengthen the QT_c interval, such as procainamide, sotalol, and amiodarone, should also be avoided (165–166). In the absence of these considerations and with the use of clinically recommended doses, neither astemizole nor terfenadine have an adverse effect on the QT_c interval and are considered to be safe. Cetirizine is not metabolized to any extent in the liver and thus is less influenced by impaired hepatic function. Loratadine, while sharing the same cytochrome P_{450} isoenzyme (CYP3A4) as astemizole and terfenadine, is also metabolized by an alternative isoenzyme (CYP2D6) (173) and no adverse cardiac effects have been substantiated with this H_1-antihistamine (174).

Topical H_1-antihistamines are associated with very low plasma levels and do not induce systemic effects. The main adverse effect of topical therapy is either nasal or ocular irritation (143, 146, 154). This affects approximately 5% of patients receiving the nasal spray and 16% of those receiving eye drops (143). This incidence of adverse reaction is similar with placebo (6% and 16%, respectively) and does not appear related to the drug per se. Detailed ophthalmological studies have not revealed any structural or functional adverse effect with regular use (150). Azelastine may alter taste sensation even when administered topically.

IX. NEW DEVELOPMENTS AND SPECIAL CONSIDERATIONS

A. H_1-Antihistamine/Decongestant Combination Therapy

The lack of effect of orally or topically administered, H_1-antihistamines on nasal obstruction has led to the development of combination products containing both H_1-antihistamines and decongestants. Two oral α-adrenergic agonists, pseudoephedrine and phenylpropanolamine, have been most commonly employed as decongestants.

Clinical trials in perennial allergic rhinitis have been undertaken with combinations of pseudoephedrine and the newer nonsedating H_1-antihistamines. Terfenadine-D is a combination of 60 mg terfenadine and 10 mg pseudoephedrine hydrochloride in an outer coat for immediate release and 110 mg pseudoephedrine in an extended release core. In chronic rhinitis, terfenadine-D twice daily was superior to placebo but not significantly better than terfenadine alone (141). It has, however, been possible with the combination product to demonstrate an objective improvement in nasal obstruction by measuring nasal peak inspiratory flow (175). Acrivastine-D, a combination of 8 mg acrivastine and 60 mg pseudoephedrine taken three times daily, has also been found to improve nasal symptoms and nasal airflow compared to placebo (176). A trial comparing astemizole-D (10 mg astemizole and 240 mg pseudoephedrine as a controlled release preparation) with an older combination product, Actifed (5 mg triprolidine and 120 mg pseudoephedrine) found that at standard doses the two had comparable effects (177) but both were associated with a high incidence of adverse effects (51% vs 53% of patients, respectively). Adverse effect reporting has been high with these combination preparations, largely due to insomnia and restlessness attributed to the pseudoephedrine which affects up to 30% of users (141, 178).

The ability to demonstrate clinical benefit of the combination product over the individual components depends on the sample size. A small study of terfenadine-D in seasonal allergic rhinitis found as in perennial disease that the combination product was more effective than terfenadine alone (179), whereas large multicenter studies of loratadine-D, involving over 800 patients (121, 180), and astemizole-D, also involving over 800 patients (181), have found that it is possible to distinguish the significant effects of the individual components as well as to demonstrate a significant effect of the combination product. The formulation may be critical to the acceptance of these preparations. In contrast to terfenadine-D, loratadine-D is a combination of 5 mg loratadine and 120 mg pseudoephedrine in a Repetab tablet formulation, in which there is an immediate-release coating of 5 mg loratadine and 60 mg pseudoephedrine and a sustained-release core of pseudoephedrine. Nervousness after administration of loratadine-D (4–7%) is still more commonly reported than after placebo (1–2%) but is less common than after astemizole-D (10–11%), and much less common than after terfenadine-D (25%).

B. Common Cold

Nasal lavage studies in the common cold have failed to demonstrate histamine release and H₁-antihistamines have not been found to be effective (182–183) (Fig. 6). One recent study of loratadine-D versus placebo, has, however, reported that loratadine-D, administered twice daily, is more effective than placebo in relieving symptoms, with improvement not only in nasal obstruction, as might be anticipated with pseudoephedrine, but also improvement in sneezing and rhinorrhea (21). Further assessment of pseudoephedrine alone administered by the same method is now required to assess whether modification of nasal obstruction alone may diminish

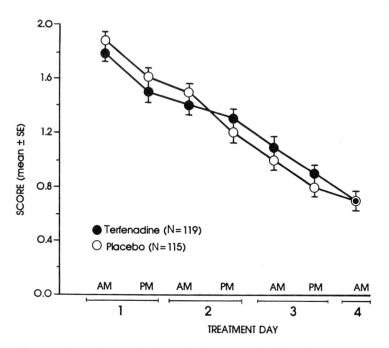

Figure 6 In a double-blind, parallel group study, 234 adults with "cold" symptoms received either terfenadine 60 mg twice daily or placebo. They recorded the severity of runny nose (shown), sniffles, sneezing, postnasal drip, cough, and sore throat on symptom cards. Terfenadine was no more effective than placebo. (From Ref. 183.)

other sequelae or whether the combination offers a valuable therapeutic option for symptom control.

C. Nonallergic Rhinitis

Some patients with rhinitis who are skin-test negative have nasal eosinophilia (NARES) and may benefit from H_1-antihistamine therapy, as do many individuals who experience cold-air-induced rhinorrhea, as in these individuals there is evidence of mast-cell degranulation (184–185). Clinical trials have thus reported some benefit of H_1-antihistamines in nonallergic rhinitic patients. Subset analysis reveals that patients with sneezing as the most prominent symptom respond best to this mode of therapy (129, 186).

D. Pregnancy

Oral H_1-antihistamines are probably best avoided in pregnancy on account of their systemic bioavailability. There is no direct evidence of teratogenicity with these agents but rodent studies have revealed an increased risk of early fetal loss, possibly through a regulatory effect of histamine on placental blood flow. Topical medications provide a safer systemic profile. There is no coordinated experience with the topical H_1-antihistamines and their safety in pregnancy.

X. SUMMARY

Antihistamines have had a long history in allergic rhinoconjunctivitis treatment. However, the discovery in the late 1970s and early 1980s of nonsedating H_1-receptor antagonists led to more widespread acceptance of this mode of therapy. This development contributed to the undertaking of well-designed clinical trials that have contributed significantly to improved understanding of rhinoconjunctivitis. Histamine is now known to contribute predominantly to nasal itch, sneeze, rhinorrhea, conjunctival itch, and lacrimation and it is these symptoms that improve most with H_1-antihistamine therapy. Oral treatment modifies both nasal and ocular symptoms and provides effective control throughout a 24-h period with only once- or twice-daily medication. The advent of topical H_1-receptor antagonists offers a wider choice of therapeutic options and provides the benefit of equal or greater efficacy with lower systemic bioavailability. Currently available medications modify predominantly histamine-regulated events despite in vitro evidence of antiallergic effects. The search for H_1-antihistamines with greater clinical antiallergic activity continues. At present, however, while having a major impact on disease it is appreciated that H_1-antihistamines can only partially modify disease, as histamine is not the only contributor to symptoms in allergic rhinoconjunctivitis.

REFERENCES

1. Staub AM, Bovet D. Action de la thymoxyethyldiethylamine (929F) et des ethers phenoliques sur le choc anaphylactique. C Rend Soc Biol 1937;125: 818–821.
2. Dale HH. Some chemical factors in the control of the circulation. Croonian lectures. Lancet 1929;1233–1237, 1285–1290.
3. Bartosch R, Feldberg W, Nagel E. Das Freiwerden eines histaminin-ahnlichen Stoffes bei der anaphylaxic des Mierschweinchens. Pfluegers Arch Physiol 1932;230:129–153.

4. Dragstedt CA and Mead FB. The role of histamine in canine anaphylactic shock. J Pharmacol Exp Ther 1936;57:419–426.
5. Blackley CH. Experimental Research on the Causes and Nature of Catarrhus Aestivus. London, Bailliere Tindall and Cox, 1873.
6. Levy L and Seabury JH. Spirometric evaluation of Benadryl in asthma. J Allergy 1947;18:244–250.
7. Herxheimer H. Antihistamines in bronchial asthma. Br Med J 1949;2: 901–905.
8. Mota I and Da Silva WD. The anti-anaphylactic and histamine releasing properties of the antihistamines. Their effect on the mast cell. Br J Pharmacol 1960;15:396–404.
9. Holgate ST, Emanuel MB, Howarth PH. Astemizole and other H_1-antihistaminic drug treatment of asthma. J Allergy Clin Immunol 1985;76: 375–380.
10. Kemp JP, Meltzer EO, Orgel HA et al. A dose-response study of the bronchodilator action of azelastine in asthma. J Allergy Clin Immunol 1987;79: 893–899.
11. Brik A, Tashkin DP, Gong H Jr, Dauphinee B, Lee E. Effect of cetirizine a new histamine H_1-antagonist on airway dynamics and responsiveness to inhaled histamine in mild asthma. J Allergy Clin Immunol 1987;80:51–56.
12. Rafferty P, Holgate ST. Terfenadine (Seldane) is a potent and selective histamine H_1-receptor antagonist in asthmatic airways. Am Rev Respir Dis 1987;135:181–184.
13. Church MK, Gradidge CF. Inhibition of histamine release from human lung in vitro by antihistamines and related drugs. Br J Pharmacol 1980;69: 663–667.
14. Lichtenstein LM and Gillespie E. The effects of the H_1 and H_2 antihistamines on allergic histamine release and its inhibition by histamine. J Pharmacol Exp Ther 1975;192:441–450.
15. Pandit PR, Kulkarni SD, Jogleker AY. Study on the relationship between antihistamine, anticholinergic and central depressant properties of various antihistamines. Indian J Med Sci 1973;27:920–928.
16. Stone CA, Wenger H, Ludden CT, Stavorski JM, Ross CA. Antiserotonin-antihistaminic properties of cyproheptadine. J Pharmacol Exp Ther 1961; 131:73–84.
17. Byck R. Drugs and the treatment of psychiatric disorders: in Goodman and Gilman (eds). The Pharmacological Basis of Therapeutics. 5th ed. New York, Macmillan, 1975:152–200.
18. Casterline CL, Evans R. Further studies on the mechanism of human histamine-induced asthma. J Allergy Clin Immunol 1977;59:420–424.
19. Moser L, Huther HJ, Koch-Weser J, Lundt PV. Effects of terfenadine and diphenhydramine alone or in combination with diazepam or alcohol on psychomotor performance and subjective feelings. Eur J Clin Pharmacol 1978;14:417–423.
20. Cheng HC, Woodward JK. Antihistaminic effect of terfenadine: A new piperidine-type antihistamine. Drug Dev Res 1982;2:181–196.

21. Janssens MML, Howarth PH. The antihistamines of the nineties. Clin Rev Allergy 1993;11:111–113.
22. Howarth PH. The cellular basis for allergic rhinitis. Allergy 1995;50(Suppl 23):6–10.
23. Howarth PH. Allergic rhinitis: a rational choice of treatment. Respir Med 1989;83:179–188.
24. Howarth PH, Wilson S, Lau L, Rajakulasingam K. The nasal mast cell and rhinitis. Clin Exp Allergy 1991;21(Suppl 2):3–8.
25. Naclerio RM, Proud D, Togias AG et al. Inflammatory mediators in late antigen induced rhinitis. N Engl J Med 1985;313:65–70.
26. Knani J, Campbell A, Enander I, Peterson CG, Michel FB, Bousquet J. Indirect evidence of nasal inflammation assessed by titration of inflammatory mediators and enumeration of cells in nasal secretions of patients with chronic rhinitis. J Allergy Clin Immunol 1992;90:880–889.
27. Devalia JL, Grady D, Harmanyeri Y, Tabaqchali S, Davies RJ. Histamine synthesis by respiratory tract microorganisms: possible role in pathogenicity. J Clin Pathol 1989;42:516–522.
28. Brown MJ, Ind PW, Barnes PJ, Jenner DA, Dollery CT. A sensitive and specific radiometric method for the measurement of plasma histamine in normal individuals. Anal Biochem 1980;109:142–146.
29. Schwartz LB. Preformed mediators of human mast cells and basophils. In: Mast Cells, Mediators and Disease. Ed: Holgate ST. Dordrecht, Kluwer Academic Publishers, 1988:129–147.
30. Bainton DF, Farquhar MG. Segregation and packaging of granule enzymes in eosinophilic leukocytes. J Cell Biol 1970;45:54–73.
31. Gomez E, Corrado OJ, Baldwin DL, Swanston AR, Davies RJ. Direct *in vivo* evidence for mast cell degranulation during allergen induced reactions in man. J Allergy Clin Immunol 1986;78:637–645.
32. Allansmith MR. The Eye and Immunology. St. Louis, CV Mosby, 1982: 47.
33. Allansmith MR, Greinder JV, Baird RS. The number of inflammatory cells in the normal conjunctiva. Am J Opthalmol 1978;86:250–259.
34. Morgan SJ, Williams JH, Walls AF, Holgate ST. Mast cell hyperplasia in atopic keratoconjunctivitis. An immunohistochemical study. Eye 1991;5: 729–735.
35. Irani AA, Butrus SI, Tabbara LF, Schwartz LB. Human conjunctival mast cells: distribution of MC$_T$ and MC$_{TC}$ in vernal conjunctivitis and giant papillary conjunctivitis. J Allergy Clin Immunol 1990;86:34–40.
36. Foster LS, Rice BA, Dutt JE. Immunopathology of atopic keratoconjunctivitis. Ophthalmology 1991;98:1190–1196.
37. Romagnani S, Maggi E, Parronchi P, Macchia D, Rici M. Increased numbers of TH$_2$-like CD$_4$ + T cells in target organs and in the allergen specific repertoire of allergic patients. Possible role of IL-4 produced by non-T cells. Int Arch Allergy Appl Immunol 1991;94:133–136.
38. Abelson M, Baird R, Allansmith M. Tear histamine levels in viral conjunctivitis and other ocular inflammations. Ophthalmology 1980;87:812–814.

39. Mygind N, Secher C, Kirkegaard J. Role of histamine and antihistamines in the nose. Eur J Respir Dis 1983;64(Suppl 28):16–20.
40. Rajakulasingam K, Mani R, Church MK, Holgate ST and Howarth PH. Changes in mucosal blood flow and airways resistance after nasal challenge with bradykinin and histamine. Thorax 1992;47:894A.
41. Raphael GD, Meredith SD, Baraniuk JN, Druce HM, Banks SM, Kaliner MA. The pathophysiology of rhinitis. II Assessment of the sources of proteins in histamine-induced nasal secretions. Am Rev Respir Dis 1989;139: 791–800.
42. McLean J, Mathews KP, Solomon WR, Brayton P, Ciarkowski A. Effect of histamine and methacholine on nasal airway resistance in atopic and nonatopic subjects. J Allergy Clin Immunol 1977;59:165–170.
43. Wong L, Hendeles L, Weinberger M. Pharmacological prophylaxis of allergic rhinitis: relative efficacy of hydroxyzine and chlorpheniramine. J Allergy Clin Immunol 1981;64:223–228.
44. Robertson J, Greaves M. Responses of human skin blood vessels to synthetic histamine analogues. Br J Clin Pharmacol 1978;5:319–322.
45. Holmberg K, Pipkorn U, Bake B, Blychert L-O. Effects of topical treatment with H_1- and H_2-antagonists on clinical symptoms and nasal vascular reactions in patients with allergic rhinitis. Allergy 1989;44:281–287.
46. Havas TE, Cole P, Parker L, Oprysk D, Ayiomamitis A. The effects of combined H_1 and H_2 histamine antagonists on alterations in nasal airflow resistance induced by topical histamine provocation. J Allergy Clin Immunol 1986;78:856–860.
47. Svensson C, Baumgarten C, Pipkorn U, Alkner U, Persson C GA. Reversibility and reproducibility of histamine induced plasma leakage in nasal airways. Thorax 1989;44:13–18.
48. Rajakulasingam K, Polosa R, Holgate ST, Howarth PH. Comparative nasal effect of bradykinin, kallidin and [des-arg^9]-bradykinin in atopic rhinitic and normal volunteers. J Physiol 1991;437:577–587.
49. Rajakulasingam K, Polosa R, Lau LCK, Church MK, Holgate ST and Howarth PH. Nasal effects of bradykinin and capsaicin: influence on plasma protein leakage and role of sensory neurons. Am Physiol Soc 1992;74: 1418–1424.
50. Ishikawa S, Sperelakis N. A novel class (H_3) of histamine receptors on perivascular nerve termination. Nature 1987;327:158–160.
51. Lorant DE, Patel KD, McIntyre TM, McEver RP, Prescott SM, Zimmerman GA. Coexpression of GMP-140 and PAF by endothelin stimulated by histamine or thrombin. A juxtacrine system for adhesion and activation of neutrophils. J Cell Biol 1990;115:223–234.
52. Symon FA, Walsh GM, Watson SR, Wardlaw AJ. Eosinophil adhesion to nasal polyp endothelium is P-selectin-dependent. J Exp Med 1994;180: 371–376.
53. Asako H, Kurose I, Wolf R et al. Role of H_1-receptors and P-selectin in histamine-induced leukocyte rolling and adhesion in postcapillary venules. J Clin Invest 1994;93:1508–1515.

54. Canonica GW, Ciprandi G, Pronzato C, Paolieri F, Scordamaglia A, Bagnasco M. Role of adhesion molecules in allergic rhinitis. In: Proc of XVI Eur Cong Allergology and Clin Immunol. A Bascomba and J Sastre (eds). Bologne, Monduzzi Editore, 1995;127–142.

55. Raible DG, Schulman ES, Di Muzio J, Cardillo R, Post TJ. Mast cell mediators prostaglandin D_2 and histamine activate human eosinophils. J Immunol 1992;148:3536–3542.

56. Kirkegaard J, Secher C, Mygind N. Effect of the H_1-antihistamine chlorpheniramine maleate on histamine-induced symptoms in the human conjunctiva. Allergy 1982;37:203–208.

57. Abelson MB, Smith LM. Levocabastine: evaluation in the histamine and compound 48/80 models of ocular allergy in humans. Ophthalmology 1988; 95:1494–1497.

58. Dechant KL, Goa KL. Levocabastine: A review of its pharmacological properties and therapeutic potential as a topical antihistamine in allergic rhinitis and conjunctivitis. Drugs 1991;41:202–224.

59. Abelson M, Udell I. H_2 receptors in the human ocular surface. Arch Ophthalmol 1981;99:302–304.

60. Bascom R, Wachs M, Naclerio RM, Pipkorn U, Galli SJ, Lichtenstein LM. Basophil influx occurs after nasal antigen challenge: effects of topical corticosteroid pretreatment. J Allergy Clin Immunol 1988;81:580–589.

61. Klementsson H, Andersson M, Pipkorn U. Allergen-induced increase in non-specific nasal reactivity is blocked by antihistamines without a clearcut relationship to eosinophil influx. J Allergy Clin Immunol 1990;86: 466–472.

62. Iliopoulos O, Proud D, Adkinson F, Norman PS, Kagey-Sobotka A, Lichtenstein LM and Naclerio RM. Relationship between the early, late, and rechallenge reaction to nasal challenge with antigen: observations on the role of inflammatory mediators and cells. J Allergy Clin Immunol 1990;86: 851–861.

63. Abelson MB, George MA, Schaefer K, Smith LM. Evaluation of the new ophthalmic antihistamine 0.05% levocabastine, in the clinical allergen challenge model of allergic conjunctivitis. J Allergy Clin Immunol 1994;94: 458–464.

64. Palma-Carlos AG, Palma-Carlos ML, Rombaut N. The effect of levocabastine nasal spray in nasal provocation tests. Int J Clin Pharmacol Res 1988; 8:25–30.

65. Horak F and Jaeger S. VCC—Vienna Challenge Chamber—The protective effect of astemizole. Allergy 1988;43(Suppl 7):110.

66. Howarth PH, Harrison K, Smith S. The influence of terfenadine and pseudoephedrine alone and in combination on allergen-induced rhinitis. Int Arch Allergy Immunol 1993;101:318–321.

67. Togias AG, Naclerio RM, Warner J et al. Demonstration of inhibition of mediator release from human mast cells by azatadine base, *in vivo* an *in vitro* evaluation. JAMA 1986;255:225–229.

68. Bousquet J-P, Lebel B, Chanal I, Morel A, Michel F-B. Antiallergic activity

of H_1-receptor antagonists assessed by nasal challenge. J Allergy Clin Immunol 1988;82:881–887.

69. Naclerio RM, Proud D, Kagey-Sobotka A et al. The effect of cetirizine on early allergic response. Laryngoscope 1989;99:596–599.

70. Naclerio RM, Kagey-Sobotka A, Lichtenstein LM et al. Terfenadine, an H_1-antihistamine inhibits histamine release *in vivo* in the human. Am Rev Respir Dis 1990;142:167–171.

71. Bousquet J-P, Campbell A, Michel F-B. Antiallergic activities of antihistamines. In: Therapeutic Index of Antihistamines. MK Church and J-P Rihoux (eds). Lewiston, Hogrefe and Huber, 1992;57–96.

72. Baumgarten CR, Lamet M, Kunkel G. The influence of topical applied loratadine upon mediator generation after nasal allergen challenge. J Allergy Clin Immunol 1990;85:164A.

73. Bradding P, Feather IH, Wilson S, Bardin P, Holgate ST, Howarth PH. Immunolocalisation of cytokines in the nasal mucosa of normal and perennial rhinitis subjects: the mast cell as a source of IL-4, IL-5 and IL-6 in human allergic inflammation. J Immunol 1993;151:3853–3865.

74. Bradding P, Mediwake R, Feather IH, Madden J, Church MK, Holgate ST, Howarth PH. TNFα is localised to nasal mucosal mast cells and is released in acute allergic rhinitis. Clin Exp Allergy 1995;25:406–415.

75. Michel L, De Vos C, Rihoux J-P, Burtin C, Benveniste J, Dubertret L. Inhibitory effect of oral cetirizine on *in vivo* antigen-induced histamine and PAF-acether release and eosinophil recruitment in human skin. J Allergy Clin Immunol 1988;82:101–109.

76. Charlesworth EN, Kagey-Sobotka A, Norman PS, Lichtenstein LM. Effect of cetirizine on mast cell mediator release and cellular traffic during the cutaneous late-phase reaction. J Allergy Clin Immunol 1989;83:905–912.

77. Howarth PH, Wilson SJ, Brewster H. The influence of cetirizine and nasal eosinophilia in seasonal allergic rhinitis. J Allergy Clin Immunol 1991; 87(Number 1, Part 2): 151.

78. Small P, Barrett D, Biskin N. Effects of azatadine, terfenadine and astemizole on allergen-induced nasal provocation. Ann Allergy 1990;64:123–131.

79. Howarth PH, Emanuel MB and Holgate ST. Astemizole, a potent histamine H_1-receptor antagonist: effect in allergic rhinoconjunctivitis, on antigen and histamine induced skin weal responses and relationship to serum levels. Br J Clin Pharmacol 1984;18:1–8.

80. Kemp JP, Buckley CE, Gershwin ME et al. Multicentre, double-blind, placebo controlled trial of terfenadine in seasonal allergic rhinitis and conjunctivitis. Ann Allergy 1985;54:502–509.

81. Falliers CJ, Brandon ML, Buchman E et al. Double-blind comparison of cetirizine and placebo in the treatment of seasonal allergic rhinitis. Ann Allergy 1991;66:257–262.

82. Dockhorn RJ, Bergner A, Connell JT et al. Safety and efficacy of loratadine (Sch-29851): a new non-sedating antihistamine in seasonal allergic rhinitis. Ann Allergy 1987;58:407–411.

83. Howarth PH, Holgate ST. Comparative trial of two non-sedative H₁-antihistamines, terfenadine and astemizole for hay fever. Thorax 1984;39:668–672.
84. Beswick KBJ, Kenyon GS, Cherry JR. A comparative study of beclomethasone dipropionate aqueous nasal spray with terfenadine tablets in seasonal allergic rhinitis. Curr Med Res Opinion 1985;9:560–567.
85. Wood SF. Oral antihistamine or nasal steroid in hay fever: a double-blind, double-dummy comparative study of once-daily oral astemizole vs twice daily nasal beclomethasone diproprionate. Clin Allergy 1986;16:195–201.
86. Pipkorn U, Bende M, Hedner J, Hedner T. A double-blind evaluation of topical levocabastine, a new specific H₁-antagonist in patients with allergic conjunctivitis. Allergy 1985;40:491–496.
87. Njaa F, Baekken T, Bjammer D et al. Levocabastine compared with sodium cromoglycate eyedrops in children with both birch and grass pollen allergy. Pediatr Allergy Immunol 1992;3:39–42.
88. The Livostin Study Group. A comparison of topical levocabastine and oral terfenadine in the treatment of allergic rhinoconjunctivitis. Allergy 1993;48:530–534.
89. Parys W, Janssens M. The efficacy and tolerability of topical levocabastine and oral terfenadine in seasonal allergic rhinitis. Allergy 1992;47(12):249.
90. Brandon ML, Weinder M. Clinical studies of terfenadine in seasonal allergic rhinitis. Arzneim-Forsch/Drug Res 1982;32:1204–1205.
91. Backhouse CI, Brewster BS, Lockhart JDF, Maneksha S, Purvis CR, Valle-Jones JC. Terfenadine in allergic rhinitis—A comparative trial of a new antihistamine versus chlorpheniramine and placebo. Practitioner 1992;226:347–351.
92. Kagan G, Dabrowski E, Huddlestone L, Kapur TR, Wolstencroft P. A double-blind trial of terfenadine and placebo in hay fever using a substitution technique for non-responders. J Int Med Res 1980;8:404–407.
93. Murphy-O'Connor JC, Renton RL, Westlake DM. Comparative trial of two dose regimens of terfenadine in patients with hay fever. J Int Med Res 1984;12:333–337.
94. Stern MA, Rosenberg RM, Smith R, Fidler C. A comparative study of terfenadine at two dose levels in the management of hay fever. Br J Clin Pract 1990;44:359–363.
95. Henauer S, Hugonot L, Hugonot R et al. Multi-centre, double-blind comparison of terfenadine once daily versus twice daily in patients with hay fever. J Int Med Res 1987;15:212–223.
96. Heykants J, Van Peer A, Woestenborghs R, Jageneau A, Vanden Bussche G. Dose-proportionality, bioavailability, and steady state kinetics of astemizole in man. Drug Dev Res 1986;8:71–78.
97. Callier J, Engelen RF, Ianniello I, Olzem R, Zeisner M, Amery WK. Astemizole (R43512) in the treatment of hay fever: an international double-blind study comparing a weekly treatment (10 mg and 25 mg) with placebo. Curr Ther Res 1981;29:24–35.
98. Brobyn R, Benoit M, Mader N, Cleaver K. Perennial rhinitis, treated with astemizole. J Allergy Clin Immunol 1982;69:110.

99. Gendreau-Reid L, Simons KJ, Simons FER. Comparison of the suppressive effect of astemizole, terfenadine and hydroxyzine on histamine-induced wheals and flares in humans. J Allergy Clin Immunol 1986;77:335–340.
100. Vanden Bussche G, Emanuel MB, Rombaut N. Clinical profile of astemizole. A survey of 50 double-blind trials. Ann Allergy 1987;58:184–188.
101. Juniper EF, White J, Dolovich J. Efficacy of continuous treatment with astemizole (Hismanal) and terfenadine (Seldane) in ragweed pollen-induced rhinoconjunctivitis. J Allergy Clin Immunol 1988;82:670–675.
102. Kaiser H, Weisberg S, Morris R. A comparison of cetirizine and chlorpheniramine vs placebo in the treatment of seasonal allergic rhinitis. J Allergy Clin Immunol 1989;83:306A.
103. Broide DH, Love S, Altman R, Wasserman SI. Evaluation of cetirizine in the treatment of patients with seasonal allergic rhinitis. J Allergy Clin Immunol 1988;81(Suppl 1):176A.
104. Panayotopoulos SM, Panayotopoulos ES. The efficacy of cetirizine in the treatment of pollinosis. A correlation with the daily pollen count of atmospheric air. A double-blind, placebo-controlled study. Acta Therapeutica 1988;14:347–353.
105. Skassa-Brociek W, Bousquet J-P, Montes F et al. Double-blind, placebo-controlled study of loratadine, mequitazine and placebo in the symptomatic treatment of seasonal allergic rhinitis. J Allergy Clin Immunol 1988;81:725–730.
106. Del Carpio J, Kabbash L, Turenne Y et al. Efficacy and safety of loratadine (10 mg once daily), terfenadine (60 mg twice daily) and placebo in the treatment of seasonal allergic rhinitis. J Allergy Clin Immunol 1989;84:741–746.
107. Clissold SP, Sorkin EM, Goa KL. Loratadine: A preliminary review of its pharmacodynamic properties and therapeutic efficacy. Drugs 1989;37:42–57.
108. Bruttmann G, Pedrali P. Loratadine (SCH 29851) 40 mg once daily vs terfenadine 60 mg twice daily in the treatment of seasonal allergic rhinitis. J Int Med Res 1987;15:63–70.
109. Gutkowski A, Del Carpio J, Gelinas B, Schulz J, Turenne Y. Comparative study of the efficacy, tolerance and side effects of dexchlorpheniramine maleate 6 mg b.i.d. with terfenadine 60 mg b.i.d. J Int Med Res 1985;13:284–288.
110. Wood SF, Emanuel MB. Astemizole in maintenance therapy of hay fever: a comparison with a long-acting formulation of pheniramine maleate. Pharmatherapeutica 1984;3:10–17.
111. Malmberg H, Holopainen E, Grahne B, Binder E, Savolainen S, Sundberg S. Astemizole in the treatment of hay fever. Allergy 1983;38:227–231.
112. Campoli-Richards DM, Buckley MM-T, Filton A. Cetirizine: A review of its pharmacological properties and clinical potential in allergic rhinitis, pollen-induced asthma and chronic urticaria. Drugs 1990;40:762–781.
113. Horak F, Bruttmann G, Pedrali P et al. A multicentric study of loratadine, terfenadine and placebo in patients with seasonal allergic rhinitis. Arzneimittelforschung 1988;38:124–128.

114. Herman D, Arnaud A, Dry J et al. Clinical effectiveness and safety of loratadine vs cetirizine in the treatment of seasonal allergic rhinitis. Clin Exp Allergy 1990;20:56.
115. Backhouse CI, Renton R, Fidler C, Rosenberg RM. Multicentre, double-blind comparison of terfenadine and cetirizine in patients with seasonal allergic rhinitis. Br J Clin Pract 1990;44:88–91.
116. Stelmach WJ, Rush DR, Pocucker PC et al. A large scale, office based study evaluates the use of a new class of non-sedating antihistamines. J Am Board Family Pract 1990;3:241–252.
117. The LASAR (Loratadine and Astemizole in Seasonal Allergic Rhinitis) Study Group. Double-blind comparison of the efficacy and safety of astemizole and loratadine in the treatment of seasonal allergic rhinitis. Drug Invest 1992;4(4):336–345.
118. Girard JP, Sommacal-Schopf D, Bigliardi P, Henauer SA. Double-blind comparison of astemizole, terfenadine and placebo in hay fever with special regard to onset of action. J Int Med Res 1985;13:102–108.
119. Oei HD. Double-blind comparison of loratadine (SCH 29851), astemizole and placebo in hay fever with special regard to onset of action. Ann Allergy 1988;61:436–439.
120. Roman IJ, Danzig MR. Loratadine: A review of recent findings in pharmacology, pharmacokinetics, efficacy and safety, with a look at its use in combination with pseudoephedrine. Clin Rev Allergy 1993;11:89–110.
121. Rijntjes E, Ghys L, Rihoux J-P. Astemizole and cetirizine in the treatment of seasonal allergic rhinitis—A comparative double-blind multicentre study. J Int Med Res 1990;18:219–224.
122. Salomonsson P, Gottberg L, Heilborn H, Norrlind K, Pegelow K-O. Efficacy of an oral antihistamine, astemizole, as compared to a nasal steroid spray in hay fever. Allergy 1988;43:214–218.
123. Dockhorn RJ, Shellenberger MK, Hassanien R, Trachelman L. Efficacy of SCH 434 (loratadine plus pseudoephedrine) vs components and placebo in seasonal allergic rhinitis. J Allergy Clin Immunol 1988;81:78.
124. Orgel HA, Meltzer EO, Kemp JP et al. Comparison of intranasal cromolyn sodium 4% and oral terfenadine, for allergic rhinitis: symptoms, nasal cytology, nasal ciliary clearance and rhinomanometry. Ann Allergy 1991;66:237–244.
125. Lindsay-Miller AJM, Chambers A. Group comparative trial of cromolyn sodium and terfenadine in the treatment of seasonal allergic rhinitis. Ann Allergy 1987;58:28–32.
126. Brostoff J and Lockhart JDF. Controlled trial of terfenadine and chlorpheniramine maleate in perennial rhinitis. Postgrad Med J 1982;58:422–423.
127. Bruttmann G, Charpin D, Germouty J, Horak F, Kunkel G, Wittmann G. Evaluation of the efficacy and safety of loratadine in perennial allergic rhinitis. J Allergy Clin Immunol 1989;83:411–416.
128. Renton R, Fidler C, Rosenberg R. Multicenter crossover study of the efficacy and tolerability of terfenadine, 120 mg, versus cetirizine 10 mg, in perennial allergic rhinitis. Ann Allergy 1991;67:416–420.

129. Wihl J-A, Nuchel-Petersen B, Nuchel Petersen I, Gundersen G, Bresson K, Myggind N. Effect of the non-sedative H_1-receptor antagonist astemizole in perennial allergic and non-allergic rhinitis. J Allergy Clin Immunol 1985; 75:720–727.

130. Berman B, Buchman E, Dockhorn R, Leese P, Mansmann H, Middleton E. Cetirizine therapy of perennial allergic rhinitis. J Allergy Clin Immunol 1988;81:177.

131. Bruttmann G, Arendt C, Bernheim J, Double-blind, placebo-controlled comparison of cetirizine, and terfenadine in atopic perennial rhinitis. Acta Therapeutica 1989;15:99–109.

132. Zetterstrom E, Halopainen E, Johnson C. Efficacy and tolerability of cetirizine and terfenadine in the treatment of perennial allergic rhinitis (PAR). Allergy 1992;47:179.

133. Lobaton P, Moreno F, Coulie P. Comparison of cetirizine with astemizole in the treatment of perennial allergic rhinitis and study of the concomitant effect on histamine- and allergen-induced wheal responses. Ann Allergy 1990;65:401–405.

134. Boniver R, Borras J, Chalmagne J, Liban F, Vanden Bussche G. Evaluation of the therapeutic effect of astemizole in perennial allergic rhinitis. A double-blind comparison with chlorpheniramine. Curr Ther Res 1986;39: 244–249.

135. Lockey RF, Fox RW. Loratadine (SCH 29851) 10 mg od vs clemastine 1mg b.i.d. in the treatment of perennial allergic rhinitis. Immunol Allergy Prac 1989;11:423–429.

136. Frolund L, Etholm B, Ivander K et al. A multicentre study of loratadine, clemastine and placebo in patients with perennial allergic rhinitis. Allergy 1990;45:254–261.

137. Boland N. A double-blind study of astemizole and terfenadine in the treatment of perennial rhinitis Ann Allergy 1988;61:18–24.

138. Robinson AC, Cherry JR, Daly S, Double-blind, cross-over trial comparing beclomethasone dipropionate and terfenadine in perennial rhinitis. Clin Exp Allergy 1989;19:569–73.

139. Harding SM, Heath S. Intranasal steroid aerosol in perennial rhinitis: comparison with an antihistamine compound. Clin Allergy 1976;6:369–372.

140. Sibbald B, Hilton S, D'Souza M. An open crossover trial comparing two doses of astemizole and beclomethasone dipropionate in the treatment of perennial rhinitis. Clin Allergy 1986;16:203–211.

141. Henauer S, Seppey M, Hugonot C, Pecoud A. Effects of terfenadine and pseudoephedrine alone and in combination in a nasal provocation test and in perennial rhinitis. Eur J Clin Pharmacol 1991;41:321–324.

142. McTavish D, Sorkin EM, Azelastine: a review of its pharmacodynamic and pharmacokinetic properties and therapeutic potential. Drugs 1989;38: 778–800.

143. Janssens MM-l, Vanden Bussche G. Levocabastine: an effective topical treatment of allergic rhinoconjunctivitis. Clin Exp Allergy 1991;21(suppl 2): 29–36.

144. Weiler JM, Meltzer EO, Dockhorn R, Widlitz MD, D'Eletto TA. Freitag JJ. A safety and efficacy evaluation of azelastine (AZ) nasal spray (NS) in seasonal allergic rhinitis (SAR). J Allergy Clin Immunol 1991;87:219.
145. Ratner PH, Findlay SR, van Bavel JH, Hampel F, Freitag JJ. Azelastine nasal spray: a two-week safety/efficacy study in seasonal allergic rhinitis. Ann Allergy 1991;66:91.
146. Schata M, Jorde W, Richarz-Barthauer U. Levocabastine nasal spray better than sodium cromoglycate and placebo in the topical treatment of seasonal allergic rhinitis. J Allergy Clin Immunol 1991;87:873–878.
147. La Force C, Dockhorn R, Prenner B et al. Safety and efficacy of azelastine (AZ) nasal spray (NS) for seasonal allergic rhinitis (SAR): A four week comparative trial. J Allergy Clin Immunol 1992;89:182.
148. Bende M, Pipkorn U. Topical levocabastine, a selective H₁-antagonist in seasonal allergic rhinoconjunctivitis. Allergy 1987;42:512–515.
149. Sohoel P, Freng BA, Kramer J et al. Topical levocabastine compared with orally administered terfenadine for the prophylaxis and treatment of seasonal allergic rhinoconjunctivitis. J Allergy Clin Immunol 1993;92:73–81.
150. Bahmer FA, Ruprecht KW. Safety and efficacy of topical levocabastine compared with oral terfenadine. Ann Allergy 1994;72:429–434.
151. Palma-Carlos AG, Chieria C, Conde TA, Cordeiro JA. Double-blind comparison of levocabastine nasal spray with sodium cromoglycate nasal spray in the treatment of seasonal allergic rhinitis. Ann Allergy 1991;67:394–398.
152. Knight A. The role of levocabastine in the treatment of allergic rhinoconjunctivitis. Br J Clin Pract 1994;48:139–143.
153. van de Hayning PH, van Haesendonck J, Creten W, Rombaut N. Effect of topical levocabastine on allergic and non-allergic perennial rhinitis: a double-blind study, levocabastine vs placebo, followed by an open prospective, single-blinded study on beclomethasone. Allergy 1988;43:386–391.
154. de Azvedo M, Castel-Branco MG, Ferraz Olivier J et al. Double-blind comparison of levocabastine eye drops with sodium cromoglycate and placebo in the treatment of seasonal allergic conjunctivitis. Clin Exp Allergy 1991;21:689–694.
155. Davies BH, Mullins J. Topical levocabastine is more effective than sodium cromoglycate for the prophylaxis and treatment of seasonal allergic conjunctivitis. Allergy 1993;48:519–524.
156. Meltzer EO. Comparative safety of H₁-antihistamines. Ann Allergy 1991;67:625–633.
157. Simons FER. H₁-receptor antagonists: comparative tolerability and safety. Drug Safety 1994;10:350–380.
158. Meador KJ, Loring DW, Thompson EE, Thompson WO. Differential cognitive effects of terfenadine and chlorpheniramine. J Allergy Clin Immunol 1989;84:322–325.
159. Bhatti JZ, Hindmarch I. The effects of terfenadine with and without alcohol on an aspect of car driving performance. Clin Exp Allergy 1989;19:609–611.
160. Hindmarch I, Bhatti JZ. Psychomotor effects of astemizole and chlorphenir-

amine alone and in combination with alcohol. Int Clin Psychopharmacol 1987;2:117–119.

161. Seidel WF, Cohen S, Bliwise NG, Dement WC. Direct measurement of daytime sleepiness after administration of cetirizine and hydroxyzine with a standardized electroencephalographic assessment. J Allergy Clin Immunol 1990;86:1029–1033.

162. Bradley CM, Nicholson AN. Studies on the central effects of the H_1-antagonist loratadine. Eur J Clin Pharmacol 1987;32:419–421.

163. Rombaut N, Bhatti JZ, Curran S, Hindmarch I. Effects of topical administration of levocabastine on psychomotor and cognitive function. Ann Allergy 1991;67:75–79.

164. Hibbert J and Howarth PH. Antihistamines and cardiac arrhythmias. Prescribers J 1994;34:31–35.

165. Zehender M, Hohnloser S, Just H. QT-interval prolonging drugs: mechanisms and clinical relevance of their arrhythmogenic hazards. Cardiovasc Drug Ther 1991;5:515–530.

166. Benedict CR. The QT-interval and drug-associated torsade de pointes. Drug Invest 1993;5:69–79.

167. Craft TM. Torsade de pointes after astemizole overdose. Br Med J 1986; 292:660–661.

168. Snook J, Boothman-Burrell D, Watkins J et al. Torsade de pointes ventricular tachycardia associated with astemizole overdose. Br J Clin Pract 1988; 42:257–259.

169. Monahan BP, Ferguson CL, Killeary ES, Lloyd BK, Troy J, Cantilena LR. Torsades de pointes occurring in association with terfenadine use. JAMA 1990;264:2788–2790.

170. Pohjola-Sintonen S, Viitasalo M, Toivonen L, Neuvonen P, Torsades de pointes after terfenadine-itraconazole interaction. Br Med J 1993;306:186.

171. Honig PK, Woosley RL, Zamani K, Connor DP, Cantilena LR. Changes in the pharmacokinetics and electrocardiographic pharmacodynamics of terfenadine with concomitant administration of erythromycin. Clin Pharmacol Ther 1992;52:231–238.

172. Yun CH, Okerholm RA, Guengerich FP. Oxidation of the antihistaminic drug terfenadine in human liver microsomes: role of cytochrome P-450 3A(4) in N-dealkylation and C-hydroxylation. Drug Metab Dispos 1993;21: 403–409.

173. Yumibe N, Huie K, Chen KJ et al. Identification of human liver cytochrome P_{450}'s involved in the microsomal metabolism of the antihistamine drug loratadine. J Allergy Clin Immunol 1994;93:234.

174. Affrime MB, Lorber R, Danzig M et al. Three month evaluation of electrocardiographic effects of loratadine in humans. J Allergy Clin Immunol 1993; 91:259.

175. Malka S, Partjdas A, Fuentes R. Estudio comparativo rhinomanometrico Y valovacion clinica de la combinacion de terfenadina and pseudofedrina vs carbinoxamina and pseudoefedrina en rhinitis obstructiva. Invest Med Int 1989;16:131–136.

176. Bronsky E, Rogers C, Altman L et al. A comparison of acrivastine and pseudoephedrine, chlorpheniramine and pseudoephedrine and placebo in the treatment of perennial allergic rhinitis. J Allergy Clin Immunol 1987; 79:

177. Blockhuys S, Janssens M. Astemizole D vs Actifed in perennial allergic rhinitis: a study performed by the Janssen research group of Belgium GP's. Allergy 1992;47(Suppl 12):319–323.

178. Stroh JE, Ayars GA, Brunstein IL et al. A comparative tolerance study of terfenadine-pseudoephedrine tablets in patients with allergic or vasomotor rhinitis. J Int Med Res 1988;16:420–427.

179. Backhouse CI, Rosenberg RM, Fidler C. Treatment of seasonal allergic rhinitis: a comparison of a combination tablet of terfenadine and pseudoephedrine with the individual ingredients. Br J Clin Pract 1990;44(7):274–279.

180. Storms WW, Bodman SF, Nathan A et al. SCH434: A new antihistamine/ decongestant for seasonal allergic rhinitis. J Allergy Clin Immunol 1989;83: 1083–1090.

181. Grant JA, Fox R, Hampel F, Harris J, Gross G. Meta-analysis of Hismanal-D clinical trials in seasonal allergic rhinitis. Allergy 1992;47(Suppl 12):242.

182. Smith MBH, Feldman W. Over the counter cold medications: a critical review of clinical trials between 1950 and 1991. JAMA 1993;269:2258–2263.

183. Gaffey MJ, Kaiser DL, Hayden FG. Ineffectiveness of oral terfenadine in natural colds: evidence against histamine as a mediator of common cold symptoms. Pediatr Infect Dis J 1988;7:223–228.

184. Togias AG, Naclerio RM, Proud D et al. Nasal challenge with cold dry air results in the production of inflammatory mediators: possible mast cell involvement. J Clin Invest 1985;76:1375–1381.

185. Proud D, Bailey GS, Naclerio RM et al. Tryptase and histamine as markers to evaluate mast cell activation during the responses to nasal challenge with allergen, cold, dry air and hyperosmolar solutions. J Allergy Clin Immunol 1992;89:1098–1110.

186. Wihl JA, Petersen BN, Mygind N. The role of histamine in non-allergic perennial rhinitis. Acta Otolaryngol (Stockh) 1984;214:99–102.

9

H_1-Receptor Antagonists in Asthma

Jonathan M. Corne and Stephen T. Holgate
University of Southampton, Southampton, England

I. INTRODUCTION

Asthma is one of the most important allergic diseases. Estimates of its prevalence vary from 7 to 14% of the population in industrialized countries (1) and the prevalence is rising for reasons as yet undetermined.

The increasing morbidity and mortality from asthma has led to a reappraisal of its treatment and development of consensus management guidelines (1). Most of these recommend inhaled corticosteroids and β_2-adrenergic agonists as the mainstay of treatment with theophylline, long-acting β_2 adrenergic agonists and oral corticosteroids to be used if control cannot be achieved using basic treatment. With the exception of Japan, where they are commonly prescribed, antihistamines have not found a major role in the treatment of asthma; yet, histamine plays an important role in the pathogenesis of asthma and they were long regarded as having great therapeutic potential in this disorder.

This chapter reviews some of the early work that identified the importance of histamine in asthma and then looks at the clinical studies examining the value of first- and second-generation antihistamines and their effect on various models of asthma.

II. HISTORICAL BACKGROUND

Following the discovery of histamine in 1907, early work concentrated on describing its effects on the cardiovascular system. Weiss et al. (2)

used histamine infusions as a way of determining blood flow velocity and performed one of the first studies in which the effects of intravenous histamine in humans are described in detail. They injected 0.001 mg kg^{-1} body weight of histamine into a peripheral vein and timed the onset of the flush and inevitable metallic taste. They observed that intravenous administration of histamine precipitated attacks of breathlessness in patients with bronchial and cardiac asthma. This was assumed to be due to pulmonary obstruction since it was associated with a reduction of vital capacity, the best measure of airway status available at the time, and a lowered position of the diaphragm.

In a subsequent study, Weiss et al. (3) measured respiratory function in 10 normal volunteers challenged with histamine infusion. They measured vital capacity before and at the peak of the histamine reaction and showed that histamine had no effect on this parameter. They also measured the rate and depth of respiration and minute volume and noted breath sounds and complaints of dyspnea. Histamine induced no changes in any of these. They concluded that as even toxic doses were unable to produce bronchial obstruction in normal subjects, the effects of histamine in asthmatics were due to an altered response to the histamine as opposed to an increase in sensitivity.

Curry (4) specifically studied the respiratory effects of histamine, confirming the earlier observations of Weiss et al. He investigated 10 normal subjects, 10 with a history of severe allergic disturbances but who were asymptomatic at the time of the study, and 9 patients with various degrees of chronic bronchitis, emphysema, and asthma. Vital capacity was used as a measure of airways obstruction. Like Weiss, he was unable to provoke bronchoconstriction in normal subjects or in asymptomatic allergic subjects. However, in the third group of 9 patients, bronchoconstriction was demonstrated following both intravenous and intramuscular histamine challenge, the response varying according to the degree of asthma. In a further study he looked at the ability of antihistamines to prevent this bronchoconstriction (5). Volunteers were initially challenged with a control injection of histamine to determine the sensitivity of their airways and then the challenge was repeated after infusion of the antihistamine β-dimethylaminoethyl-benzhydryl-ether hydrochloride. This drug was known to have potent antihistamine effects in animal experiments and to be effective for symptoms of allergic rhinitis, also thought to be a histamine-mediated disease. Intravenous antihistamine administration was shown to provide full protection against bronchoconstriction.

It is clear that histamine causes bronchoconstriction in asthmatic subjects and indeed bronchoconstriction induced by inhaled histamine is now widely used as a measure of airway reactivity in the histamine challenge test in which the concentration of histamine required to produce a 20% fall (PC_{20}) in the forced expiratory volume in the first second (FEV_1) is determined. To determine whether naturally released histamine has a role in asthma, as opposed to simply being a substance to which asthmatics react, one needs to demonstrate its presence within anatomically and physiologically relevant cells and also demonstrate increases in histamine following experimentally induced or naturally occurring attacks of asthma.

III. CELLULAR LOCALIZATION OF HISTAMINE

Riley and West (6), in an important series of experiments, localized histamine to the mast cell. Heparin had previously been shown to be an important component of this cell and in animal tissues they demonstrated a positive correlation between extractable heparin and extractable histamine. They also compared the histamine content of fetal tissue with tissue from fully developed animals and showed there were far higher histamine levels in mature tissue, in which mast cells are completely developed, than in fetal tissue, in which they are not. In a further series of experiments, they demonstrated that maneuvers that led to destruction of mast cells, such as injection of stilbalemoine, liberated histamine, and maneuvers that increased numbers of mast cells led to a raised level of extractable histamine.

Graham et al. (7) attempted to identify histamine in blood. Following the observation that most histamine was found in the buffy coat, blood elements were separated by centrifugation and the individual portions were analyzed for histamine. Blood was obtained from several sources including patients with polycythemia rubra vera and chronic myelocytic leukemia. Most histamine was located within the granulocyte fraction, with multiple regression analysis suggesting that the basophil was the major source.

Two independent groups of investigators discovered reaginic antibody (IgE) (8, 9) in 1967 and the presence of the high affinity IgE receptor on human mast cells was confirmed shortly afterward (10). This allowed a mechanistic link to be developed between allergen and histamine release. Early models of asthma emphasized the importance of allergen causing the cross-linking of IgE on mast cells leading to degranulation and histamine release and subsequent bronchoconstriction.

Ishizaka et al. (11) investigated whether leukocytes other than basophils were involved in histamine release by anti-immunoglobulin E (anti-IgE). Blood was separated by differential centrifugation followed by chromatography. The subfractions so obtained were challenged with anti-IgE, which resulted in histamine release from basophil granulocytes but *not* from any of the other fractions.

This has important implications in asthma since the role of histamine-containing mast cells and basophils in the pathogenesis of asthma has been clearly demonstrated. Flint et al. (12) performed bronchoalveolar lavage (BAL) in patients with atopic asthma and in healthy controls. Compared to control subjects, BAL fluid from the asthmatic subjects contained a greater proportion of mast cells and eosinophils and the histamine content of lavaged cells was greater. Furthermore, there was a significant inverse relationship between FEV_1 and levels of histamine and proportion of mast cells. The cells obtained by BAL were incubated with antihuman IgE and those from asthmatic subjects released a significantly higher proportion of cellular histamine than cells from control subjects. Electron microscopy has shown mast-cell degranulation to be an important feature of asthma.

Basophils have been identified in the sputum of patients during attacks of asthma and have been shown to disappear on resolution of the attack (13). Guo et al. (14) studied cells found during the late phase of antigen challenge using BAL following subsegmental allergen challenge. Asthmatic volunteers were challenged with ragweed antigen. BAL was performed before and 17–22 h after challenge. Cells were identified by morphology, flow cytometry, functional studies, and pharmacological studies. Following antigen challenge there was a 20- to 200-fold increase in IgE-bearing, histamine-containing, cells, of which 95% were classified as basophils using the above criteria.

Measurements of histamine in blood are beset with problems and the results need to be interpreted carefully. It has a half-life of approximately 20 s. The histamine released from mast cells, the local supplier in asthma, accounts for only 2% of the circulating amount, the rest being derived from basophils. Nevertheless, elevations of plasma histamine have been described during acute exacerbations of asthma and also following allergen challenge. Bruce et al. (15) compared histamine levels in plasma, blood, and urine in 64 normal subjects, 10 hospitalized patients with severe asthma attacks and 10 control patients with other respiratory illnesses. Histamine levels were increased in asthmatics on the day of admission compared to levels in normal subjects. Interestingly, levels in the control group (i.e., those with other respiratory illnesses) were higher than those in the asthmatics on day 3 (in whole blood) and on day 14 (in plasma).

Simon et al. (16) monitored plasma histamine levels in 60 asthmatics; 20 were experiencing acute bronchoconstriction, 20 were in partial remission, and 20 were in complete remission. Histamine was significantly ele-

vated in the group experiencing acute bronchoconstriction with the
amount correlating with the severity of the attack.

Barnes et al. (17) measured plasma histamine in stable asthmatic subjects employing a more sensitive radioenzymatic assay than used in previous studies. Plasma histamine was significantly higher in asthmatics than
in their age-matched controls, with no significant correlation between
plasma histamine concentration and severity of the asthma. Interestingly,
patients with airway obstruction due to chronic bronchitis did not show
the same elevation of plasma histamine as the asthmatics did (Fig. 1).

A more sensitive measure is local histamine release. Histamine levels
have been shown to be higher in BAL fluid from allergic asthmatics than
from normal controls, with a correlation found between levels of histamine
release and bronchial hyperreactivity (18). Other work has demonstrated

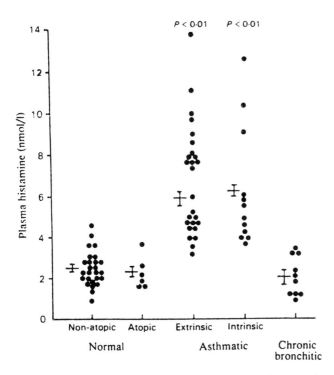

Figure 1 Venous plasma histamine in nonatopic normal, atopic nonasthmatic,
extrinsic asthmatic, intrinsic asthmatic, and chronic bronchitic subjects. Means
± SEM for each asthmatic group are significantly higher ($p < 0.01$) than those
for normal or chronic bronchitic groups. (From Ref. 17.)

histamine release following hyperosmolar challenge, thought to an important stimulus for release of chemical mediators of inflammation in exercise-induced asthma.

Although histamine has potent bronchoconstrictive properties, it is not the only mediator released from mast cells. Many other mediators may also be involved in the inflammatory process underlying the asthmatic state. Chief among these are the prostaglandins: PGD2, PGD9, PGD11, and PGF2 have been shown to be increased in asthma (19). All of these have proinflammatory properties. PGD2, in particular, has potent spasmogenic properties (20) of its own and also potentiates the bronchoconstrictor properties of histamine. Recently IL-4 has been found in mast cells (21), thus giving this cell a potential role for releasing cytokines.

An interesting link between histamine and asthma has recently been described. In childhood asthma, over 78% of exacerbations have been attributed to upper respiratory tract viral infections (22, 23). Time course analysis studies show that this association is also found in adults. A study carried out in Leicester, England (24) demonstrated that 50% of asthma exacerbations in adults were linked to upper respiratory tract virus infections and this is probably an underestimate (25). A number of respiratory tract viruses have been shown to enhance basophil histamine releasability. Influenza A enhances basophil histamine release in vitro as does respiratory syncytial virus (RSV). Paramyxovirus has been shown to cause histamine release when incubated with peripheral blood basophils. Increased levels of histamine have been found in the nasal secretions of patients with upper respiratory tract infections. If viruses are common precipitants of exacerbations of asthma, then histamine may be the important link.

IV. OTHER ACTIONS OF HISTAMINE AND ANTIHISTAMINES IN ASTHMA

In asthma, the effects of histamine are more widespread than the simple bronchoconstriction first described by Weiss. The bronchoconstriction itself is not only due to a direct effect on smooth muscle but also occurs through stimulation of airway afferent sensory nerves. Both of these effects are H_1-receptor mediated. There is an increase in vascular permeability caused by venular dilatation and creation of interendothelial pores (26), again mediated through the H_1-receptor. Eosinophil chemoattractant properties have been described, though these are generally not inhibited by blocking either the H_1- or H_2-receptor. Histamine also enhances the secretion of a viscid mucus rich in glycoproteins.

Antihistamines have been shown to have effects in addition to the antagonism of H_1-mediated bronchoconstriction as discussed in Chapters 5 and 6 of this book. Many of these effects might be beneficial in asthma. They have been shown to prevent mast-cell and basophil mediator release at low concentrations (27, 28). Ketotifen inhibits platelet migration and platelet mediator release (29) and also prevents the down-regulation of β_2-receptors that occurs with long-term β_2-agonist treatment (30).

V. EFFECTS OF ANTIHISTAMINES

With histamine playing such an important role in the pathogenesis of asthma, and animal experiments appearing so promising, the therapeutic potential of antihistamines in asthma would seem to be very great. Herxheimer (31) was the first to propose the use of antihistamines in asthma. In an open, uncontrolled study he administered the antihistamines antisan, promethazine, and diphenhydramine to 26 patients with mild and moderate asthma, all of whom showed definite clinical improvement. Since then, many other studies have examined the efficacy of antihistamines in asthma. These have focused on different aspects of asthma: effect on resting bronchial tone; effect on bronchial hyperresponsiveness to methacholine, histamine, and exercise; effect on the early and late reactions to allergen; and effects on clinical outcome. We will review each of these in turn.

VI. EFFECTS ON BRONCHIAL TONE

A medication that causes reduction of resting bronchial tone and therefore reduction of airway obstruction is likely to be beneficial in the treatment of asthma. Early studies concentrated on this aspect of antihistamines. Popa (32) measured forced expiratory flow in 10 asthmatic volunteers before, and up to 5 h after, administration of oral chlorpheniramine (8 mg), intravenous (IV) chlorpheniramine (10 mg given twice), IV aminophylline (5.5 mg/kg), and butabarbital (30 mg). There was a reproducible improvement following intravenous chlorpheniramine starting 15 min postdose and reaching a peak at 120 min. The average improvement was 15%, comparable to the effect of aminophylline. A significant, but smaller improvement occurred with oral chlorpheniramine and no improvement was noted with butabarbital or the placebo control.

Azelastine in a single dose of 4 mg was found to improve the FEV_1 within 3 h of dosing in a double-blind, randomized, placebo-controlled study (33). A single dose of terfenadine was also shown to cause a small, but nevertheless significant, improvement in FEV_1 2 h post dosing, lasting for 4 h. Interestingly, there was no dose–response relationship suggesting that the effects of the antihistamine were more dependent on local endogenous histamine release than on the degree of H_1-receptor blockade.

In a study of ten children 8–14 years of age, Groggins et al. (34) looked at the bronchodilating effects of inhaled chlorpheniramine. Patients had measurements of peak expiratory flow (PEF), forced vital capacity (FVC) and forced expiratory volume in 0.75 s ($FEV_{0.75}$) at baseline and then inhaled nebulized chlorpheniramine, with respiratory function tests repeated regularly up to 2 h after inhalation. The children were studied on three occasions with doses of 2, 4, and 8 mg of chlorpheniramine. All treatment groups showed improvement in PEF and $FEV_{0.75}$ at 2 min, which progressed over the following 30–45 min. The chlorpheniramine was, however, poorly tolerated, causing cough on 13 of the 30 occasions. In three patients, this cough was associated with a fall in PEF.

The older antihistamines could not be given in high doses because of their sedative side effects. In an attempt to overcome this, Norgrady (35) compared the bronchodilator effect of inhaled nebulized clemastine with that of inhaled salbutamol and placebo. Twelve patients hospitalized following an asthma exacerbation were treated with either nebulized clemastine, nebulized salbutamol, or placebo, with PEF and FEV_1 measured before and after treatment. There was no significant difference in the mean maximum increases in either PEF or FEV_1 brought about by either salbutamol or clemastine, though both were significantly better than placebo. Clemastine did, however, have a longer duration of action than salbutamol. This effect was not found in the study by Partridge et al. (36). Eleven stable asthmatics received nebulized clemastine as a single dose and six received it for 2 weeks, the latter in a double-blind, cross-over study. No improvement was noted in FEV_1 following a single dose of clemastine and no improvement in PEF or symptom scores was noted during maintenance treatment.

Hodges (37) examined the ability of inhaled clemastine and inhaled chlorpheniramine to cause bronchodilation in schoolchildren. Fourteen asthmatics 6–14 years of age inhaled 4 mg chlorpheniramine, 1 mg clemastine, or placebo each made up to a 2-ml solution. Lung function was assessed by measuring PEF and $FEV_{0.75}$ before inhalation and at 30-min intervals afterward. The purpose of this study was to compare the bronchodilating properties of clemastine, a nonirritant, nonsedating H_1-blocker, with the established bronchodilator properties of inhaled chlorpheniramine. Both chlorpheniramine and clemastine resulted in significant improvement in both PEF and $FEV_{0.75}$ when compared with placebo.

Wood-Baker (38) examined the effects of oral chlorpheniramine, terfenadine, brompheniramine, cetirizine, cyproheptadine, clemastine, and as-

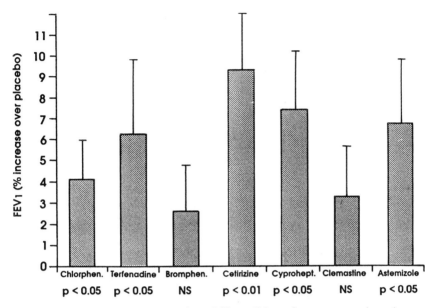

Figure 2 The bronchodilator action of H₁-antihistamines expressed as the percentage increase in baseline FEV_1 after administration of active drug compared with that after administration of placebo. Bars represent the mean ± standard error of the mean. (From Ref. 38.)

temizole in a single-dose study. Administration of all of these antihistamines increased baseline FEV_1 from 2.58 to 9.28%, which was significant for all drugs except brompheniramine and clemastine (Fig. 2). A study has been carried out looking at the effect of 2 weeks' treatment with astemizole (Howarth, in press). There was no significant change in FEV_1 and no relationship between final FEV_1 and first-dose bronchodilating effect. The implication is that in asthma, single-dose studies do not correlate with the efficacy of prolonged treatment, perhaps due to the development of tachyphylaxis in the lower airways.

VII. EFFECTS ON METHACHOLINE/ HISTAMINE CHALLENGE

Asthma is a variable disease, so studies looking at baseline spirometry are not a particularly sensitive means of determining drug efficacy. Bronchial hyperreactivity is a reasonable measure of airway "lability" and therefore a useful marker of the underlying inflammation. It is usually measured by challenging the patient with increasing doses of either histamine or methacholine to determine the concentration that causes a 20% fall in FEV_1 (PC_{20}). A number of studies have used bronchial hyperreactivity as a measure of efficacy.

One study (39) examined the effects of nebulized chlorpheniramine and ipratropium bromide in nine asthmatic children challenged with methacholine and histamine. Patients were all atopic, perennial asthmatics requiring only β_2 agonists or no therapy at all. Patients had bronchial hyperreactivity tested with histamine and methacholine before and after inhalation of chlorpheniramine and ipratropium bromide. Ipratropium bromide shifted the PD_{20} to methacholine whereas chlorpheniramine shifted the PD_{20} to histamine but not to methacholine. Wood-Baker (38) also looked at the effects of many different antihistamines on histamine-provoked bronchoconstriction. He studied chlorpheniramine, terfenadine, astemizole, cyproheptadine, brompheniramine, clemastine, and cetirizine and found significant improvements in PC_{20} histamine with all of these (Fig. 3).

Rafferty (40) looked at the effect of three doses of terfenadine—60 mg, 120 mg, and 180 mg—on histamine and methacholine challenge. In a

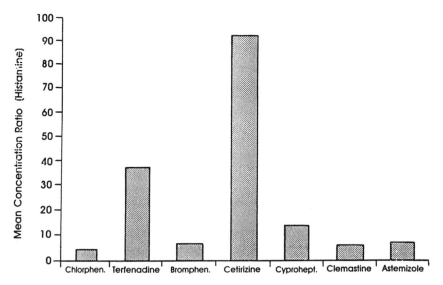

Figure 3 The protective efficacy of antihistamines against the airway response to histamine expressed as a mean concentration ratio for nine subjects dosed with each drug. (From Ref. 38.)

double-blind, single-dose study, nine asthmatic patients were given either terfenadine or placebo and bronchial challenge was carried out 3 h later with histamine or methacholine. Once again, terfenadine was shown to cause a small, but nevertheless significant, improvement (9 to 10%) in FEV_1. There was also a shift to the right of the dose–response curve for histamine challenge with a dose-dependent effect, but there was no effect on methacholine challenge.

It should be noted that the antihistamine doses used in these studies have been comparatively high and continuous use of these doses may, in the clinical situation, expose some patients to risks from side effects such as cardiac arrhythmias. This must be taken into consideration when interpreting these studies and the potential risks as well as benefits of the medication should be compared to currently used asthma medication.

VIII. EFFECTS ON EXERCISE-INDUCED ASTHMA

Exercise-induced asthma is noted in at least 60% of asthmatics and elevated levels of histamine after exercise have been reported by a number of investigators, although it has never been certain whether this comes from lung mast cells or circulating basophils. Early studies failed to show any benefit of antihistamines in its treatment but it is possible that the doses given were too low, limited by the sedative effects of the early drugs. Astemizole, a second-generation antihistamine, is less sedating and so can be given in higher doses. In a study of nine patients (41), the mean fall in FEV_1 following exercise was reduced by astemizole, although this effect was not consistent for all patients. Patel (42) investigated the effect of terfenadine at doses of 60, 120, and 180 mg on ten patients with exercise-induced asthma. FEV_1 was recorded before and after 6 to 8 min of steady-state running with the exercise carried out 4 h after the treatment with drug or placebo when plasma levels of terfenadine acid metabolite would be at their peak. The degree of protection from the exercise-induced fall in FEV_1 was noted. Terfenadine provided significant protection compared to placebo. There was a considerable variation in response, with three patients showing no response at all. In the seven patients who did respond, there was a dose–response relationship with 60 mg of terfenadine providing 25% protection and 180 mg providing 59% protection.

Ghosh studied the effect of oral and inhaled cetirizine in patients with exercise-induced asthma (43). In the first part of the study, 12 patients were given either placebo or cetirizine 10 mg twice daily for 1 week and exercised for 6 to 8 min on a treadmill 2 h after the final dose. Oral cetirizine failed to inhibit exercise-induced bronchoconstriction. In the second phase, 8 patients were given either 5 mg or 10 mg cetirizine or placebo through a nebulizer, again in a double-blind, randomized fashion and then exercised. Nebulized cetirizine was effective in reducing exercise-induced bronchoconstriction at both doses, perhaps due to its increased concentration locally.

The effect of prolonged treatment on exercise-induced asthma has also been investigated. In a study of eight patients following 2 weeks' treatment with astemizole, the degree of exercise-induced bronchoconstriction fell from a drop of FEV_1 of 20.9% $\pm$ 5.7% before treatment to 10.8% $\pm$ 5.7% after treatment, this being significant 2–30 min postchallenge.

IX. EFFECTS ON THE EARLY REACTION TO ALLERGEN

Experimental models of asthma have focused on the early and late reactions following antigen challenge (44). The early reaction occurs within 10 to 30 min of challenge, and resolves within 2 h. This is thought to be predominantly mast-cell-mediated, since mast-cell stabilizers such as sodium cromoglycate have a protective effect and mast cells recovered from bronchoalveolar lavage show signs of degranulation.

Several studies have investigated the effects of antihistamines on the early phase reaction. Howarth et al. (45) studied eight mild, seasonal grass pollen-sensitive asthmatics, challenging them on two separate occasions with grass allergen. The challenges were separated by a 2-week period during which the patients received astemizole 10 mg daily. At the first challenge the mean maximal fall in FEV_1 was 27% whereas following treatment with astemizole FEV_1 fell by only 15.4%, a significant protective effect. The greatest protection occurred at the earlier time points with 72% protection at 2 min and 50% protection at 5 min. Astemizole failed to have any effect on increments of plasma histamine, therefore suggesting that its H_1-receptor-blocking properties rather than any possible mast-cell stabilizing properties, were responsible for this effect.

Curzen et al. (46) investigated the action of terfenadine on allergen-induced bronchoconstriction in seven atopic, nonasthmatic subjects. Volunteers were tested before and 3 h after 180 mg of terfenadine and also before and after treatment with flurbiprofen (a cyclo-oxygenase inhibitor) and a combination of terfenadine and flurbiprofen. Following terfenadine, the decline in FEV_1 after inhaled allergen was slower than after placebo ($p < 0.01$) and the maximum fall was inhibited by 50%, also significant. The maximum inhibitory effect occurred at 15 min.

Chan et al. (47) studied the effect of three doses of terfenadine (60 mg, 120 mg, 180 mg) on the early- and late-phase response to antigen. Terfenadine caused a shift of the antigen dose–response curve to the right, but, although this was a statistically significant effect, the shift was small and unlikely to be clinically important.

In a study by Rafferty et al. (48) the effect of terfenadine on adenosine monophosphate (AMP), histamine, and antigen challenge was determined. Terfenadine inhibited histamine-induced bronchoconstriction completely, AMP-induced bronchoconstriction by 86%, and allergen-induced bronchoconstriction by only 50% thus suggesting a smaller role for histamine in allergen-induced, as opposed to histamine- or AMP-induced bronchoconstriction.

Eiser et al. performed a double-blind, placebo-controlled, cross-over study of terfenadine—60 mg, 120 mg, and 180 mg—on the early phase response to antigen challenge and showed significant shift of the antigen dose–response curve to the right.

Cistero et al. (49) examined the effects of treatment with astemizole on the early phase reaction. Patients were age 18–45 and RAST-positive to *Dermatophagoides pteronyssinus*. In a double-blind, placebo-controlled, cross-over trial either placebo, astemizole 10 mg/day, or astemizole 30 mg/day was administered. Allergen challenge with *Der p* I was carried out 7 and 28 days into treatment. A diary card was used to assess tolerance and side effects. Twelve patients were studied. With a dose of 10 mg/day, there was a significant reduction in bronchial response at ten days whereas with a 30 mg/day dose there was a reduction in response at both 7 and 28 days. There was a significant improvement in bronchial reactivity between 7 and 28 days at both doses. No late response was detected in any of these patients, so the effect of astemizole on this response could not be determined.

X. EFFECTS ON THE LATE-PHASE REACTION TO ALLERGEN

Six to 8 h following allergen challenge, 60% of patients develop a late-phase reaction characterized by infiltration of the airways with eosinophils, neutrophils, and macrophages. This response is considered to be a model of clinical asthma as it is accompanied by an increase in nonspecific bronchial hyperresponsiveness.

Very few studies have examined the effects of antihistamines on the late-phase reaction, presumably because it has not been thought of as a mast-cell-mediated phenomenon and therefore histamine has not been felt to play an important role in its generation. One double-blind study did, however, produce encouraging results. The effect of azelastine on the late-phase reaction was examined in 11 atopic asthmatics. In six of these patients no late reaction occurred and, in the remaining subjects, the late-phase bronchoconstriction was reduced by 70%.

XI. THE EFFECTS OF ANTIHISTAMINES IN CLINICAL ASTHMA

The ultimate test of a drug's efficacy is whether it works in the clinical situation—whether it makes the patient feel better. Herxheimer was the first to suggest that antihistamines could play a role in the treatment of clinical asthma and since then a number of studies have looked directly at clinical efficacy. Herxheimer's measures were crude, using only vital capacity, number of inhalations the patient needed, and a global score of "physical efficiency." Most clinical studies nowadays would involve the patient keeping diary cards of outcome measures including symptom scores, PEF variability, and diurnal variation, all sensitive markers of asthma control. Many studies have been carried out and it is best to consider these according to the individual drugs.

A. Clemastine

Partridge et al. (36) examined the effect of inhaled clemastine as both an acute and maintenance treatment. In the maintenance arm of the study six asthmatic patients continued with their normal treatments (inhaled steroids in all and oral steroids in one) but, in addition, were given either two puffs of clemastine or placebo four times a day. No difference was found between the clemastine and placebo group.

B. Terfenadine

Taytard et al. (50), in a much larger trial, studied terfenadine, a drug noted previously in this chapter to have a significant bronchodilator effect. Fifty-two patients with atopic, perennial asthma were evaluated. A 2-week run-in period on placebo was followed by two periods of 2 weeks on either terfenadine 120 mg twice daily or placebo. Patients were not taking inhaled or oral corticosteroids but were allowed theophylline, sodium cromoglycate, or salbutamol. There were significant differences between the terfenadine and placebo group in *patient*-recorded parameters of symptom score ($p < 0.01$), salbutamol use ($p < 0.05$), morning PEF ($p < 0.05$), and evening PEF ($p < 0.05$). However, there were no significant differences between the *physician*-assessed variables of symptom score, degree of discomfort, or PEF after salbutamol, although there was a significant difference in PEF before salbutamol.

Rafferty et al. (51) looked at the effect of terfenadine on seasonal asthma, studying 18 mild grass pollen-sensitive asthmatics in a 9-week double-blind cross-over study (Figs. 4, 5). There was a run-in period of 1 week followed by a 4-week treatment period with either terfenadine

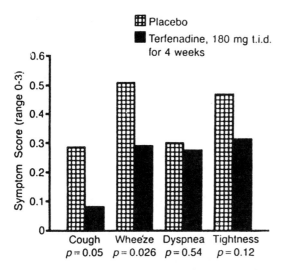

Figure 4 Mean symptom scores in grass-pollen-sensitive patients with asthma following 4 weeks' treatment with terfenadine 180 mg three times daily (From Ref. 51.)

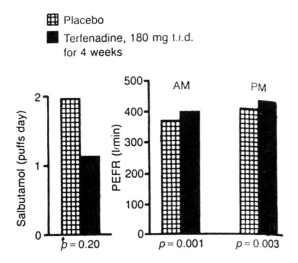

Figure 5 Mean salbutamol use and peak expiratory flow rate in grass-pollen-sensitive patients with asthma following 4 weeks' treatment with terfenadine 180 mg three times daily (From Ref. 51.)

180 mg or placebo 3 times daily followed by 4 weeks of the alternative. Treatment with terfenadine resulted in a significant increase in morning PEF from 379 L/min to 400 L/min ($p < 0.001$) and evening PEF from 405 L/min to 430 L/min ($p < 0.003$). Terfenadine also resulted in a significant reduction in cough (76.9%; $p < 0.05$) and wheeze (46.9%; $p < 0.026$).

C. Azelastine

A number of studies have looked at the effects of oral azelastine, but many of these have been uncontrolled and have used vague measures of disease outcome. Takishima et al. reported that 4 mg/day of oral azelastine led to "fair to excellent overall improvement" in 52% of asthmatics after 2 weeks and 74% after 4 weeks, but there was no control group for comparison. Ohno et al. (52) found that 2 weeks' treatment with azelastine in children led to a reduction in frequency and severity of asthma attacks in 33–86% of patients.

Gould et al. (53) showed that treatment for 7 weeks with oral azelastine, when compared to placebo, improved morning and evening PEF, wheeze score, and number of inhalations of salbutamol in a group of 21 atopic asthmatics. Tinkelman et al. (54) studied 220 patients treated with azelastine for 12 weeks and showed significant improvement in FEV_1, and decreased use of bronchodilator therapy.

D. Cetirizine

Bruttmann et al. (55) carried out a double-blind study of 57 grass pollen-sensitive patients with severe asthma, that is, a minimum of three to five attacks per week. During the pollen season twenty-nine patients were treated with placebo and 28 with cetirizine 15 mg daily for 14 days. In the cetirizine group, respiratory function tests and PEF remained constant, whereas in the placebo group these deteriorated considerably. Furthermore, a majority of the placebo group withdrew due to declining PEF despite treatment with fenoterol.

Kurzeja et al. (56), in a small study of 20 patients, compared oral cetirizine 10 mg twice daily with terfenadine 60 mg twice daily, the latter acting as a control since in this dose it was felt to have minimal antiasthmatic activity. Cetirizine treatment resulted in a reduction in the need for β_2-agonists compared to the control treatment. Interestingly the difference was more marked in the third week ($p < 0.005$) than in the second week ($p < 0.04$). Patients' and doctors' assessments also favored the cetirizine-treated group.

Bousquet (57) performed a double-blind, multicenter study of cetirizine in grass pollen-induced asthma. Ninety-seven grass pollen-sensitive asthmatics were recruited. All had been symptom-free for 2 months and all had a history of allergic rhinoconjunctivitis, so it seemed clear that their symptoms were the result of grass pollen exposure. Patients were randomized to either cetirizine 10 mg twice daily, cetirizine 15 mg twice daily, terfenadine 60 mg twice daily, or placebo. The results were not as encouraging as the previous studies described. There was a significant improvement in symptoms but only at the 15 mg/twice daily dose of cetirizine and only when the total symptom scores were pooled. There was no significant difference in PEF or reduction in β_2-agonist use between active treatment and placebo groups.

E. Ketotifen

A number of studies have investigated the effects of ketotifen. Lane (58) looked at its steroid-sparing effect in a multicenter, double-blind study in 86 corticosteroid-dependent asthmatics. The patients recruited had been on maintenance treatment with oral corticosteroids for at least 1 month and had airflow obstruction with proven response to bronchodilators. After a run-in period of 2 weeks, they were allocated to treatment with either ketotifen or placebo and then their oral corticosteroids were reduced by 1 mg a week until either they had been off corticosteroids for 1 week or they became symptomatic. Ten patients receiving ketotifen were weaned off oral corticosteroids completely (24%) compared to three patients receiving placebo, a significant difference in favor of ketotifen ($p < 0.04$), and the daily requirement for oral prednisolone was reduced from 8.4 mg to 4.4 mg in those on ketotifen. The major adverse effect was drowsiness reported by 5% of patients on ketotifen.

A multicenter Canadian study was performed in 138 children with chronic asthma (59). This study lasted for 7 months. After a 1-month baseline patients were given either ketotifen or placebo in a double-blind manner for 6 months. After 10 weeks of treatment, an attempt was made to reduce their other medications. In the ketotifen group 60% of patients were able to stop theophylline, compared to 34% taking placebo. Significant differences ($p < 0.05$) were also found for patient evaluation of global well-being.

Dyson et al. (60) came to less favorable conclusions. Fifty atopic asthmatic patients received ketotifen 1 mg, ketotifen 2 mg, or placebo for 1 month in a random order. PEF was recorded 3 times daily as were symptoms of rhinorrhea, itchy eyes and skin. There was no significant difference found in mean PEF in either active treatment group. There was a significant reduction in salbutamol use (5 puffs per week, $p < 0.005$) in those patients *not* on inhaled corticosteroids but no difference was apparent in those taking inhaled corticosteroids. Again, the major adverse effect was drowsiness with 14% of patients either withdrawing from the study or reducing their ketotifen dose as a consequence.

XII. SUMMARY

Although a number of clinical studies have shown beneficial effects of antihistamines in the treatment of asthma, these benefits have fallen far short of original expectations. The effects are small compared to more established treatments and the side effects are potentially more troublesome; thus the indications, if any, for antihistamines are unclear. The use of antihistamines has, however, helped us to understand the role of histamine in the pathogenesis of asthma.

REFERENCES

1. British Thoracic Society. Guidelines for the treatment and management of asthma. Thorax 1993;48:S1–S24.
2. Weiss S, Robb G, Blumgart H. The velocity of blood flow in health and disease as measured by the effect of histamine on the minute vessels. Am Heart J 1929;4:664–691.
3. Weiss S, Robb G, Ellis C. The systemic effects of histamine in man. Arch Intern Med 1932;49:360–396.
4. Curry J. The action of histamine on the respiratory tract in normal and asthmatic subjects. J Clin Invest 1946;25:785–794.
5. Curry J. The effect of antihistamine substances and other drugs on histamine bronchoconstriction in asthmatic subjects. J Clin Invest 1946;25:792–99.
6. Riley JF, West GB. The presence of histamine in tissue mast cells. J Physiol 1953;120:528–537.
7. Graham H, Leway O, Wheelwright F, Lenz N, Parish H. Distribution of histamine among leucocytes and platelets. Blood 1955;10:467–481.
8. Ishizaka K, Ishizaka T, Terry W. Antigenic structure of E globulin and reaginic antibody. J Immun 1967;99:849–858.
9. Johansson S, Bennich H. Immunological studies of an autosomal myeloma protein. Immunology 1967;73:381.
10. Tomioka H, Ishizaka K. Mechanisms of passive sensitization—Presence of receptors for IgE on monkey mast cells. J Immunol 1971;107:971–978.
11. Ishizaka T, DeBernado R, Tomioka H, Lichtenstein L, Ishizaka K. Identification of basophil granulocytes as a site of allergic histamine release. J Immunol 1972;108:1000–1008.
12. Flint KC, Leung KB, Hudspith BN, Brostoff J, Pearce F, McJohnson N. Bronchoalveolar mast cells in extrinsic asthma: a mechanism for the initiation of antigen-specific bronchoconstriction. Br Med J 1985;291:923–926.
13. Kimura I, Tanizaki Y, Saito K, Takahashi K, Ueda N, Sato S. Appearance of basophils in the sputum of patients with bronchial asthma. Clin Allergy 1975;5:95–98.
14. Guo C, Liu M, Galli S, Bochner B, Kagey-Sobotka A, Lichtenstein L. Identification of IgE-bearing cells in the late phase response to antigen in the lung as basophils. Am J Resp Cell Mol Biol 1994;10:384–390.

15. Bruce C, Weatherstone R, Seaton A, Taylor W. Histamine levels in plasma, blood and urine in severe asthma and the effect of corticosteroid treatment. Thorax 1976;31:724–729.

16. Simon R, Stevenson D, Arrogane C, Tan E. The relationship of plasma histamine to the activity of bronchial asthma. J Allergy Clin Immunol 1977; 60:312–316.

17. Barnes P, Ind P, Brown M. Plasma histamine and catecholamines in stable asthmatic subjects. Clin Sci 1982;62:661–665.

18. Casale TB, Wood D, Richerson HB, et al. Elevated bronchoalveolar lavage fluid histamine levels in allergic asthmatics are associated with methacholine bronchial hyperresponsiveness. J Clin Invest 1987;79:1197–1203.

19. Liu M, Bleeker E, Lichtenstein L, et al. Evidence for elevated levels of histamine, prostaglandin D2 and other bronchoconstricting prostaglandins in the airways of subjects with mild asthma. Am Rev Respir Dis 1990;142: 126–132.

20. Fuller R, Dixon C, Dollery C, Barnes P. Prostaglandin D2 potentiates airway responsiveness to histamine and methacholine. Am Rev Respir Dis 1986; 133:252–254.

21. Bradding P, Feather IH, Wilson S, Bardin PG, Heusser CH, Holgate ST, Howarth PH. Interleukin 4 is localised to and released by human mast cells. J Exp Med 1992;176:1381–1386.

22. Johnston S, Pattermore P, Sanderson G, Lampe F, Josephs L, Myint S, Tyrell D, Holgate ST. Role of virus infection in exacerbations in children with recurrent wheeze and cough. Thorax 1993;48:1055.

23. Johnston S, Campbell M, Pattermore P, Smith S, Sanderson G, Symington P, O'Toole S, Myint S, Tyrell D, Holgate S. Community status of role of viral infections in exacerbations of asthma in 9–11 year-old children. Br Med J 1995;36:1225–1229.

24. Kent J, Ireland D. Respiratory viruses and exacerbations of asthma in adults. Br Med J 1993;307:982–986.

25. Corne J, Chanarin N. Respiratory viruses and asthma—importance of infection underestimated. Br Med J 1994;308:57.

26. Marjo G, Gilmore V, Leventhal M. On the mechanism of vascular leakage caused by histamine type mediators. Circ Res 1967;21:833–847.

27. Arunlakshana O, Schild H. Histamine release by antihistamines. J Physiol 1953;119:47P–48P.

28. Rimmer SJ, Church MK. The pharmacology and mechanisms of action of histamine H₁-antagonists. Clin Exp Allergy 1990;20:3–17.

29. Page C, Tomiak R, Morley J, Saunders R. Susceptibility of platelet dependent bronchoconstriction by antihistamine drugs. Am Rev Respir Dis 1984; 129:A25.

30. Koshimo T, Agrawal D, Townley R. Effect of ketotifen on the down-regulation of beta-adrenoreceptors in guinea pig lung and spleen. Am Rev Respir Dis 1988;138:127–137.

31. Herxheimer H. Antihistamines in bronchial asthma. Br Med J 1949;2:901.

32. Popa V. Bronchodilating activity of an H_1-blocker, chlorpheniramine. J Allergy Clin Immunol 1977;59:54–63.
33. Kemp JP, Meltzer EO, Orgel HA, et al. A dose response study of the bronchodilator action of azelastine in asthma. J Allergy Clin Immunol 1987;79: 893–899.
34. Groggins RC, Milner AD, Stokes GM. The bronchodilator effects of chlorpheniramine in childhood asthma. Br J Dis Chest 1979;73:297–301.
35. Nogrady S, Bevan C. Inhaled antihistamines—bronchodilation and effects on histamine- and methacholine-induced bronchoconstriction. Thorax 1978; 33:700–704.
36. Partridge M, Saunders K. Effect of an inhaled antihistamine (clemastine) as a bronchodilator and as a maintenance treatment in asthma. Thorax 1979; 34:771–776.
37. Hodges IGC, Milner AO, Stokes GM. Bronchodilator effect of two inhaled H_1-receptor antagonists clemastine and chlorpheniramine in wheezy schoolchildren. Br J Dis Chest 1983;77:270–275.
38. Wood-Baker R and Holgate ST. The comparative actions and adverse effects profile of single doses of H_1-receptor antihistamines in the airways and skin of subjects with asthma. J Allergy Clin Immunol 1993;91:1005–1014.
39. Woenne R, Kattan M. Orange RP, Levison H. Bronchial hyperreactivity to histamine and methacholine in asthmatic children after inhalation of SCH 1000 and chlorpheniramine maleate. J Allergy Clin Immunol 1978;62: 119–123.
40. Rafferty P, Holgate ST. Terfenadine (Seldane) is a potent and selective histamine H_1-receptor antagonist in asthmatic airways. Am Rev Respir Dis 1987; 135:181–184.
41. Clee MD, Ingram CG, Reid PC, Robertson AS. The effect of astemizole on exercise-induced asthma. Br J Dis Chest 1984;78:180–183.
42. Patel K. Histamine- and exercise-induced asthma. Eur J Clin Res 1991;2: 89–96.
43. Ghosh SK, de Vos C, McIlroy I, Patel KR. Effect of cetirizine on exercise-induced asthma. Thorax 1991;46:242–244.
44. Cartier A, Thomson N, Frith P, Roberts R, Hargreave F. Allergen-induced increase in bronchial responsiveness to histamine: relationship to the late asthmatic response and change in airway caliber. J Allergy Clin Immunol 1982;70:170–7.
45. Howarth P. Histamine and asthma: an appraisal based on specific H_1-receptor antagonism. Clin Exp Allergy 1990;20:31–41.
46. Curzen N, Rafferty P, Holgate S. Effects of a cyclo-oxygenase inhibitor, flurbiprofen, and an H_1-histamine receptor antagonist, terfenadine, alone and in combination on allergen-induced, immediate bronchoconstriction in man. Thorax 1987;42:946–952.
47. Chan TB, Shelton DM, Eiser NM. Effect of an oral H_1-receptor antagonist terfenadine on antigen-induced asthma. Br J Dis Chest 1986;80:375–384.
48. Rafferty P, Beasley R, Southgate P, Holgate S. The role of histamine in

allergen- and adenosine-induced bronchoconstriction. Int Arch Allergy Appl Immun 1987;82:292–4.

49. Cistero A, Abadias M, Lleonart R, et al. Effect of astemizole on allergic asthma. Ann Allergy 1992;69:123–7.

50. Taytard A, Beaumont D, Pujet J, Saphe M, Lewis P. Treatment of bronchial asthma with terfenadine; a randomised controlled trial. Br J Clin Pharmacol 1987;24:743–6.

51. Rafferty P. Antihistamines in the treatment of clinical asthma. J Allergy Clin Immunol 1990;88:647–650.

52. Ohno H, Moriguchi N, Matsumota K. Clinical studies of azelastine in bronchial asthma in children. Med Consult N Remed 1988;:1003–1010.

53. Gould CAL, Ollier S, Aurich R, Davies RJ. A study of the clinical efficacy of azelastine in patients with extrinsic asthma and its effect on airway responsiveness. Br J Clin Pharmacol 1988;26:515–525.

54. Tinkelman D, Bucholtz G, Kemp J, Repsher L, Spector S. The long term use of azelastine in asthmatics. N Engl Reg Allergy Proc 1988;9:365.

55. Bruttmann G, Pedrali P, Arendt C, Rihoux JP. Protective effect of cetirizine in patients suffering from pollen asthma. Ann Allergy 1990;64:224–228.

56. Kurzeja A, Riedelsheimer B, Hulhoven R, Bernheim J. Cetirizine in pollen associated asthma. Lancet 1989;i:556.

57. Bousquet J, Emonst A, Germoutz J, et al. Double-blind multicentre study of cetirizine in grass-pollen-induced asthma. Ann Allergy 1990;65:504–508.

58. Lane DJ. A steroid-sparing effect of ketotifen in steroid-dependent asthmatics. Clin Allergy; 1980;10:519–525.

59. Rackham A, Brown C, Chandra R, et al. A Canadian multicentre study with Zaditen (ketotifen) in the treatment of bronchial asthma in children aged 5 to 17 years. J Allergy Clin Immunol 1989;84:286–295.

60. Dyson AJ, Mackay AD. Ketotifen in adult asthma. Br Med J 1980;280:360–361.

10

H₁-Receptor Antagonists in Chronic Urticaria

Tommy C. Sim and J. Andrew Grant
University of Texas Medical Branch
at Galveston, Galveston, Texas

I. INTRODUCTION

Approximately 10 to 20% of people experience one or more episodes of urticaria during their lifetime. Urticaria is characterized by the appearance of transient pruritic, edematous, erythematous papules, plaques, or wheals in the skin. The condition results from localized vasodilation and transudation of fluid from small cutaneous blood vessels. The ideal treatment for urticaria is identification and removal of its trigger. However, although the cause of acute urticaria is usually detectable, the cause of chronic urticaria (i.e., persisting longer than 6 weeks), remains a mystery in more than 75% of cases. Therefore, therapy must focus on measures that provide relief of symptoms. Although several different chemical mediators of inflammation can increase vascular permeability and cause wheals, histamine is the best documented mediator of urticaria. Antihistamines are thus an important pharmacotherapeutic option in this disorder.

II. HISTAMINE AND URTICARIA

Although many mediators may play a role in urticaria, histamine appears to be the most important one involved; indeed, clinical and pharmacological studies suggest a primary role for histamine-mediated reactions in urticaria (1, 2). Histologically, urticarial lesions exhibit dermal edema with perivascular mononuclear cell infiltrate that may contain lymphocytes, neutrophils, and eosinophils (3). Under the electron microscope, degranulated mast cells and eosinophils are seen. A histological study done on 43 patients with chronic urticaria revealed that their lesional skin contained ten times the normal number of mast cells and monocytes (4). No explanation for the accumulation of these cells was apparent, but it has long been known that the mediators of urticaria are mast-cell products.

Although the precise etiology and pathogenesis of chronic urticaria are unknown, an immunological basis has been suggested (Table 1). A large body of evidence supports the role of histamine in the pathophysiology of chronic urticaria. It is known that the amounts of histamine recovered from skin blister fluids are elevated in patients with chronic urticaria (1, 5). In a study by Cohen and Rosenstreich (6), patients with urticaria were found to react to codeine at concentrations almost 100 times lower than normal subjects; thus, patients with chronic urticaria exhibit increased histamine releasability. In addition, they are reported to have increased reactivity to histamine itself (7). Patients with chronic urticaria have increased dermal mast-cell histamine release, which occurs only when urticaria is active and disappears when the disease is in remission (8). It is therefore apparent that blockade of the effects of histamine is important in the treatment of chronic urticaria.

III. CHRONIC URTICARIA AND HISTAMINE-RELEASING FACTORS

Mast-cell activation and subsequent mediator release have been reported in a variety of inflammatory disorders, including chronic urticaria, in which an IgE-dependent mechanism is not demonstrable. Several pathogenic mechanisms have been proposed regarding the origin of chronic urticaria. Grattan et al. (9) demonstrated that the serum of some patients

Table 1   Immunological Abnormalities Associated with Chronic Urticaria

1. Increase in the number of mast cells and monocytes in the skin
2. Increased skin reactivity to histamine and codeine
3. Increased skin histamine content
4. Elevated number of CD4$^+$ T-cells and monocytes in skin lesions
5. Significant recovery of histamine releasing factors (HRF) in blister fluids from skin lesions
6. Demonstration of histamine-releasing IgG autoantibodies against the high-affinity IgE receptor in some patients

with chronic urticaria contains a mediator or factor that will produce a wheal reaction on reinjection into their skin, while causing no response in normal subjects. Alam and Rozniecki (10) reported an immediate cutaneous allergic reaction in atopic patients after intradermal injection of supernatants from cultures of mononuclear cells of allergic donors. Some of these individuals also exhibited a late-phase cutaneous reaction. These observations raise the possibility that a circulating histamine releasing factor (HRF) might be present, the actions of which might lead to histamine release from mast cells and basophils, possibly through cell-bound IgE (Fig. 1). Of interest, Elias et al. (4) first speculated that HRF could have a seminal role in producing hyperreactive mast cells in chronic urticaria, on the basis of their observation of increased numbers of CD4⁺ T-cells and monocytes in the biopsy specimens of skin lesions. Recently, using a skin blister model, Claveau et al. (11) demonstrated significant recovery of HRF in blister fluids from lesional skin of patients with chronic urticaria as compared with both nonlesional skin of patients with chronic urticaria and skin of normal individuals. Moreover, histamine content in blister fluids from affected skin of patients was also significantly higher

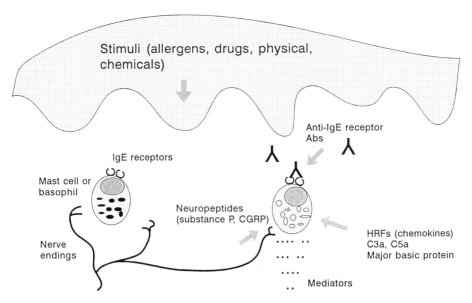

Figure 1 Mechanisms of chronic urticaria. Abbreviations: IgE, immunoglobulin E; CGRP, calcitonin gene-related peptide; Abs, antibodies; HRFs, histamine-releasing factors; C = complement.

and correlated somewhat with the degree of HRF activity. During the past 3 years, studies by Alam et al. (12, 13) and others (14) indicated that a newly described distinct group of chemotactic cytokines (chemokines) may represent HRF. Chemokines such as monocyte chemotactic and activating factor (MCAF)/monocyte chemotactic protein-1 (MCP-1), MCP-2, MCP-3, and macrophage inflammatory protein-1α (MIP-1α) were found to be potent secretagogues for basophils and mast cells.

Hide et al. (15) recently demonstrated that histamine-releasing IgG autoantibodies against the α subunit of the high-affinity IgE receptor (FcϵRI) are present in the circulation of some patients with chronic urticaria. They also reported that the mechanism of histamine release induced by the autoantibody against FcϵRI is unusual; unlike histamine release induced by allergens, anti-IgE antibodies, and some of the histamine-releasing cytokines, this action of the autoantibody is not dependent on cell-surface IgE. These new findings suggest that autoantibody-mediated cross-linking of IgE receptors may be an important pathogenic mechanism of chronic urticaria.

IV. CLINICAL EFFICACY OF H$_1$-RECEPTOR ANTAGONISTS IN CHRONIC URTICARIA

During the past 50 years, numerous H$_1$-antihistamines have been marketed (16). For many years, the use of H$_1$-antihistamines in dermatological conditions has been closely linked to the treatment of urticarias and to the symptomatic relief of pruritus. H$_1$-antihistamines are generally accepted as the standard first-line therapy for treatment of chronic urticaria. Although these drugs do not always decrease the number of urticarial lesions or the frequency of eruptions, they usually diminish pruritus significantly.

In addition to being pharmacological antagonists of histamine, many of the H$_1$-antihistamines have properties beyond receptor blockade (16, 17) (Fig. 2). The capacity of these agents to inhibit mediator release from mast cells and basophils seems unrelated to their ability to antagonize the biological activities of histamine at H$_1$-receptor sites. The cellular mechanism of inhibition of mediator release by H$_1$-antihistamines is not fully understood at present (Chap. 5). Although the older antihistamines such as hydroxyzine, chlorpheniramine, and diphenhydramine also decrease mediator release, this process has been best studied for the newer, specific H$_1$-receptor blocking agents. There are numerous reports of in vitro studies of the modulatory actions of newer antihistamines such as terfenadine, astemizole, azelastine, loratadine, and cetirizine on mediator

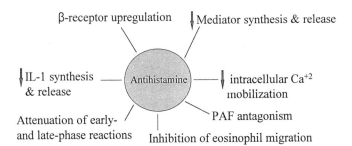

Figure 2 Non-H₁-receptor actions of antihistamines. At present, the clinical relevance of the above activities of some antihistamines is still unclear and further studies of these effects are necessary. Abbreviations: IL-1, interleukin-1; PAF, platelet activating factor.

release from murine mast cells, human tissue mast cells, and peripheral blood basophils using a range of immunological (antigen and anti-IgE) and nonimmunological (compound 48/80, substance P, and calcium ionophore) stimuli (16). Some of the newer antihistamines also significantly reduce the generation and release of leukotrienes and prostaglandins (18). H₁-antihistamines have various degrees of antiallergic effects, depending on the type and source of the cell, the type and concentration of the secretagogue or H₁-receptor antagonist used, and the mediator being measured.

A. First-Generation/Sedating H₁-Antihistamines

1. General Considerations

Older H₁-antihistamines are generally effective in controlling the signs and symptoms of urticaria. Because many of these drugs have similar therapeutic efficacy, central nervous system and anticholinergic side effects and pharmacokinetic considerations are important in the selection of a specific antihistamine for an individual patient (19). Moreover, the less convenient dosing schedules (two- to four-times daily) of some older H₁-antihistamines may also compromise their efficacy.

2. Therapeutic Efficacy of First-Generation H₁-Antihistamines

Antihistamines such as chlorpheniramine, brompheniramine, and triprolidine have repeatedly been found superior to placebo in the treatment of chronic urticaria (17). However, some of these drugs require frequent dosing and often lead to incomplete or limited relief of urticarial symptoms. For these reasons, the longer-acting traditional antihistamine, hydroxyzine, is probably the agent most frequently used for the treatment of chronic urticaria. It acts both as an antihistamine and as a serotonin antagonist and its combined properties as an antipruritic, sedative, and tranquilizer make it an efficacious drug for this vexing cutaneous disorder. Hydroxyzine has been shown to suppress wheal and flare reactions and pruritus induced by intradermally injected histamine for up to 36 h, with maximal effects occurring 6 to 12 h after taking the drug (20, 21). This prolonged duration of action is thought to be the result of fixation to the skin in addition to a relatively long serum elimination half-life. It has been suggested that hydroxyzine is especially useful in cases where emotional factors are involved both in the induction of symptoms or where symptoms produce anxiety. When compared in a double-blind manner with recommended doses of diphenhydramine, cyproheptadine, or placebo, its inhibition of exogenous histamine-mediated pruritus was 75- to 150-fold better (22). In addition, it was found clinically to be more effective than brompheniramine in patients with chronic urticaria (23).

Selection of particular H_1-antihistamines must be made on an individual basis and tailored to each patient situation. It is usually based on side effects, rather than efficacy, since most H_1-antihistamines will reduce symptoms of urticaria to some degree. Diphenhydramine is reported to be effective in at least 50% of patients with urticaria and causes fewer gastrointestinal side effects than hydroxyzine (16). Both these medications have a pronounced sedative effect that may preclude daytime use or even necessitate discontinuation. Clemastine, used twice daily, seems to control acute urticaria better than chronic urticaria. It is also reported anecdotally that cyproheptadine may add to the effect of hydroxyzine when the two drugs are used together. In current practice, the use of specific H_1-antihistamines is recommended over others in the treatment of some dermatological conditions. For example, cyproheptadine is reported to be preferred for symptomatic relief of cold-induced urticaria (24, 25); and hydroxyzine is preferred for the treatment of cholinergic urticaria (26).

Although doxepin is mainly used as a tricyclic antidepressant, it has both H_1- and H_2-blocking activities (16). Doxepin has been used in patients with chronic urticaria who were resistant to some conventional H_1-antihistamines. In several controlled clinical trials it was superior to other H_1-antihistamines. The efficacy of doxepin has been linked to its wide spectrum of activity. Besides antimuscarinic, antiserotoninergic, and antiadrenergic properties, its H_1-antihistaminic effect is approximately 750 times greater than that of diphenhydramine, and 50 times greater than hydroxyzine (27). These properties may also explain its adverse effects, which commonly include lethargy, dry mouth, and constipation. In addition, its effectiveness in cold urticaria may derive from a specific inhibition of the release of platelet-activating factor (PAF) (28). Regular dosage of 25 to 50 mg 1–3 times daily is usually effective in controlling symptoms of urticaria, but is still considered to be "low," since antidepressant dosages may reach 250 to 500 mg daily.

B. Second-Generation/Nonsedating H₁-Antihistamines

1. General Considerations

Current experience with first-generation H₁-antihistamines in the treatment of chronic urticaria often suggests that they are only mildly effective in dosages recommended for allergic rhinitis and other allergic disorders. Unfortunately, the profound anticholinergic and sedative side effects have often restricted the dosages that could be tolerated and may in part explain the poor compliance often encountered (16). A significant advance has been the development of more specific and highly potent H₁-antihistamines in which these properties are greatly reduced or apparently absent. A number of the second-generation H₁-antihistamines (e.g., terfenadine, astemizole, loratadine, azelastine, and cetirizine) possess additional antiallergic properties that may be of value. These agents have been shown to act on effector cells, such as mast cells and leukocytes, preventing their attraction to the reaction site (chemotaxis) or their functional bioactivity (e.g., their capacity to synthesize and secrete proinflammatory mediators). In clinical studies of chronic urticaria, the relatively nonsedating H₁-antihistamines have proven to be significantly better than placebo, and comparable to most of the first-generation H₁-antihistamines (17).

2. Therapeutic Efficacy of Second-Generation H₁-Antihistamines

Terfenadine   Terfenadine was the first of the newer H₁-antihistamines. It was introduced worldwide in 1981 and became available in the United States in 1985 (Table 2). Terfenadine has a unique structure and is not a derivative of a first-generation H₁-antihistamine. Following a single dose of terfenadine, maximum suppression of the wheal response to intradermal histamine in healthy volunteers is evident after 4 h (29) and persists for more than 12 h (30). Skin prick and nasal and bronchial challenge studies in humans generally confirm the antiallergic activity of terfenadine (31). Terfenadine 60 mg twice daily produced marked or complete relief of pruritus, and size and number of wheals after 1 week in patients with chronic urticaria and was equally effective when administered as a single

Table 2 Terfenadine: Double-Blind Clinical Studies in Chronic Urticaria

Study	Patients (No.)	Dosage	Results (overall efficacy)	Statistical significance
Ormerod et al. (33)	16 (cross-over)	T 60 mg b.i.d. B 12 mg b.i.d.	T = B	N/A
Fredriksson et al. (37)	60	T 60 mg b.i.d. Cl 1 mg b.i.d. P	T/Cl > P T/Cl > P	Yes Yes
Go et al. (35)	28 (cross-over)	T 60 mg b.i.d. Cet 10 mg q.d. P	T = Cet T/Cet > P	N/A Yes
Grant et al. (36)	121	T 60 mg b.i.d. Chl 4 mg t.i.d. P	T > Chl T/Chl > P	Yes Yes
Boggs et al. (34)	37	T 60 mg b.i.d. H 25 mg q.i.d. P	T = H T/H > P	N/A Yes
Paul and Bodeker (38)	40	T 60 mg b.i.d. A 10 mg q.d.	A > T	No

Abbreviations: T = terfenadine; B = brompheniramine; Cl = clemastine; Cet = cetirizine; Chl = chlorpheniramine; A = astemizole; H = hydroxyzine; P = placebo; q.d. = once daily; b.i.d. = twice daily; t.i.d. = three times daily; q.i.d. = four times daily; and N/A = not applicable. Modified from Ref. 32.

daily 120-mg dose (32). Based on global assessment of efficacy, terfenadine was at least as effective as brompheniramine 24 mg/day (33), hydroxyzine 100 mg/day (34), cetirizine 10 mg/day (35), chlorpheniramine 12 mg/day (36), or clemastine 2 mg/day (37), but had a less persistent effect than astemizole 10 mg/day (38). Combined treatment with terfenadine and ranitidine, an H_2-antihistamine, produced greater relief of pruritus than terfenadine or ranitidine alone in patients with chronic idiopathic urticaria (39). In one study, improvement in intractable urticarial symptoms was sufficient that it permitted several patients to discontinue concomitant corticosteroid therapy (40). In a multicenter study, terfenadine 60 mg twice a day was as effective as chlorpheniramine 4 mg 3 times a day, and both medications were superior to placebo (36). Terfenadine and chlorpheniramine were rated as good or excellent by 83% and 71% of patients, respectively.

Astemizole Astemizole has been available worldwide since the mid-1980s and became available in the United States in 1989 (Table 3). It has

Table 3 Astemizole: Double-Blind Clinical Studies in Chronic Urticaria

Study	Patients (No.)	Dosage	Results (overall efficacy)	Statistical significance
Sussman and Jancelewicz (46)	36	A 10 mg q.d. D 25 mg t.i.d. H 25 mg t.i.d. P	A = H A/H > D A/H/D > P	N/A Yes Yes
Cainelli et al. (44)	42	A 10 mg q.d. T 60 mg b.i.d.	A > T	No
Mobacken et al. (45)	53	A 10 mg q.d. Chl 4 mg t.i.d.	A > Chl	Yes
Knight (47)	35	A 10 mg q.d. P	A > P	Yes

Abbreviations: A = astemizole; D = diphenhydramine; H = hydroxyzine; T = terfenadine; Chl = chlorpheniramine; P = placebo; q.d. = daily; b.i.d. = twice daily; t.i.d. = three times daily; and N/A = not applicable. Modified from Ref. 41.

a long duration of action, permitting once-daily administration (41). The efficacy of astemizole and its long half-life make this H₁-antihistamine ideal for the treatment of chronic urticaria. In previous placebo-controlled clinical studies and in comparative trials, astemizole has been shown to be effective for the relief of urticarial symptoms. Astemizole and hydroxyzine have similar efficacy with regard to suppression of wheals produced by histamine in patients with chronic urticaria (42). A review by Monroe (43) reported the results of six published double-blind trials on the efficacy of astemizole in the treatment of chronic urticaria. These studies, which involved a total of 279 patients, showed astemizole to be more effective than placebo in controlling the signs and symptoms of chronic urticaria (38, 44–47). In a double-blind trial of 51 patients with chronic idiopathic urticaria, astemizole caused a significant improvement in symptoms and lesions in 75% of treated patients versus 20% in the placebo group (14). In addition, clinically benefited patients had persistent suppression of their urticarial symptoms and lesions for an average of 38 days beyond the treatment period. Another study demonstrated astemizole to be superior to chlorpheniramine in chronic urticaria and dermatographism (45). Astemizole was also reported to be more effective in the treatment of chronic urticaria than terfenadine (38).

Loratadine Loratadine became available on prescription in the United States in 1993, although it had been used in other countries since the late

1980s. It is related to the first-generation H_1-antihistamine azatadine (48, 49). Loratadine offers the convenience of once-daily administration. The inhibitory effect of loratadine on histamine-induced wheal formation is significantly greater than that of placebo, and in general has exceeded the effect of chlorpheniramine in comparative studies (49, 50). Hammarlund et al. (51) found that pretreatment with loratadine significantly inhibited both histamine- and allergen-induced blood flow increase to the skin as well as the late-phase response (LPR); pretreatment with topical corticosteroids only mildly inhibited the LPR after allergen challenge. Loratadine has also been evaluated in a number of single- and multicenter double-blind clinical trials conducted in patients with chronic idiopathic urticaria (52–60) (Table 4). In a 4-week clinical trial involving 153 patients with chronic idiopathic urticaria, it produced marked or complete relief of symptoms in significantly more patients than placebo did (52). Similar results in favor of loratadine compared with placebo were recorded in a 4-week double-blind clinical trial involving 30 patients with chronic urticaria (53). Furthermore, Belaich et al. (54) found that loratadine provided significantly greater relief from symptoms of chronic idiopathic urticaria and had a much lower treatment failure rate than terfenadine and placebo in a 28-day study involving 44 patients. The overall therapeutic response at the endpoint of treatment was rated as marked clinical improvement of symptoms in 64%, 50%, and 21% of the patients in the loratadine, terfenadine, and placebo treatment groups, respectively. However, in two studies involving 172 patients by Pleskow (55) and 100 patients by Soto et al. (56), loratadine and terfenadine at recommended doses were similarly effective over a 4-week period and both drugs were significantly superior to placebo. Monroe recently reported that loratadine was as effective as hydroxyzine in the treatment of chronic urticaria (47% and 43%, respectively) and also demonstrated a significant antipruritic effect in atopic dermatitis (57% in the loratadine group and 38% in the hydroxyzine group) (57).

   Cetirizine   Cetirizine is a piperazine derivative and the carboxylic acid metabolite of hydroxyzine (61). Although some studies have shown greater evidence of sedation in patients who received cetirizine than placebo, cetirizine is significantly less sedating than its parent compound hydroxyzine (62). Of interest, Simons et al. (63) performed a single-dose, placebo-controlled, double-blind, cross-over comparative study of the antihistaminic effects of the second-generation H_1-antihistamines, including controls such as placebo and chlorpheniramine. Using a computerized wheal and flare measurement, these investigators were able to show significant differences among the study drugs with regard to time of onset of action, degree of inhibition of the histamine-induced wheal and flare reac-

Table 4   Loratadine: Double-Blind Clinical Studies in Chronic Urticaria

Study	Patients (No.)	Dosage	Results (overall efficacy)	Statistical significance
Monroe (52)	153	L 10 mg q.d. P	L > P	Yes
Monroe (57)	59	L 10 mg q.d. H 25 mg t.i.d. P	L = H L/H > P	N/A Yes
Hamerlinck et al. (58)	Not available	L 10 mg q.d. A 10 mg q.d. P	L = A L/A > P	N/A Yes
Sayag et al. (59)	168	L 10 mg q.d. Cet 10 mg q.d. P	L = Cet L/Cet > P	N/A Yes
Pleskow et al. (55)	172	L 10 mg q.d. T 60 mg b.i.d. P	L = T L/T > P	N/A Yes
Belaich et al. (54)	44	L 10 mg q.d. T 60 mg b.i.d. P	L > T L/T > P	No Yes
Bruttmann et al. (60)	172	L 10 mg q.d. T 60 mg b.i.d. P	L > T L/T > P	No Yes
Bernstein et al. (53)	30	L 10 mg q.d. P	L > P	Yes

Abbreviations: L = loratadine; H = hydroxyzine; T = terfenadine; A = astemizole; Cet = cetirizine; P = placebo; q.d. = daily; b.i.d. = twice daily; t.i.d. = three times daily; and N/A = not applicable. Modified from Ref. 49.

tion, and duration of effect. Overall, cetirizine was found to be superior to terfenadine, astemizole, loratadine, chlorpheniramine, and placebo. In a series of experiments, Charlesworth and colleagues (64, 65) studied cetirizine to determine its effects on mediators and cellular infiltration during the cutaneous LPR. Using a heat/suction skin blister model, they demonstrated the inhibitory effects of cetirizine on mediator (e.g., histamine, prostaglandin D_2) release and also the attenuation of inflammatory cell (e.g., eosinophils, basophils) migration into the skin chamber. Interestingly, Fadel et al. (66), in a double-blind, randomized, cross-over study, investigated the inhibitory effects of single doses of cetirizine and loratadine on histamine- and grass pollen-induced skin reactions. Cetirizine significantly inhibited histamine- and allergen-induced skin responses, as well as the accumulation of eosinophils measured by using a modified Rebuck window technique 24 h after allergen challenge. Loratadine, however, did not significantly inhibit the skin responses nor the eosinophil accumulation. Juhlin et al. (67) found that cetirizine significantly decreased the wheal and flare reaction in patients with chronic idiopathic and cold urticaria induced by intradermal administration of several stimuli, including histamine, compound 48/80, platelet-activating factor (PAF), and kallikrein. In a separate study in patients with chronic urticaria, Townley et al. (68) observed that cetirizine markedly reduced both the immediate wheal and flare reaction caused by PAF and the delayed skin reaction at 6 h. Clinically, the first indication of an unusual and potent activity was obtained by Juhlin et al. (61) when some patients with chronic idiopathic urticaria who were resistant to other H_1-antihistamines responded to cetirizine. The efficacy of cetirizine in the treatment of chronic idiopathic urticaria was also demonstrated in several double-blind, randomized comparative studies with other H_1-antihistamines (Table 5). Cetirizine had

Table 5 Cetirizine: Double-Blind Clinical Studies in Chronic Urticaria

Study	Patients (No.)	Dosage	Results (overall efficacy)	Statistical significance
Kalivas et al. (71)	211	Cet 5–20 mg q.d. H 25 mg t.i.d. P	Cet = H Cet/H > P	N/A Yes
Alomar et al. (70)	36	Cet 10 mg q.d. A 10 mg q.d.	Cet = A	N/A
Andri et al. (73)	30	Cet 10 mg q.d. T 60 mg b.i.d.	Cet > T Cet > T	Yes No
Dockx et al. (72)	115	Cet 10 mg q.d. T 60 mg b.i.d. P	Cet/T > P	Yes
Breneman et al. (69)	187	Cet 10 mg q.d. A 10 mg b.i.d. P	Cet > A Cet/A > P	No Yes

Abbreviations: Cet = cetirizine; H = hydroxyzine; T = terfenadine; A = astemizole; P = placebo; q.d. = daily; b.i.d. = twice daily; t.i.d. = three times daily; and N/A = not applicable. Modified from Ref. 62.

overall efficacy equivalent to that of astemizole and hydroxyzine (69–71), and was similar to or, in one study, significantly more effective than, terfenadine (72, 73). Results from a study by Paul et al. (74) in 500 patients with chronic idiopathic urticaria indicated that cetirizine 10 mg daily for 30 weeks significantly reduced the pruritus and the urticarial lesion scores to 25 and 17% of baseline levels, respectively.

Cetirizine has also shown some benefit in the treatment of urticaria triggered by delayed-pressure, light, or cold. In 11 patients with chronic urticaria, the occurrence of pressure-induced hives was significantly attenuated by pretreatment with cetirizine 10 mg 3 times daily for 1 week when compared with placebo (75). In six patients with solar urticaria, cetirizine 10 mg daily or terfenadine 60 mg twice daily for 2 days significantly increased the threshold for urticaria induced by light (76). In a study by Soter (19) involving 12 patients with cold urticaria and positive ice cube tests, cetirizine demonstrated an overall therapeutic efficacy in most subjects. After cetirizine treatment, the urticarial reaction to an ice cube disappeared in five patients and was markedly reduced in the remaining seven patients. In two patients whose baseline reaction time to cold stimulation was 30 s, urticaria was delayed until 6 min.

Acrivastine Acrivastine is a relatively nonsedating analogue of triprolidine, with a rapid onset of action (77, 78). Due to its short elimination half-life, the recommended dosing schedule is every 8 h. Acrivastine is effective in the management of chronic idiopathic urticaria, symptomatic dermatographism, cholinergic urticaria, idiopathic acquired cold urticaria, and also the itching of atopic dermatitis. Gibson et al. (79) have reported that acrivastine possesses a high level of efficacy in the treatment of chronic idiopathic urticaria when given at a dose of 8 mg 3 times daily. In double-blind cross-over trials in patients with chronic idiopathic urticaria or physical urticaria, treatment for 5 days with acrivastine 8 mg 3 times daily was of similar efficacy to clemastine 1 mg, hydroxyzine 20 mg, cyproheptadine 4 mg, or terfenadine 60 mg, each administered 3 times daily, and tended to be more effective than chlorpheniramine 4 mg 3 times daily (80–85) (Table 6). Leynadier et al. (86) have reported the comparative effects of acrivastine and terfenadine on skin reactivity to histamine. Both H₁-antihistamines were consistently superior to placebo as early as 1 h after administration. Acrivastine had an earlier onset of action than terfenadine and was effective in less than 30 min. The action of the two H₁-antihistamines was similar 2 h after drug administration, but terfenadine proved to be more active than acrivastine at later times (>4 h). Since mid-1994 acrivastine has been available in the United States but is only marketed for rhinitis in a combined formulation with pseudoephedrine.

Table 6   Acrivastine: Double-Blind Clinical Studies in Chronic Urticaria

Study	Patients (No.)	Dosage	Results (overall efficacy)	Statistical significance
Gale et al. (80)	20 (cross-over)	Ac 8 mg t.i.d. Chl 4 mg t.i.d.	Ac > Chl	No
Kobza-Black et al. (84)	10	Ac 8 mg t.i.d. H 20 mg t.i.d. P	Ac = H Ac/H > P	N/A Yes
Neittaanmaki et al. (85)	18	Ac 8 mg t.i.d. Cyp 4 mg t.i.d. P	Ac > Cyp Ac/Cyp > P	Yes Yes
Gibson et al. (79)	20 (cross-over)	Ac 4 mg t.i.d. Ac 8 mg t.i.d. P	Ac 8 mg > 4 mg Ac > P	No Yes
Salo et al. (82)	21 (cross-over)	Ac 8 mg t.i.d. H 20 mg b.i.d. P	Ac = H Ac/H > P	N/A Yes
Leyh et al. (81)	20 (cross-over)	Ac 4/8 mg t.i.d. Cl 1 g t.i.d. P	Ac = Cl Ac/Cl > P	N/A Yes
van Joost et al. (83)	56 (cross-over)	Ac 4/8 mg t.i.d. T 60 mg t.i.d. P	Ac = T Ac/T > P	N/A Yes

Abbreviations: Ac = acrivastine; T = terfenadine; Cyp = cyproheptadine; Cl = clemastine; Chl = chlorpheniramine; H = hydroxyzine; P = placebo; b.i.d. = twice daily; t.i.d. = three times daily; and N/A = not applicable. Modified from Ref. 77.

Ketotifen   Ketotifen has been reported to have multiple pharmacological properties (87), including antihistaminic, mast-cell stabilizing, calcium channel blocking, and antianaphylactic actions. It also increases the intracellular concentrations of cyclic adenosine monophosphate (cAMP) (88, 89).

Ketotifen is effective in the management of severe allergic skin disorders. In chronic urticaria its efficacy has been demonstrated in large double-blind comparative trials (90–93) (Table 7). In one study, ketotifen was significantly more effective than clemastine at decreasing the severity of the wheal and flare and the intensity of pruritus experienced by patients (90). A marked or moderate global improvement rating was recorded in almost 65% of ketotifen-treated patients. Similarly, in other studies, significant symptomatic improvement was achieved in 60 to 80% of subjects,

Table 7 Ketotifen: Double-Blind Clinical Studies in Chronic Urticaria

Study	Patients (No.)	Dosage	Results (overall efficacy)	Statistical significance
Kamide et al. (90)	305	K 1 mg b.i.d. Cl 1 mg b.i.d.	K > Cl	Yes
Phanuphak et al. (91)	30	K 1 mg b.i.d. P	K > P	Yes
Kuokkanen (93)	48	K 1 mg b.i.d. K 2 mg b.i.d. Cyp 1 mg b.i.d.	K 4 mg > 2 mg K 4 mg > Cyp K 2 mg = Cyp	Yes Yes N/A
St. Pierre et al. (92)	11 (cross-over)	K 1 mg b.i.d. P	K > P	Yes

Abbreviations: K = ketotifen; Cl = clemastine; Cyp = cyproheptadine; P = placebo; b.i.d. = twice daily; and N/A = not applicable. Modified from Ref. 88.

some of whom had complete resolution of lesions. Ketotifen was also clearly superior to placebo in a double-blind comparison, producing more than 50% reduction in symptom scores in 75% of treated subjects as compared with 14% of placebo recipients. Ketotifen has also been investigated in selected patients exhibiting severe physical urticarias (e.g., cold- and/ or exercise-induced urticarias). In one study, ketotifen significantly increased the minimum cold exposure time required to induce wheal formation in comparison with placebo (92). Moreover, Czerniawska-Mysik et al. (94) reported the duration of cold-induced wheal was decreased by ketotifen administered for a minimum of 2 weeks, with a greater than 50% reduction or even complete disappearance of the skin reaction in 11 of 14 patients after 4 weeks of treatment. Indeed, a pronounced rise in plasma histamine level was initially demonstrated in patients with severe physical urticaria by Huston et al. (95). The three patients studied by them were rechallenged after 1 month of treatment with ketotifen, at which time two patients exhibited minimal flare without wheals and one patient was asymptomatic. Two patients had reduced peak plasma histamine levels after physical challenge (18,250 and 10,650 ng/L reduced to 970 and 410 ng/L, respectively) while the plasma histamine level of the third patient remained within the normal range of <300 ng/L. In one report, we described four consecutive cases of severe cholinergic urticaria with systemic manifestations refractory to conventional therapy and subsequently treated with ketotifen (96). Plasma histamine levels which had increased from < 900 ng/L at baseline to postexercise peaks of 1700 to 9000 ng/L prior to treatment remained at preexercise levels in the three asymptomatic subjects, and the peak level in the patient with residual mild urticaria was significantly lower. In our experience with long-term administration of ketotifen in patients with chronic refractory urticaria, 82% had an average of almost 4 years of continuous significant positive clinical responses to ketotifen, and used fewer concomitant medications during ketotifen therapy (97). Furthermore, cessation or significant reduction of concurrent corticosteroid treatment was observed in all 16 of our patients who were initially on a concurrent steroid regimen. Benefit of ketotifen in patients with other histamine-mediated skin disorders such as atopic dermatitis and urticaria pigmentosa/systemic mastocytosis has also been demonstrated (88).

Azelastine Azelastine, like ketotifen, has a potent inhibitory activity against mediator production and release induced by antigen and nonantigen stimuli (98). Azelastine pretreatment reduces the histamine or allergen-induced wheal and flare reactions by 30 to 86% (99). Although data are limited, azelastine appears to have some beneficial effects on several histamine-mediated skin disorders (97–99). In small studies, azelastine therapy was judged to be moderately to markedly beneficial in about 85% of patients with eczema and in 54% of patients with atopic dermatitis. Recently, Friedman et al. (100) reported that azelastine was superior to chlorpheniramine at conventional doses in suppressing skin reactions to

Table 8 Guidelines for the Use of H_1-Antihistamines in the Treatment of Chronic Urticaria

---

Chronic Idiopathic Urticaria
1. Prescribe an H_1-antihistamine with little or no sedative effect (e.g., loratadine 10 mg/day or terfenadine 60 mg twice daily)
2. If treatment fails or is inadequate, either try another H_1-antihistamine from a different group; or use an H_1- and H_2-antihistamine combination
3. If this fails, consider the possibility of another type of treatment (e.g., corticosteroids)

Physical Urticaria
1. Persuade the patient to avoid the causative agent
2. If this is impossible, prescribe an H_1-antihistamine; special cases—cold urticaria: use cyproheptadine; cholinergic urticaria: use hydroxyzine

---

It is recommended that the above guidelines be implemented with other current acceptable preventive modalities.

intradermal histamine and morphine, and in suppressing pruritus in patients with mastocytosis.

Other Investigational H₁-Antihistamines   Preliminary evidence suggests the potential usefulness of other second-generation H₁-antihistamines such as ebastine and flezelastine in the treatment of allergic dermatoses (102, 103). Most of them are found to have special properties similar to the above second-generation H₁-antihistamines, including antiallergic activities.

V. SUMMARY

Unquestionably, the primary indication for H₁-antihistamines in dermatological diseases is urticaria, but they are also prescribed for the symptom of pruritus, regardless of its cause (104). As long as a precipitating cause or triggering factor is not identified, the treatment of chronic urticaria focuses on symptom relief. At present, H₁-antihistamines are the principal therapy for chronic urticaria. Sedative effects limit the daytime use of first-generation H₁-antihistamines. The introduction of the second-generation, low-sedating H₁-antihistamines has been a major step forward. One may start by prescribing an H₁-antihistamine, bearing in mind the information that there are considerable variations in individual responses to different H₁-antihistamines (Table 8). In general, the usefulness of many H₁-antihistamines, especially the older ones, is compromised by undesirable central nervous system side effects such as daytime sedation and anticholinergic side effects such as dry mouth. Newer H₁-antihistamines devoid of the above side effects are the drugs of first choice in this instance. In addition, these antihistamines have a long duration of action and can be conveniently taken once or twice a day. They are therefore a major therapeutic advance in the treatment of allergic and other histamine-mediated disorders.

There is strong evidence to support the view that human skin possesses H₂- as well as H₁-receptors (105). An H₂-antihistamine, when given together with an H₁-antihistamine, may bring about greater suppression of histamine-mediated reactions in the skin than the H₁-antihistamine given alone. If H₁-antihistamines, alone or in combination with H₂-antihistamines, have no effect in the treatment of severe or refractory urticaria, the addition of other types of pharmacotherapy is an appropriate consideration (e.g., corticosteroids or other immunosuppressive drugs; calcium channel blocking agents, or β_2-adrenergic agonists) (26,106).

REFERENCES

1. Kaplan AP, Horakawa Z, Katz SI. Assessment of tissue fluid histamine levels in patients with urticaria. J Allergy Clin Immunol 1978;61:350–354.
2. Phanuphak P, Schocket AL, Arroyave CM, Kohler PF. Skin histamine in chronic urticaria. J Allergy Clin Immunol 1980;65:371–375.
3. Natbony SF, Phillips M, Elias JM, Godfrey HP, Kaplan AP. Histologic studies of chronic idiopathic urticaria. J Allergy Clin Immunol 1983;71: 179–183.
4. Elias JM, Boss E, Kaplan AP. Studies of the cellular infiltrate of chronic idiopathic urticaria: prominence of T-lymphocytes, monocytes, and mast cells. J Allergy Clin Immunol 1986;78:914–918.
5. Bedard PM, Brunet C, Pelletier G, Hebert J. Increased compound 48/80 induced local histamine release from nonlesional skin of patients with chronic urticaria. J Allergy Clin Immunol 1986;78:1121–1125.
6. Cohen RW, Rosenstreich DL. Discrimination between urticaria-prone and other allergic patients by intradermal skin testing with codeine. J Allergy Clin Immunol 1986;77:802–807.
7. Krause LB, Shuster S. Enhanced wheal and flare response to histamine in chronic idiopathic urticaria. Br J Clin Pharmacol 1985;20:486–488.
8. Jacques P, Lavoie A, Bedard PM, Brunet C, Hebert J. Chronic idiopathic urticaria: profiles of skin mast cell histamine release during active disease and remission. J Allergy Clin Immunol 1992;89:1139–1143.
9. Grattan CEH, Wallington TB, Warin RP, Kennedy CTC, Bradfield JW. A serological mediator in chronic idiopathic urticaria: a clinical, immunological, and histological evaluation. Br J Dermatol 1986;114:583–590.
10. Alam R, Rozniecki J. A mononuclear cell-derived histamine releasing factor in asthmatic patients. II. Activity in vivo. Allergy 1985;40:124–129.
11. Claveau J, Lavoie A, Brunet C, Bedard PM, Hebert J. Chronic idiopathic urticaria: possible contribution of histamine releasing factor to pathogenesis. J Allergy Clin Immunol 1993;92:132–137.
12. Alam R, Forsythe PA, Stafford S, Kormos C, Kenamore K, Lett-Brown MA, Grant JA. Monocyte chemotactic and activating factor (MCAF) is a potent histamine-releasing factor for basophils. J Clin Invest 1992;89:723–8.
13. Alam R, Forsythe PA, Stafford S, Lett-Brown MA, Grant JA. Macrophage inflammatory protein-1α activates basophils and mast cells. J Exp Med 1992;176:781–786.
14. Baggiolini M, Dahinden CA. CC chemokines in allergic inflammation. Immunol Today 1994;15:127–133.
15. Hide M, Francis DM, Grattan CEH, Hakimi J, Kochan JP, Greaves MW. Autoantibodies against the high-affinity IgE receptor as a cause of histamine release in chronic urticaria. N Engl J Med 1993;328:1599–1604.
16. Simons FER, Simons KJ. Antihistamines. In: Allergy: Principles and Practice (Middleton E, Reed CE, Ellis EF, Adkinson NF, Yunginger JW, Busse WW, eds.), 4th ed. St. Louis: CV Mosby, 1993:856–892.

17. Simons FER, Simons KJ. The pharmacology and use of H₁-receptor-antago-
     nist drugs. N Engl J Med 1994;330:1663–1670.
18. Rimmer SJ, Church MK. The pharmacology and mechanisms of action of
     histamine H₁-antagonists. Clin Exp Allergy 1990;20:3–17.
19. Soter NA. Treatment of urticaria and angioedema: low-sedating H₁-type
     antihistamines. J Am Acad Dermatol 1991;24:1084–1087.
20. Galant SP, Bullock J, Wong D, Maibach HI. The inhibitory effect of antial-
     lergy drugs on allergen- and histamine-induced wheal and flare response.
     J Allergy 1973;51:11–21.
21. Simons FER, Simons KJ, Frith EM. The pharmacokinetics and antihista-
     minic effects of the H₁-receptor antagonist hydroxyzine. J Allergy Clin Im-
     munol 1984;73:69–75.
22. Rhoades RB, Leifer KN, Cohan R, Wittig HJ. Suppression of histamine-
     induced pruritus by three antihistaminic drugs. J Allergy Clin Immunol 1975;
     55:180–185.
23. Moore-Robinson M, Warin RP. Some clinical aspects of cholinergic urti-
     caria. Br J Dermatol 1968;80:794–799.
24. Wanderer AA, Ellis EF. Treatment of cold urticaria with cyproheptadine.
     J Allergy Clin Immunol 1971;48:366–371.
25. Wanderer AA. Cold urticaria syndromes: historical background, diagnostic
     classification, clinical and laboratory characteristics, pathogenesis, and
     management. J Allergy Clin Immunol 1990;85:965–981.
26. Kaplan AP. Urticaria and angioedema. In: Allergy (Kaplan AP, ed.). New
     York; Churchill Livingstone, 1985;439–471.
27. Richelson E. Antimuscarinic and other receptor-blocking properties of anti-
     depressants. Mayo Clin Proc 1983;58:40–46.
28. Grandel KE, Farr RS, Wanderer AA, Eisenstadt TC, Wasserman SI. Asso-
     ciation of platelet-activating factor with primary acquired cold urticaria. N
     Engl J Med 1985;313:405–409.
29. Huther KJ, Renftle G, Barraud N, Burke JT, Koch-Weser J. Inhibitory
     activity of terfenadine on histamine-induced skin wheals in man. Eur J Clin
     Pharmacol 1977;12:195–199.
30. Van Landeghem VH, Burke JT, Thebault J. The use of a human bioassay
     in determining the bioequivalence of two formulations of the antihistamine
     terfenadine. Clin Pharmacol Ther 1980;27:290–291.
31. Akagi M, Mio M, Miyoshi K, Tasaka K. Antiallergic effects of terfenadine
     on immediate type hypersensitivity reactions. Immunopharmacol Immuno-
     toxicol 1987;9:257–259.
32. McTavish D, Goa KL, Ferrill M. Terfenadine: an updated review of its
     pharmacological properties and therapeutic efficacy. Drugs 1990;39:
     552–574.
33. Ormerod AD, Baker R, Watt J, White MI. Terfenadine and brompheni-
     ramine maleate in urticaria and dermographism. Dermatologica 1986;173:5–8.
34. Boggs PB, Ellis CN, Grossman J, Washburne WF, Gupta AK, Ball R,
     Shulan D. Double-blind, placebo-controlled study of terfenadine and hy-

droxyzine in patients with chronic idiopathic urticaria. Ann Allergy 1989; 63:616–620.

35. Go MJ, White J, Arendt C, Bernheim J. Double-blind, placebo-controlled comparison of cetirizine and terfenadine in chronic idiopathic urticaria. Acta Therapeutica 1989;15:77–85.

36. Grant JA, Bernstein DI, Buckley CE, Chu T, Fox RW et al. Double-blind comparison of terfenadine, chlorpheniramine, and placebo in the treatment of chronic idiopathic urticaria. J Allergy Clin Immunol 1988;81:574–579.

37. Fredriksson T, Hersle K, Hjorth N, Mobacken H, Persson T. Terfenadine in chronic urticaria: a comparison with clemastine and placebo. Cutis 1986; 38:128–130.

38. Paul E, Bodeker RH. Comparative study of astemizole and terfenadine in the treatment of chronic idiopathic urticaria. Ann Allergy 1989;62:318–320.

39. Paul E, Bodeker RH. Treatment of chronic urticaria with terfenadine and ranitidine: a randomized double-blind study in 45 patients. Eur J Clin Pharmacol 1986;31:277–280.

40. Cerio R, Lessof MH. Treatment of chronic idiopathic urticaria with terfenadine. Clin Allergy 1984;14:139–141.

41. Richards DM, Brogden RN, Heel RC, Speight TM, Avery GS. Astemizole: a review of its pharmacodynamic properties and therapeutic efficacy. Drugs 1984;28:38–61.

42. Bateman DN, Chapman PH, Rawlins MD. The effects of astemizole on histamine-induced wheal and flare. Eur J Clin Pharmacol 1983;25:547–551.

43. Monroe EW. Chronic urticaria: a review of nonsedating H_1-antihistamines in treatment. J Am Acad Dermatol 1988;19:842–849.

44. Cainelli T, Seidenari S, Valsecchi R, Mosca M. Double-blind comparison of astemizole and terfenadine in the treatment of chronic urticaria. Pharmatherapeutica 1986;4:679–686.

45. Mobacken H, Faergeman J, Gisslen H. Astemizole in chronic idiopathic urticaria: a comparison with chlorpheniramine [Abstract]. Ann Allergy 1985;55:254.

46. Sussman G, Jancelewicz Z. Controlled trial of H_1-antagonists in the treatment of chronic idiopathic urticaria. Ann Allergy 1991;67:433–439.

47. Knight A. Astemizole in the management of chronic urticaria. In: Astemizole: A New Non-Sedative, Long-Acting H_1-Antagonist. Oxford: Medical Education Services, 1984:95–100.

48. Roman IJ, Danzig MR. Loratadine: a review of recent findings in pharmacology, pharmacokinetics, efficacy, and safety, with a look at its use in combination with pseudoephedrine. Clin Rev Allergy 1993;11:89–110.

49. Clissold SP, Sorkin EM, Goa KL. Loratadine: a preliminary review of its pharmacodynamic properties and therapeutic efficacy. Drugs 1989;37: 42–57.

50. Batenhorst RL, Batenhorst AS, Graves DA, Foster TS, Kung M. Pharmacologic evaluation of loratadine (SCH29851), chlorpheniramine and placebo. Eur J Clin Pharmacol 1986;31:247–250.

51. Hammarlund A, Olsson P, Pipkorn U. Blood flow in histamine- and allergen-induced wheal and flare responses, effects of an H_1 antagonist, alpha-adrenoceptor agonist and a topical glucocorticoid. Allergy 1990;45:64–70.

52. Monroe EW, Fox RW, Green AW, Izuno GT, Bernstein DI, Pleskow WW, Willis I, Brigante JR. Efficacy and safety of loratadine in the management of idiopathic chronic urticaria. J Am Acad Dermatol 1988;19:138–139.

53. Bernstein IL, Bernstein DI. The efficacy and safety of loratadine in the management of chronic idiopathic urticaria [Abstract]. J Allergy Clin Immunol 1988;21:211.

54. Belaich S, Bruttmann G, DeGreef H, Lachapelle JM, Paul E, Pedrali P, Tennstedt D. Comparative effects of loratadine and terfenadine in the treatment of chronic idiopathic urticaria. Ann Allergy 1990;64:191–194.

55. Pleskow WW. Studies of a new antihistamine in the management of perennial rhinitis and idiopathic chronic urticaria. 44th Annual Congress, American College of Allergists, Boston, Nov, 1987.

56. Soto LD, Falabella R, Hojyo T, Muzquiz D. Multicenter study comparing two H_1 antihistamines in the management of chronic idiopathic urticaria. Dermatol Rev Mex 1987;31:9–14.

57. Monroe EW. Relative efficacy and safety of loratadine, hydroxyzine, and placebo in chronic idiopathic urticaria and atopic dermatitis. Clin Ther 1992; 14:17–21.

58. Hamerlinck FFV, Boyden B, Oel HD, Niordson AM, Avrach W, Ottevanger V, Danzig MR. Presented at European Academy of Dermatology and Venereology, Athens, Oct, 1991.

59. Sayag J, Guillet G, Leroy D, Wessel F, Guillot B, Moulin G, Bonerandi JJ, Amblard P, Weber M, Czarlewski W, LeHalpere A, Cougnard J. Evaluation of the efficacy and safety of loratadine in chronic idiopathic urticaria and atopic dermatitis [Abstract]. Clin Exp Allergy 1990;20:55.

60. Bruttmann G, Charpin D, Germouty J, Horak F, Kunkel G, Wittmann G. Comparative effects of loratadine and terfenadine in the treatment of chronic idiopathic urticaria. J Allergy Clin Immunol 1989;83:411–416.

61. Juhlin L, Arendt C. Treatment of chronic urticaria with cetirizine dihydrochloride, a nonsedating antihistamine. Br J Dermatol 1988;119:67–72.

62. Spencer CM, Faulds D, Peters DH. Cetirizine: a reappraisal of its pharmacological properties and therapeutic use in selected allergic disorders. Drugs 1993;46:1055–1080.

63. Simons FER, McMillan JL, Simons KJ. A double-blind, single-dose, crossover comparison of cetirizine, terfenadine, loratadine, astemizole, and chlorpheniramine versus placebo: suppressive effects on histamine-induced wheals and flares during 24 hours in normal subjects. J Allergy Clin Immunol 1990;86:540–547.

64. Charlesworth EN, Kagey-Sobotka A, Norman PS, Lichtenstein LM. Effect of cetirizine on mast cell-mediator release and cellular traffic during the cutaneous late-phase reaction. J Allergy Clin Immunol 1989;83:905–912.

65. Charlesworth EN, Massey WA, Kagey-Sobotka A, Norman PS, Lich-

tenstein LM. Effect of H_1-receptor blockade on the early and late response to cutaneous allergen challenge. J Pharmacol Exp Ther 1992;262:964–970.

66. Fadel R, Herpin-Richard N, Dufresne F, Rihoux JP. Pharmacological modulation by cetirizine and loratadine of antigen- and histamine-induced skin wheals and flares, and late accumulation of eosinophils. J Int Med Res 1990; 18:366–371.

67. Juhlin L, De Vos C, Rihoux JP. Inhibiting effect of cetirizine on histamine-induced and 48/80-induced wheals and flares, experimental dermographism, and cold-induced urticaria. J Allergy Clin Immunol 1987;80:599–602.

68. Townley RG, Okada C. Use of cetirizine to investigate non-H_1 effects of second-generation antihistamines. Ann Allergy 1992;68:190–196.

69. Breneman D, Bergstresser P, Boggs P, Bronsky E, Bruce S et al. A multisite comparative study of the efficacy and safety of cetirizine versus astemizole and placebo in the treatment of chronic idiopathic urticaria [Abstract]. J Allergy Clin Immunol 1992;89:249.

70. Alomar A, de la Cuadra, Fernandez J. Cetirizine versus astemizole in the treatment of chronic idiopathic urticaria. J Int Med Res 1990;18:358–365.

71. Kalivas J, Breneman D, Tharp M, Bruce S, Bigby M et al. Urticaria: clinical efficacy of cetirizine in comparison with hydroxyzine and placebo. J Allergy Clin Immunol 1990;86:1014–1018.

72. Dockx P, Lambert J, Arendt C, Darte D. Comparison of cetirizine and terfenadine in idiopathic urticaria. Presented at the Cetirizine Symposium Proceedings, Brussels, March 1987.

73. Andri L, Senna GE, Betteli C, Givanni S, Andri G, Lombardi C, Mezzelani P. A comparison of the efficacy of cetirizine and terfenadine. A double-blind, controlled study of chronic idiopathic urticaria. Allergy 1993;48:358–365.

74. Paul E, Paul C. Longterm therapy with cetirizine in chronic urticaria. Results of a multicenter study. Allergologie 1993;16:56–60.

75. Kontou-Fili K, Maniatakou G, Demaka P, Gonianakis M, Palaiologos G, Aroni K. Therapeutic effects of cetirizine in delayed pressure urticaria: clinicopathologic findings. J Am Acad Dermatol 1991;24:1090–1093.

76. Bilsland D, Ferguson J. A comparison of cetirizine and terfenadine in the management of solar urticaria. Photodermatol Photoimmunol Photomed 1991;8:62–64.

77. Brogden RN, McTavish D. Acrivastine: a review of its pharmacological properties and therapeutic efficacy in allergic rhinitis, urticaria and related disorders. Drugs 1991;41:927–940.

78. Gibson JR, Manna VK, Salisbury J. Acrivastine: a review of its dermato-pharmacology and clinical activity. J Int Med Res 1989;17:28B–34B.

79. Gibson JR, Harvey SG, Barth JH, Moss MY, Burke CA. An assessment of the novel antihistamine BW 825C in the treatment of chronic idiopathic urticaria. Dermatologica 1984;169:179–183.

80. Gale AE, Harvey SG, Calthrop JG, Gibson JR. A comparison of acrivastine versus chlorpheniramine in the treatment of chronic idiopathic urticaria. J Int Med Res 1989;17:25B–27B.

81. Leyh F, Harvey SG, Gibson JR, Manna VK. A comparison of acrivastine versus clemastine and placebo in the treatment of patients with chronic idiopathic urticaria. J Int Med Res 1989;17:22B–24B.

82. Salo OP, Harvey SG, Calthrop JG, Gibson JR. A comparison of acrivastine versus hydroxyzine and placebo in the treatment of chronic idiopathic urticaria. J Int Med Res 1989;17:18B–21B.

83. van Joost T, Blog FB, Westerhof W, Jansen FC, Starink TM, den Boer MS, Kuneman JJ, Harvey SG, Gibson JR. A comparison of acrivastine versus terfenadine and placebo in the treatment of chronic idiopathic urticaria. J Int Med Res 1989;17:14B–17B.

84. Kobza-Black A, Aboobaker J, Gibson JR, Harvey SG. Acrivastine versus hydroxyzine in the treatment of cholinergic urticaria. A placebo-controlled study. Acta Derm Venereol 1988;68:541–544.

85. Neittaanmaki H, Fraki JE, Gibson JR. Comparison of the new antihistamine acrivastine (BW825C) versus cyproheptadine in the treatment of idiopathic cold urticaria. Dermatologica 1988;177:98–103.

86. Leynadier F, Murrieta M, Dry J, Colin JN, Gillotin C, Steru D. Effects of acrivastine and terfenadine on skin reactivity to histamine. Ann Allergy 1994;72:520–524.

87. Simons FER, Luciuk GH, Becker AB, Gillespie CA. Ketotifen: a new drug for prophylaxis of asthma in children. Ann Allergy 1982;48:145–150.

88. Grant SM, Goa KL, Fitton A, Sorkin EM. Ketotifen: a review of its pharmacodynamic and pharmacokinetic properties, and therapeutic use in asthma and allergic disorders. Drugs 1990;40:412–448.

89. Morley J, Page CP, Mazzoni L, et al. Effects of ketotifen upon responses to platelet activating factor: a basis for asthma prophylaxis. Ann Allergy 1986;56:335–340.

90. Kamide R, Niimura M, Ueda H, Imamura S, Yamamoto S, Yoshida H, Kukita A. Clinical evaluation of ketotifen for chronic urticaria: multicenter double-blind comparative study with clemastine. Ann Allergy 1989;62:322–325.

91. Phanuphak P, Locharernkul C. Double-blind, placebo-controlled study of ketotifen in chronic urticaria [Abstract]. J Allergy Clin Immunol 1986;77:187.

92. St. Pierre JP, Kobric M, Rackham A. Effect of ketotifen treatment on cold-induced urticaria. Ann Allergy 1985;55:840–843.

93. Kuokkanen K. Comparison of a new antihistamine HC 20-511 with cyproheptadine (Periactin) in chronic urticaria. Acta Allergol 1977;32:316–320.

94. Czerniawska-Mysik G, Woloszynski J, Adamek-Guzik T, Koterba A, Prochowska K. Proceedings, annual meeting of the European Academy of Allergology and Clinical Immunology, Vol. 2, pp. 828–830, Clermont-Ferrand, Sept, 1981.

95. Huston DP, Bressler RB, Kaliner M, Sowell LK, Baylor MW. Prevention of mast-cell degranulation by ketotifen in patients with physical urticarias. Ann Intern Med 1986;104:507–510.

96. McClean SP, Arreaza EE, Lett-Brown MA, Grant JA. Refractory choliner-

gic urticaria successfully treated with ketotifen. J Allergy Clin Immunol 1989;83:738–741.

97.   Sim TC, Bush RK, McClean SP, Grant JA. Long-term efficacy and safety of ketotifen in the treatment of chronic refractory urticaria [Abstract]. Ann Allergy 1991;66:68.

98.   McTavish D, Sorkin EM. Azelastine: a review of its pharmacodynamic and pharmacokinetic properties, and therapeutic potential. Drugs 1989;38: 778–800.

99.   Hinogaki Y, Hidano N. Efficacy of azelastine in atopic dermatitis, eczema and contact dermatitis. Med Consult N Remed 1988;25:581–586.

100. Yoshida M, Tezuka T. Clinical efficacy of azelastine in eczema and dermatitis. J N Remed Clin 1988;37:171–174.

101. Friedman BS, Santiago ML, Berkebile C, Metcalfe DD. Comparison of azelastine and chlorpheniramine in the treatment of mastocytosis. J Allergy Clin Immunol 1993;92:520–526.

102. Rivest J, Despontin K, Ghys L, Rihoux JP, Lachapelle JM. Pharmacological modulation by cetirizine and ebastine of the cutaneous reactivity to histamine. Dermatologica 1991;183:208–211.

103. Chand N, Sofia RD, Szelenyi I, Schmidt J. Flezelastine: inhibition of synthesis and release of chemical mediators [Abstract]. Am J Respir Crit Care Med 1994;149:A770.

104. Advenier C, Queille-Roussel C. Rational use of antihistamines in allergic dermatological conditions. Drugs 1989;38:634–644.

105. Theoharides TC. Histamine-2 (H_2)-receptor antagonists in the treatment of urticaria. Drugs 1989;37:345–355.

106. Champion RH. A practical approach to the urticarial syndromes. Clin Exp Allergy 1990;20:221–224.

11

Anaphylaxis and Histamine Antagonists

Stephen L. Winbery
Methodist Central Hospital Teaching Practice,
Memphis, Tennessee

Philip L. Lieberman
University of Tennessee, Cordova, Tennessee

I. INTRODUCTION

The incidence of anaphylaxis seems to be increasing. Perhaps this rise is due to increased environmental and medical exposure to antigens such as antibiotics and other drugs, latex, and food additives. A review of the incidence of anaphylaxis is shown in Tables 1 and 2.

In its strictest sense, the term anaphylaxis refers to an immediate systemic reaction mediated by IgE antibody-antigen-induced degranulation of mast cells and basophils. The classic allergen is a multivalent, physiological cross-link between receptor-bound IgE molecules on the surface of mast cells and basophils.

The term "anaphylactoid reaction" refers to an event that is clinically indistinguishable from anaphylaxis, but not due to IgE antibody-antigen-induced mast-cell and basophil degranulation. A pathophysiological classification of anaphylaxis and anaphylactoid reactions is described in Table 3. Both clinical events may be caused by the same vasoactive and inflammatory mediators, including histamine, which probably plays a major role.

II. MEDIATORS OF ANAPHYLAXIS

Anaphylaxis and many anaphylactoid reactions are initiated by the release of preformed chemical mediators of inflammation from granules in mast

Table 1 Approximate Overall Incidence of Anaphylaxis

Overall incidence	Ref.
41 fatal cases reported 1895–1923	(1)
68 fatal cases reported 1923–1935	(2)
9 cases per 20,064 admissions 1990	(3)
140 cases, 41 fatal, in Great Britain 1966–1975	(4)
6 reactions, 0.87 fatalities per 10,000 patients in Boston Collaborative Drug Surveillance Program in 1973	(5)
12 cases in a series of 32,812 with 2 deaths in 1977	(6)
4 cases per 10 million population with 2 deaths in Province of Ontario	(7)
As high as 1 in every 3000 patients, more than 500 deaths annually in USA	(8)

cells and basophils. Besides preformed products, activated mast cells also synthesize and release mediators including arachidonic acid metabolites and platelet activating factor (PAF). Leukotrienes are synthesized by mast cells after stimulation. Slow-reacting substance of anaphylaxis (SRS-A) is a mixture of leukotrienes, LTB4, LTC4, LTD4, and LTE4. The mixture is chemotactic and can call forth other inflammatory cells. The reaction is amplified as chemotactic agents recruit inflammatory cells and their mediators.

Histamine is only one of many mediators involved in the pathophysiology of anaphylaxis (Table 4). The multiplicity of mediators offers an explanation as to why antihistamines alone do not control anaphylactic episodes. In addition to its direct effects, histamine seems to have a regulatory role in the subsequent release and production of mediators after stimulation of mast cells by antigen, since antihistamines can block allergen-induced increases in histamine and several other mediators. Also, there appears to be an interdependence of PAF and histamine in anaphylactic reactions (24).

A. The Late Phase of the Anaphylactic Response

After resolution of an anaphylactic event, symptoms can reappear 3 to 12 h later. The cause for this recurrence has not been established, but it

Table 2 Approximate Incidence of Anaphylaxis to
Selected Agents

Incidence for specific agents	Ref.
Penicillin	
1–5 reactions per 10,000 patient treatments	(9)
Cephalosporins	
3–7% of patients with a history of penicillin allergy and a positive skin test	(10)
Radiocontrast media	
Estimated 2667 reactions to hyperosmolar agents, with 500 deaths	(11)
Severe reactions in 0.04% of patients receiving lower osmolar agents	(12)
Drugs used during perioperative period	
10 of 336 anaphylaxis to drugs—penicillin, analgesics, nonsteroidal anti-inflammatory drugs	(13)
Australia—frequency between 1/5000 and 1/25,000 with mortality of 3.4%	(14)
France—1 reaction per 4500 cases	(15)
MMR	
28 patients reported since 1980	(16)
Insect stings	
0.4 to 3% of population is sensitive	(17)
Estimated 25 to 50 deaths per year	(18)
39 of 138 patients with a previous history upon rechallenge	(19)
Aspirin and nonsteroidal anti-inflammatory agents	
Anaphylactoid reactions occur in as many as 0.9% of patients taking aspirin.	(20)
35 of 51,797 patients taking NSAIDS experienced shock	(21)
Most common agent causing anaphylaxis in a series of 267 adult cases	(22)

is reasonable to assume that the late response may be due to recruitment of cells and their mediators by chemotactic factors released from mast cells and/or basophils in the initial reaction. Chemotactic factors promote adhesion, diapedesis, migration, and activation of inflammatory cells (25).

Table 3 Pathophysiological Classification of Anaphylaxis and Anaphylactoid Reactions

---

I. Anaphylaxis–IgE-mediated reaction
  A. Food
    1. Peanut
    2. Treenuts
    3. Fish
    4. Seafood (shellfish)
    5. Egg
    6. Milk
    7. Grains
  B. Drugs
    1. Penicillins
    2. Cephalosporins
    3. Sulfonamides
  C. Venoms
    1. Hymenoptera
    2. Fire ant
    3. Snakes
  D. Human proteins
    1. Insulin
    2. Corticotropin
    3. Vasopressin
    4. Semen and seminal proteins
  E. Exercise (some cases)
II. Anaphylactoid
  A. Direct release of mediators from mast cells and basophils
    1. Drugs
      a. opiates
      b. paralytic agents
      c. vancomycin
      e. fluorescein
      f. dextran
    2. Hyperosmolar solutions
    3. Idiopathic
    4. Exercise
    5. Physical factors such as cold, sunlight
  B. Disturbances in arachidonic acid metabolism
    1. Aspirin
    2. Other nonsteroidal anti-inflammatory drugs
  C. Immune aggregates
    1. Gamma globulin
    2. IgG- anti-IgA
  D. Cytotoxic
    Transfusion reactions to cellular elements
  E. Miscellaneous and multimediator activity
    1. Nonantigen-antibody-mediated complement activation
      a. radiocontrast material

Table 3 (*Continued*)

> b. protamine reactions (some cases)
> c. dialysis membranes
> 2. Activation of contact system
> a. dialysis membranes
> c. radiocontrast material

B. Amplification of the Anaphylactic Response

Amplification of the acute anaphylactic event occurs through recruitment of other inflammatory pathways. Anaphylaxis can result in activation and recruitment of inflammatory cells, neuropeptides, the contact (kallikrein) system, complement, clotting, and the clot lysis systems.

Neuropeptides can be detected in nasal secretions of patients with allergic symptoms. Substance P and calcitonin gene-related peptide (CGRP) are released by sensory nerves while vasoactive intestinal peptide (VIP) is released from parasympathetic nerve endings. Both sets of nerves are probably stimulated in a nonspecific manner by inflammation associated with allergic reactions (26). Substance P causes increases in vascular permeability and blood flow, which are mediated in part by histamine acting on H_1-receptors (27).

Mast-cell kininogenase and basophil kallikrein can activate the contact system (28, 29). Tryptase has kallikreinlike activity and can activate complement and cleave fibrinogen (30).

Circumstantial evidence for activation of complement as part of anaphylactic reactions comes from the relationship between decreased levels of C3a and severity of the reaction in eight patients rechallenged with wasp stings (19).

C. Modulation by Histamine

Histamine may have a role in modulating anaphylaxis. Histamine binds to H_2-receptors on basophils and mast cells leading to inhibition of further histamine release by a cAMP-dependent mechanism (31–33). H_2-receptor agonists cause a dose-dependent inhibition of antigen-induced histamine release from sensitized guinea pig hearts. This suggests that histamine

Table 4 Mediators of Anaphylaxis

Mediator	Action	Signs and symptoms
Histamine	Smooth muscle relaxation and contraction	Vasodilation, hypotension, bronchospasm, coronary artery spasm, increased gastrointestinal (GI) motility, diarrhea
	Increase in capillary permeability	Angioedema, urticaria, flush
	Positive inotropy	Increased contractility
	Positive chronotropy	Tachycardia
	Exocrine gland secretion	Increased respiratory secretions
	Irritation of sensory nerves	Pruritus
	Stimulation of release or synthesis of most other mediators	Potentiation of reaction, late-phase reaction
Heparin	Anticoagulant	Coagulopathy, could have anti-inflammatory effect, possible complement inhibition
Prostaglandins (D2 and F2α)	Smooth muscle relaxation and contraction, peripheral vasodilation	Vasodilation, flush, hypotension, bronchospasm, coronary artery spasm, increased GI motility, diarrhea
	Mucus secretion	Rhinorrhea, increased respiratory secretions
	Enhanced basophil mediator release	Potentiation
Leukotrienes (B4, C4, D4)	Chemotaxis	Potentiation
	Contraction of airway smooth muscle	Bronchospasm
	Increased vascular permeability	Vasodilation, efflux of inflammatory cells
	Negative inotropic effect	Hypotension, myocardial depression

Table 4 (*Continued*)

Mediator	Action	Signs and symptoms
	Goblet and mucosal cell secretion	Increased respiratory secretions
Platelet activating factor (PAF)	Contraction of airway smooth muscle	Bronchospasm
	Increased vascular permeability	Hypotension, flush, urticaria, angioedema
	Neutrophil aggregation	Inflammation
	Platelet aggregation	Platelet activation
Lymphokines (IL-3, IL-5, TNF)	Adherence, degranulation, and chemotaxis of inflammatory cells	Potentiation, inflammation
Eosinophilic chemotactic factor (ECF)	Eosinophil chemotaxis	Amplification, inflammation
Neutrophilic chemotactic factor (NCF)	Neutrophil chemotaxis	Amplification, inflammation
Neutral proteases	Proteolysis	Inflammation
Tryptase	May cleave C3 to activate complement	Amplification
Major basic protein (MBP)	Stimulates histamine release	Amplification
Arachidonic acid stimulating factors	Production of lipoxygenase and cyclooxygenase products	Amplification
Neuropeptides, substance P, and vasoactive intestinal protein (VIP)	Peripheral nerve stimulation	Itching, pain
	Vasodilation	Flush, hypotension
	Possible mast-cell degranulation	Amplification

feedback inhibition of antigenic histamine release via H_2-receptors occurs (34). Theoretically, if histamine modulates the anaphylactic response via H_2-receptors, then H_2-receptor antagonists could have a detrimental effect upon the treatment of anaphylaxis. Practically, this does not appear to occur in the clinical setting and, as discussed later, H_2-antagonists may be of benefit in the treatment or prevention of anaphylactic episodes.

III. HISTAMINE AND ANAPHYLAXIS

Histamine is an endogenous imidazole compound that is synthesized, stored, and released primarily by mast cells and circulating basophils. In 1949, MacIntosh and Paton showed that histamine can be released by basic substances such as diamines, diamides, and diquanides. Histamine is the mediator responsible for many of the symptoms early in the course of anaphylaxis and may be indirectly responsible for many symptoms during the late response. Most of the signs and symptoms of anaphylaxis can be produced by histamine infusion, and many can be blocked by H_1 and/or H_2-histamine antagonists (36). Histamine infusion produces increased vascular permeability of the postcapillary venules, vasodilation, decreased total peripheral resistance and hence blood pressure, myocardial depression, and warm extremities.

The considerable species variation in response to histamine has led to confusion in interpretation of experimental results. For instance, guinea pigs and rabbits are extremely sensitive to both the respiratory and cardiovascular effects of histamine and require a much lower lethal dose than the relatively insensitive rat. There is strong species variability in the histamine-releasing ability of the anaphylatoxins C3a and C5a (33). In most animals, histamine-induced increases in microvascular permeability are mediated by H_1-receptors, except in the hamster where these effects are mediated by H_2-receptors (37). The human response tends to be somewhat intermediate, and sensitivity varies from person to person.

Serum histamine levels of less than 1 ng/ml are associated with metallic taste, headache, and nasal congestion. Higher levels cause generalized skin reactions, gastrointestinal stimulation, flushing, tachycardia, cardiac arrhythmias, and hypotension. Life-threatening hypotension, ventricular fibrillation, bronchospasm, and cardiopulmonary arrest generally result from histamine levels above 12 ng/ml (38).

It has been suggested that the clinical manifestations of histamine release during anesthesia are different from those of classic systemic anaphylaxis due to the effects of anesthesia and other drugs used perioperatively. Because of this, increased histamine levels may have been neglected as a contributing factor to perioperative complications. Elevated levels of histamine may produce untoward effects in the perioperative patient even though the classic symptoms of anaphylaxis are not always present. Elevated levels of plasma histamine are associated with arrhythmias, increased thrombosis, stress ulceration, increased intrapulmonary shunt, increased risk of death from shock, and even adult respiratory distress syndrome. Plasma histamine concentrations ranging from 0.2 ng/ml to 1.0 ng/ml must be accepted as elevated, and their significance cannot be dismissed because of the absence of the classical signs and symptoms of anaphylaxis during anesthesia (39).

A. Histamine Receptors

Histamine stimulates H_1-, H_2-, and H_3-receptors to produce its characteristic responses (Table 5). Many of the actions of histamine are local and tissue-dependent. H_1- and H_2-receptors are better characterized than the H_3-receptor. In many instances, H_1-receptor stimulation results in the breakdown of the second messenger inositol phosphate and calcium mobilization while H_2-mediated responses are mostly due to activation of adenylate cyclase. However, there is considerable "cross-talk" between second-messenger systems (40). The mechanisms responsible for many actions of histamine such as vascular smooth muscle dilation are not understood and may involve release of endothelial factors. Histamine can stimulate endothelial cells to produce nitric oxide, a smooth muscle relax-

Table 5 Actions of Histamine Related to Anaphylaxis

H_1-receptor-mediated actions	H_2-receptor-mediated actions
Elevation of cyclic GMP	Increased intracellular cyclic AMP
Smooth-muscle contraction	Increased gastric acid and pepsin
Increased endothelial permeability	secretion
Stimulation of nerve endings	Decreased fibrillation threshold of
Pruritus	cardiac muscle
Vagal irritant receptors (cough and	Positive inotropy
bronchospasm)	Positive chronotropy
Increased viscosity of mucus	Increased mucus secretion
Vasodilation	
Direct	H_1- plus H_2-mediated actions
Production of endothelium-derived	
relaxing factors	Vasodilation and flush
Release of neuropeptides from nerve	Increased vascular permeability
endings	Hypotension
Epinephrine secretion from adrenal	Eosinophil chemotaxis
medulla	H_3-receptor-mediated actions
Increased rate of depolarization of the	
sinoatrial node	
Slowed rate of atrioventricular	Modulation of peripheral and
conduction	bronchial neurotransmission
	Possible bronchospasm

ant (41, 42). In guinea pig lung tissue, histamine binds to H_1-receptors to produce a phospholipase-C-dependent calcium mobilization that stimulates the conversion of L-arginine to nitric oxide. Nitric oxide activates guanylate cyclase leading to production of cyclic GMP (43).

H_3-receptors have been located on presynaptic nerve endings both peripherally and in the central nervous system. H_3-receptors control acetylcholine release at the level of the myenteric plexus in the guinea pig intestine (44), modify peripheral neuropeptide release, modulate airway reactivity (45), and control histamine synthesis in brain and lung tissue. A role for H_3-histamine receptors in anaphylaxis has not been established.

B. Anaphylaxis-Related Actions of Histamine

1. Vascular System

Histamine exerts a complex action on the vascular system. The predominant action of histamine is a net vasodilation from smooth muscle relaxation in arterioles, precapillary sphincters, and venules. This leads to a decrease in total peripheral resistance and a fall in blood pressure. Intravenous histamine infusion produces a marked decrease in diastolic blood pressure and a widening of pulse pressure. Histamine contracts postcapillary venules, which exposes permeable capillary membranes to hydrostatic forces leading to edema. The initial flush from intradermally administered histamine is due to local cutaneous vasodilation while wheal formation is due to local edema. The vascular effects of histamine are mediated through both H_1- and H_2-receptor subtypes, but predominantly through H_1-receptors.

2. Nerves

Histamine directly stimulates nerve endings to produce cutaneous pain or pruritus. Local inflammation nonspecifically stimulates nerve endings to release neurotransmitters and neuropeptides that can perpetuate and potentiate the actions of histamine and other inflammatory mediators.

3. Heart and Coronary Arteries

Mast cells are present in cardiac muscle and coronary arteries. Histamine acts on the heart to increase inotropy and chronotropy, probably through increasing calcium influx in cardiac myocytes. H_2-receptors directly mediate most of the increase in contractile force and heart rate. H_1-receptors appear to mediate effects on the conducting system and produce coronary artery vasospasm in patients with variant angina. H_1-receptor stimulation results in negative chronotropic effects secondary to atrioventricular conduction delay. In experimental animal models, histamine acting on H_1-receptors slows atrioventricular nodal conduction and decreases fibrillation threshold. Dysrhythmia may result from a combination of effects on conduction and ischemia from coronary artery vasospasm (46). Anaphylactic reactions of isolated perfused guinea pig hearts are characterized by a short initial cardiac stimulation, followed by precipitous constriction of coronary arteries and long-lasting impaired myocardial function. Vasoactive anaphylactic mediators other than histamine are also involved in cardiac malfunction occurring during the later phase of systemic and cardiac anaphylaxis. Generally, H_1-receptor antagonist pretreatment prevents myocardial depression and cardiac shock in experimental models of anaphylaxis. It has been thought that H_2-receptor-mediated effects are of minor importance in cardiovascular manifestation of anaphylaxis (47). However, histamine acting on H_2-receptors causes pronounced stimulation of spontaneously active cardiac Purkinje fibers in sheep hearts (48). Theoretically, H_2-antagonists could have detrimental effects by potentiating coronary artery vasoconstriction (49) or by reversing H_2-receptor-mediated inhibition of cardiac histamine release (34). Thus far, investigators have not shown detrimental effects of H_2-antagonists on cardiovascular function (47).

A prominent feature of antigen-induced mediator release in isolated rat and guinea pig hearts is coronary artery vasoconstriction and ischemic myocardial damage (49). In other experimental models, histamine has a biphasic response on coronary arteries. Vasodilation is produced in part by H_2-receptor-stimulated synthesis of an endothelium-derived relaxing factor, which has been postulated to be nitric oxide. An inhibitor of nitric oxide synthesis, N-methyl-L-arginine, causes a more pronounced coronary constriction in response to histamine. This suggests that dysfunctions in the formation of nitric oxide could precipitate histamine-induced coronary artery vasospasm and myocardial infarction (50). Several studies have demonstrated that H_1-receptor stimulation of epicardial coronary arteries causes constriction, whereas H_2-receptor activation induces dilation. In patients with normal coronary arteries, direct coronary artery infusion of histamine acts via H_1-receptors located in the endothelium to cause coronary artery dilation and decreased coronary artery vascular resistance. H_2-receptors in vascular smooth muscle also mediate coronary vasodilation, whereas stimulation of H_1-receptors in vascular smooth muscle induces vasoconstriction (51). Thus, the action of histamine on coronary arteries is complex and shows considerable species variation.

4. Respiratory System

Histamine acts via H_1-receptors in the respiratory tract to contract bronchial smooth muscle, dilate or constrict vascular smooth muscle, cause microvascular leak, and activate sensory nerves. In healthy individuals, histamine does not cause significant bronchoconstriction, but patients with asthma and other pulmonary diseases are hypersensitive to its bronchoconstrictive effects. H_1-receptors also mediate an increase in airway fluid and electrolyte secretions and increased mucus viscosity. In awake sheep, bronchial vasodilation is mediated primarily by histamine acting on H_1-receptors (52). In experimental models, H_2-receptor stimulation may actually relax constricted bronchial smooth muscle. It is doubtful that this is of clinical significance, since patients with reactive airway disease tolerate H_2-antagonists (53). Most reports indicate that cimetidine and ranitidine do not clinically cause or potentiate bronchoconstriction in normal and asthmatic patients. However, in one study, exacerbation of bronchoconstriction occurred in 4 of 24 asthmatic patients treated with cimetidine, and isolated basophils from H_2-antagonist-treated asthmatic patients showed enhanced histamine release (54). H_3-receptors in the lungs modulate cholinergic neurotransmission, neuropeptide release, and bronchoconstriction (55, 45).

5. Gastrointestinal System

Histamine plays a physiological role in stimulating gastric acid secretion through H_2-receptors. There has been widespread clinical use of H_2-antagonists to decrease gastric acid secretion. Histamine produces gastrointestinal smooth muscle contraction and relaxation. These responses are caused by direct actions on smooth muscle H_1- and H_2-receptors, respectively, and indirect actions to release several active substances from the gut's own intrinsic nervous system.

6. Histamine Levels in the Clinical Diagnosis of Anaphylaxis

Anaphylaxis is largely a clinical diagnosis based upon clinical signs and possible identification of the offending agent, but atypical presentations of anaphylaxis may be more common than previously thought. One may not be able to identify the offending agent in the majority of cases (22). Improved techniques allow histamine levels to be measured more accurately and quickly than ever before. Histamine begins to rise in 5 to 10 min and may remain elevated for up to 60 min. Urine histamine and histamine metabolites may stay elevated longer and are more stable (56). Analysis of the histamine metabolite methylhistamine in urine can be helpful in diagnosing anaphylaxis (57).

IV. ANTIHISTAMINES

Binding of the H_1-antagonist to the H_1-receptor does not produce a response, but rather blocks the actions of endogenous histamine at H_1-receptors. For the most part, H_1-antagonist binding is competitive and reversible, with the notable exceptions of astemizole, terfenadine, and loratadine.

A. Newer H_1-Antagonists

The newer H_1-antagonists such as astemizole, terfenadine, loratadine and cetirizine are much more selective for the H_1-receptor and are generally well tolerated at doses that produce high levels of antihistamine activity. At the present time, none of the second-generation H_1-antagonists are available for injection, but several of the agents have been given intravenously to squirrel monkeys (58). These newer agents have not yet been widely used in the treatment or prevention of anaphylaxis in humans but, like their predecessors, they prevent anaphylactic death in experimental models. The improved selectivity and additional antiallergic properties of these drugs may be of benefit in the treatment of anaphylaxis (58–61).

B. Nonhistamine-Receptor Antiallergic Properties of Antihistamines

Many of the first-generation antihistamines have antimuscarinic, antidopaminergic, anti-α-adrenergic and antiserotoninergic properties. The contribution of these properties to the treatment of anaphylaxis is unknown but is probably of little consequence. The H_1-antagonists not only block histamine interaction with the H_1-receptor, but also inhibit cell-mediator release (62, 63), basophil migration (64), and eosinophil recruitment (65). H_1-antagonists, such as chlorpheniramine, mepyramine, ketotifen, promethazine, diphenhydramine, cyclizine, and oxatomide, in therapeutic concentrations, can inhibit IgE-induced histamine release (66). At supratherapeutic concentrations, the H_1-antagonists can release histamine by an antigen-independent direct cytotoxic effect on mast cells. Inhibition of histamine release from mast cells and basophils has been shown with the second-generation H_1-antagonists as well, including terfenadine, loratadine, cetirizine, azelastine, ketotifen, and oxatomide (67–69). Their antiallergic effects vary depending on the source of the mast cells and the stimulus used (compound 48/80, anti-IgE, antigen, substance P, Con A, or a calcium ionophore). The exact mechanism of inhibition of histamine release is unknown, but may be due in part to the lipophilic and cationic nature of the H_1-antagonists. Many H_1-antagonists are lipophilic, cationic drugs and may dissolve into the cell membrane and produce stabilization of sodium and calcium ion flux. However, there is imprecise correlation between lipophilicity and inhibition of histamine release (67).

Several other nonhistamine antagonist effects have been observed with the second-generation antihistamines, including inhibition of allergen-induced eosinophil, basophil, and neutrophil migration and inhibition of PAF-induced eosinophil accumulation in the skin. There is a definite need for standardization and characterization of the antiallergic properties of H_1-antagonists using double-blind, placebo-controlled human studies with strict criteria for serum concentrations, standardization of mediator assays, and reproducibility.

Terfenadine inhibits histamine release and production of the sulfidoleukotrienes (LTC4,LTD4,LTE4) from basophils and inhibits both immediate and late-phase cutaneous reactions to histamine (70). Oral terfenadine reduces release of prostaglandin D2 (PGD2) and leukotrienes C4 and D4 from mast cells isolated from human lung, but histamine release was not affected (71). Terfenadine pretreatment reduces nasal lavage levels of histamine, kinins, and albumin (72).

Azelastine inhibits several actions of leukotrienes, inhibits leukotriene synthesis, and may inhibit nonallergic histamine secretion from lung tissue (73). Azelastine also inhibits the effects of PAF, including platelet aggregation and bronchoconstriction (74). In studies of early and late phases of bronchoconstriction in humans, azelastine had no antiallergic effect on the late-phase response (75). Once again, the importance of these antiallergic effects of antihistamines in the treatment of anaphylaxis has not been established and further clinical investigation is warranted.

V. ANTIHISTAMINES IN ANAPHYLAXIS

A. Treatment of Acute Anaphylaxis with Antihistamines

Before the discovery of the H_1- and H_2-receptor subtypes, histamine was thought to be the mediator of anaphylaxis because agents like diphenhydramine blocked or reversed histamine-induced hypotension and bronchospasm. However, antihistamines alone were clinically ineffective in reversing all the symptoms of anaphylaxis and were of little benefit in severe anaphylaxis. We now know that although histamine mediates anaphylactic signs and symptoms through both H_1- and H_2-subtypes of receptors, there are many other mediators of the anaphylactic response. Antihistamines are not the major therapy for the anaphylactic event but are helpful as adjunctive treatment after epinephrine has been injected. To treat anaphylaxis, intravenous or intramuscular diphenhydramine 1 to 2 mg/kg, up to 75 mg, should be given early and repeated every 4 to 6 h as long as anaphylactic symptoms persist (76). Ranitidine, up to 50 mg, or cimetidine 200 to 300 mg administered intramuscularly or intravenously have been used adjunctively. Orally administered H_2-antagonists may be of benefit in less serious anaphylactic reactions. An important principle regarding the activity of antihistamines is that their effect is delayed compared to peak serum levels and lasts longer than the serum elimination half-life of these drugs (77). For example, with hydroxyzine, maximal suppression of wheal and flare does not occur until 7 h after peak serum concentrations have been reached (78). This may indicate that the therapeutic effect of these agents is also delayed. Another disadvantage of antihistamine therapy in acute anaphylaxis is that H_1-antagonists cannot reverse the consequences of H_1-receptor activation after the fact. They are most effective if they block receptors before histamine binds. This offers a tenable explanation as to why these agents are more effective as prophylactic drugs for the prevention of anaphylaxis than to treat the acute event. As noted previously, histamine is only one of a number of anaphylactic mediators that contribute to the pathophysiology (60) and H_1-antagonists should not be used as single agents for the therapy of anaphylaxis.

B. Prevention of Anaphylaxis with Antihistamines

Pretreatment with antihistamines can prevent or attenuate anaphylaxis in experimental models, whether the reaction is triggered by antigen, exogenous histamine, or compound 48/80. Pretreatment with either first- or second-generation antihistamines can prevent anaphylactic death and pulmonary, cardiac, hemodynamic, and cutaneous manifestations of anaphylaxis in experimental animals (79, 80). In patients, pretreatment with H_1-antagonists prevents bronchoconstriction to histamine challenge, nonisotonic aerosols, and exercise (81–83), and decreases microvascular leak in the skin wheal response to histamine. Cetirizine and terfenadine have been reported to protect against late-phase reactions to allergen provocation tests (84, 70). In a cross-over trial with six healthy adult volunteers, 8 mg of dimetindene and 400 mg of cimetidine administered intravenously reversed hypotension and tachycardia caused by intravenous histamine infusion (85). Clinically, pretreatment with antihistamines has been employed to prevent anaphylaxis and anaphylactoid reactions in patients at risk for reactions to radiocontrast media, volume expanders, plasma exchange, fluorescein and other drugs, and anesthesia.

1. Reactions to Radiocontrast Materials

Radiocontrast materials (RCM) were first used in 1929. Reactions to RCM are heterogeneous and involve nonallergic mast-cell degranulation and histamine release as well as complement activation. In pigs, RCM causes release of histamine from cardiac mast cells (86). However, studies with canine mastocytoma cells suggest that direct release of histamine from mast cells does not completely explain the mechanism of RCM reactions. Although antihistamines are not the primary medications for treatment of anaphylaxis, they have played an important role in prevention of anaphylaxis and anaphylactoid reactions. The prototype for the clinical efficacy of antihistamines in this regard is their use in the prevention of anaphylactoid reactions to radiocontrast material in patients who have experienced previous reactions.

The first use of an antihistamine for prevention of RCM reactions in previously reactive subjects was by Zweiman and associates (87). Diphenhydramine 50 mg was given intramuscularly prior to the administration of radiocontrast media. They reported a significant reduction in recurrence rates in previous reactors, which, in the untreated patient, range from 16% to a high of 44% (88). Since then the pretreatment protocol (Table 6) has been refined. Patterson and associates showed that the addition of prednisone enhanced clinical efficacy of the H_1-antagonist (90).

Ephedrine (25 mg) added empirically because of the propensity for hypotension with anaphylaxis produced a further reduction in reaction rates (91). Ephedrine should not be used in patients with conditions where sympathomimetics are contraindicated, such as ischemic heart disease, moderate to severe congestive heart failure, and treatment with monoamine oxidase inhibitors.

An H_2-antagonist has also been employed in this setting. Greenberger and associates found that the addition of cimetidine reduced the efficacy of the pretreatment protocol. However, in another study, patients undergoing intravenous urography were stratified into four treatment groups. Group 1 received intravenous prednisolone alone; group 2 received clemastine; group 3 received clemastine and cimetidine; and group 4 received only saline. There was significant reduction in anaphylactic

Table 6 Prophylactic Regimen for Patients with History of Reaction to RCM

Prednisone 50 mg by mouth, 13 h, 7 h, and 1 h before
Diphenhydramine 50 mg, intramuscularly, 1 h before
Ephedrine 25 mg, by mouth, 1 h before[a]
Ranitidine 150 mg or cimetidine 300 mg, by mouth, 3 h before[b]

RCM = radiocontrast materials.
[a] Omit if patient has contraindication to sympathomimetics, including ischemic heart disease, angina, and cardiac arrhythmias.
[b] Optional, see text.
Source: Ref. 89.

signs and symptoms for patients in group 3 compared with the other groups. The combination of an H_1- and H_2-antagonist was superior to an H_1-antagonist alone. Importantly, not all anaphylactic reactions were suppressed, but there were no severe reactions (92). A third study evaluated 100 patients with previous reactions to RCM. The addition of an H_2-antagonist did not alter the results of prophylaxis, but, in combination with glucocorticoid-antihistamine treatment, prevented the majority of patients from having any reactions and none had severe reactions (93). At best, the results are inconclusive as to whether or not an H_2-antagonist is an effective adjunctive agent to diphenhydramine, prednisone, and ephedrine (94). Therefore, the use of an H_2-antagonist should be left to the discretion of the physician managing the patient.

It should be noted that the protocol should be used in patients at risk regardless of the route of administration of the RCM or the procedure being performed. Reactions have been reported after hysterosalpingo-grams, myelograms, and retrograde pyelograms (95).

The risk of reactions to RCM in patients with previous reactions decreases to 1% with the prophylaxis regimen plus the use of a low osmolar agent (94). This antihistamine and prednisone prophylaxis regimen has been so successful that it has been adopted for prevention of other anaphylactic reactions in patients at risk. It has been used successfully to prevent reactions due to the administration of volume expanders (96); during plasma exchanges (97); due to the administration of fluorescein (98); during general anesthesia (99); due to the administration of specific drugs such as morphine (100), vancomycin (101), protamine (102), and chymopapain (103); and in subjects with cold urticaria who must undergo bypass surgery (104).

The efficacy of this regimen has been demonstrated only to prevent anaphylaxis or anaphylactoid episodes. In two reported instances, the pretreatment regimen failed to prevent recurrence of adult respiratory distress syndrome and noncardiac edema associated with RCM administration (105, 106).

2. Volume Expanders

A combination of H_1- and H_2-antagonists with a corticosteroid effectively prevented reactions to urea-linked gelatin solutions used as volume expanders in two studies. In one study volunteers received intravenous infusions of the plasma expander. Fifteen of 50 controls had anaphylactic symptoms compared to 0 of 50 in the treated group (96). In the other study 27 of 150 orthopedic patients had significant reactions when pretreated with only intravenous saline. When pretreated with dimetidene and cimetidine, 4 of 150 patients had reactions, and with chlorpheniramine plus cimetidine 9 of 150 patients had reactions (96).

3. Plasma Exchange

Patients undergoing plasma exchange may experience anaphylactic reactions due to multiple causes. Pretreatment using a modification of the regimen established for the prevention of RCM reactions has been successful in preventing repeat reactions (97).

4. Fluorescein

Fluorescein is commonly employed as an intravenous microvascular contrast agent by ophthalmologists and optometrists. In patients with previous reactions to fluorescein, the incidence of reaction with subsequent exposure can be nearly 50% (107). A modification of the treatment regimen for RCM reactions has been used to prevent recurrent reactions. The modification includes a 3-day regimen of oral prednisone followed by 50 mg of diphenhydramine with 400 mg of cimetidine 1 h before the study (98).

5. Idiopathic Anaphylaxis

Even after exhaustive searches for an avoidable cause of anaphylaxis, some triggers remain undetected. Patients with infrequent mild episodes of idiopathic anaphylaxis occurring 3 or 4 times per year probably do not need preventive therapy. When mild to moderate episodes occur more frequently, concurrent administration of an H_1- and H_2-antagonist should be considered on a daily basis. The nonsedating H_1-antagonists should be used in preference to potentially sedating drugs. When patients are refractory to antihistamine therapy, glucocorticoids, salbutamol (albuterol), cromolyn, or ephedrine may be empirically added (108).

6. Anaphylaxis to Latex

Exposure to latex is not limited to gloves, but includes rubber tubing used for bladder catheterization, intravenous injection ports and tubing, ventilator tubing, electrocardiogram pads, condoms, balloons, and latex-cuffed enema tubes (109). In patients with latex allergy, perioperative prophylaxis with anti-inflammatory corticosteroids, H_1- and H_2-antagonists, and ephedrine has been recommended (109, 110). However, such a prophylactic regimen, while useful, has not been as successful in preventing reactions to latex as it has been in preventing reactions to RCM in sensitive individuals. Perhaps the disparity is due to differences in the inducing or causative mechanisms between RCM and latex reactions.

7. *Exercise-Induced Anaphylaxis*

Exercise-induced anaphylaxis was first described by Sheffer and Austen in 1980 (111). The mechanism responsible for anaphylaxis appears to be non-IgE mediated mast-cell activation. Successful prophylaxis with antihistamine therapy is possible (112, 113).

8. *Drug and Anesthesia-Related Anaphylaxis*

Drugs used perioperatively and for anesthesia are frequently implicated as causes of anaphylaxis. Even in asymptomatic patients undergoing surgery and general anesthesia there is evidence of mediator release from mast cells (114). Histamine is liberated from mast cells in the lung vasculature and the heart during pediatric cardiopulmonary bypass (115). Basophil and mast-cell release of histamine appears to be the mechanism responsible for adverse reactions to many anesthetic agents. Propanidid, althesin, chemaphor, and opiates are examples of perioperative substances that directly release histamine (33). Muscle relaxants, especially atricurium, can release histamine from human lung and skin mast cells, but the effect shows considerable patient to patient variability (116). With the exception of buprenorphine, the majority of these drugs stimulate release of preformed mediators but do not generally stimulate synthesis of late-phase mediators. In a study of 11 subjects with previous reactions to anesthesia, patients were premedicated with prednisone and diphenhydramine. In addition, in seven of these patients, substances to which they had positive skin test reactions were avoided. Only one of the 11 patients had any repeat reaction and none had anaphylaxis (99). Other studies have demonstrated the clinical efficacy of antihistamine prophylaxis in perioperative patients (117, 118). In general, histamine antagonists can decrease the occurrence and severity of anaphylactoid reactions to anesthesia.

When considering much of the evidence from patients undergoing anesthesia, especially the fact that histamine may be elevated in the majority of asymptomatic patients, it is tempting to recommend universal prophylactic antihistamine therapy (33, 119, 120). In a single study of 240 general surgery patients not known to be at risk for anesthesia-related reactions, pretreatment with combination H_1- and H_2-antagonists was associated with a significantly lower incidence of histamine-related cardiorespiratory disturbances (120). Universal prophylaxis with antihistamine in the perioperative patient is controversial (121). Full endorsement of such a policy should come only after large multicenter trials show unequivocal benefit. For now, prophylactic regimens are indicated for those individuals with previous adverse reactions to procedures that cannot be avoided.

9. Morphine

Morphine is used for preoperative analgesia and induction. Philbin and associates (100) investigated the relationship between plasma histamine levels, systemic vascular resistance, and diastolic blood pressure in patients given 1 mg/kg of intravenous morphine before cardiac bypass surgery. In patients undergoing bypass, intravenous morphine causes significant histamine release. Four groups of ten patients each were studied. Group 1 received placebo; group 2 received cimetidine; group 3 received diphenhydramine; and group 4 received both the H_1- and H_2-histamine antagonists. Patients in both groups 3 and 4 had significant decreases in morphine-related reactions, but the combination therapy was more effective than treatment with the H_1-antagonist alone. Thus, pretreatment with antihistamines can attenuate the hemodynamic complications of intravenous morphine administered to cardiac bypass patients (100).

10. Vancomycin

Intravenous injection of vancomycin or polymyxin causes nonimmune dose-dependent degranulation of mast cells and basophils leading to anaphylactoid reactions. The reaction to vancomycin can be prevented by pretreatment with antihistamines (101).

11. Protamine

Protamine is being used increasingly in cardiac catheterization and cardiothoracic and vascular surgical procedures. Protamine is contained in some insulin preparations and is used in dialysis and leukophoresis. Reactions to protamine can be IgE-mediated (122). The incidence of reaction to protamine in cardiac bypass can be as high as 10.7% (102). Patients with known sensitivity to protamine who must undergo procedures where hexadimethrine bromide cannot be substituted should be pretreated with the prophylactic regimen used for reactions to RCM.

12. Chymopapain

Chymopapain is used topically for enzymatic debridement of surface lesions. It is known to induce anaphylactoid reactions. Antihistamines have been shown to be effective in the prophylaxis of chymopapain-induced anaphylaxis. In an uncontrolled trial, a combination H_1- and H_2-antihistamines was more effective than H_1-antagonists alone for prevention of reaction to chymopapain (103).

VI. THE CASE FOR COMBINATION H_1- AND H_2-ANTAGONIST THERAPY

In 1966 Ash and Schild (123) noticed that the classic antihistamines did not block all of the actions of histamine, especially gastric acid secretion.

This led to the discovery of the H_2-histamine receptor subtype by Black et al. in 1972 (124). Generally, the combination of an H_1- and H_2-receptor antagonist is needed to maximally reduce histamine-induced peripheral vasodilation, hypotension, and mucus secretion (53). Combinations of antagonists are more effective in preventing the decrease in diastolic blood pressure and the widening of pulse pressure by histamine (Fig. 1).

A. Experimental Evidence

Compound 48/80 produces an anaphylactoid response in pentabarbitone-anesthetized rats that includes decreased mean arterial blood pressure, decreased left ventricular pressure, and increased frequency of ventricular tachycardia and fibrillation. Either cimetidine or diphenhydramine reduces the frequency of ventricular tachycardia and fibrillation, but only

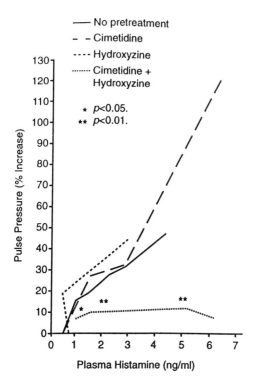

Figure 1 Effects of pretreatment with histamine receptor antagonists on widening of pulse pressure in response to increasing concentrations of plasma histamine. (From Ref. 36.)

the combination of antagonists completely inhibits hypotension and decreased left ventricular pressure and ventricular arrhythmia (125). Numerous studies have shown that H_1- plus H_2-receptor antagonists produce a greater inhibition of histamine-induced cardiovascular reactions (126, 100). In a controlled trial using dog and piglet models of anaphylaxis, the combination of H_1- and H_2-antagonists dimetindene and cimetidine was more effective than dimetindene alone and in combination with famotidine or ranitidine (127). In a cross-over study with six human volunteers, dimetindene and cimetidine inhibited histamine-induced decrements in mean arterial pressure and cutaneous reactions (85). There is increased serum histamine, decreased diastolic blood pressure, and decreased vascular resistance when 1 mg/kg morphine is given intravenously before bypass surgery. Cimetidine plus diphenhydramine was more effective in preventing hemodynamic effects during reaction to morphine than either antagonist alone (100). In a dorsal hand vein compliance technique in human subjects, venodilatory response to histamine was mediated by both H_1- and H_2-receptor subtypes. These studies provide compelling evidence that combined H_1- and H_2-antagonist therapy may prevent hypotension better than an H_1-antagonist alone during anaphylaxis (128).

In several reports, the addition of an H_2-antagonist reversed symptoms in patients that were unresponsive to H_1-antagonists and/or ephedrine (129–131). The combination of H_1- and H_2-antagonists is more effective than either alone for the prevention of anaphylactic reactions to chymopapain, perioperative agents, plasma expanders, and morphine (94). The combination of H_1- and H_2-antagonists for the prophylaxis of anaphylaxis may have additional effects other than blocking the action of histamine at its receptors. The combination may exert a mast-cell-stabilizing effect or a suppressive effect on immune cell function (33).

VII. SUMMARY

1. Anaphylaxis is a clinically significant problem that increases with increased exposure to drugs, synthetic substances, and medical procedures.
2. Histamine is a primary mediator of anaphylaxis and anaphylaxis can be reproduced by histamine infusion.
3. Antihistamines are adjunctive therapy for anaphylaxis. Antihistamines alone have been disappointing for the treatment of anaphylaxis because anaphylaxis is mediated by many other chemical substances from mast cells and basophils, besides histamine. Also, compared to epinephrine, antihistamines have a slow onset of action, and they cannot block events that occur subsequent to histamine binding to its receptors.

4. Antihistamines are an important component of regimens for the prevention of anaphylaxis in patients at risk and may even have more universal application in the perioperative setting. In some patients with exercise-induced anaphylaxis or anaphylactic reactions to latex, prophylaxis regimens are not always effective.

5. H_2-antagonists are not detrimental in the therapy of anaphylaxis and many studies show a favorable outcome when H_1- and H_2-antagonists are combined for prophylaxis. H_2-antagonists should be added to H_1-antagonist therapy at the discretion of the treating physician.

6. Because of decreased antimuscarinic and central nervous system side effects, the newer antihistamines can be given at doses which provide greater blockade of histamine receptors. These agents should lead to a reevaluation of the usefulness of antihistamines in both anaphylaxis treatment and in prophylactic regimens. The unavailability of parenterally administered second-generation H_1-antagonists may limit their usefulness in anaphylaxis treatment and perioperative prophylaxis.

REFERENCES

1. Lamson RW. Sudden death associated with the injection of foreign substances. JAMA 1924;82:1091.
2. Vaughn WT, Pipes DM. On the probable frequency of allergic shock. Am J Dig Dis 1936;3:558.
3. Amornmarn L, Bernard L, Kumar N, Bielory L. Anaphylaxis admissions to a university hospital. J Allergy Clin Immunol 1992;89:349.
4. Davies DM. Anaphylaxis and the community nurse. In: Davies DM, ed. Adverse Drug Reaction Bulletin. Newcastle Upon Tyne: Regional Postgraduate Institute of Medicine and Dentistry, 1977:228.
5. Boston Collaborative Drug Surveillance Program: Brief reports: Drug induced anaphylaxis. JAMA 1973;224:613.
6. Porter J, Jick H. Boston Collaborative Drug Surveillance Programs: Drug induced anaphylaxis, convulsions, deafness, and extrapyramidal symptoms. Lancet 1977;1:587.
7. Orange RP, Donsky GJ. Anaphylaxis. In: Middleton Jr, E, Reed CE, Ellis EF eds. Allergy Principles and Practice. St. Louis: Mosby, 1978:564.
8. Bochner BS, Lichtenstein LM. Anaphylaxis. N Eng J Med 1991;324(25): 1785–1790.
9. Idsoe O, Guthe T, Wilcox RR, et al. Nature and extent of penicillin side-reactions with particular reference to fatalities from anaphylactic shock. Bull World Health Organ 1968;38:159–188.
10. Saxon A. Immediate hypersensitivity reactions to beta-lactam antibiotics. Ann Intern Med 1987;107:204–215.
11. Cohan R, Dunnick N, Bashore T. Treatment of reactions to radiographic contrast material. AJR 1988;151:263.

12. Katayama H, Yamaguchi K, Kozuka T, et al. Adverse reactions to ionic and nonionic contrast media; A report from the Japanese Committee on the safety of contrast media. Radiology 1990;175:621.

13. van der Klauw MM, Stricker BHCH, Herings RMC, et al. A population based case-cohort study of drug-induced anaphylaxis. Br J Clin Pharm 1993; 35:400–408.

14. Fisher MMcD, More DG. Epidemiology and clinical features of anaphylactic reactions in anaesthesia. Anaesth Intens Care 1981;9:226–234.

15. Hatton F, Tiret L, Maujol L, et al. Enquete epidemiologique sur les anesthesies. Ann Fr Anesth Reanim 1983;2:333–385.

16. Kelso JM, Jones RT, Yunginger JW. Anaphylaxis to measles, mumps, and rubella vaccine mediated by IgE to gelatin. J Allergy Clin Immunol 1993; 91:867–872.

17. Golden DBK. Epidemiology of allergy to insect venoms and stings. Allergy Proc 1989;16:103.

18. Barnard JH. Studies of 400 Hymenoptera sting deaths in the United States. J Allergy Clin Immunol 1973;52:259.

19. van der Linden PWG, Hack CE, Poortman J, et al. Insect-sting challenge in 138 patients: Relationship between clinical severity of anaphylaxis and mast cell activation. J Allergy Clin Immunol 1992;90:110–118.

20. Settipane GA, Chafee FH, Klein DE. Aspirin intolerance: II. A prospective study in an atopic and normal population. J Allergy Clin Immunol 1974;53: 200–204.

21. Strom BL, Carson JUL, Morse ML, et al. The effect of indication on hypersensitivity reactions associated with zomepirac sodium and other nonsteroidal antiinflammatory drugs. Arthritis Rheum 1987;30:1142.

22. Kemp S, Lieberman P, Wolf B. A review of 267 cases of anaphylaxis in clinical practice. J Allergy Clin Immunol 1993;91:153.

23. Lieberman PL. Specific and idiopathic anaphylaxis: pathophysiology and treatment. In Bierman CW, Pearlman DS, Shapiro GG, Busse WW (eds.): Allergy, Asthma, and Immunology From Infancy to Adulthood (3rd edition). WB Saunders Co., Philadelphia, 1996;297–319.

24. Lohman IC, Halonen M. The effects of combined histamine and platelet-activating factor antagonism on systemic anaphylaxis induced by immunoglobulin E in the rabbit. Am Rev Respir Dis 1993;147:1223–1228.

25. Pradalier A. Late-phase reaction in asthma: basic mechanisms. Int Arch Allergy Immunol 1993;101:322–325.

26. Mosimann BL, White MV, Hohman RJ, et al. Substance P, calcitonin gene-related peptide, and vasoactive intestinal peptide increase in nasal secretions after allergen challenge in atopic patients. J Allergy Clin Immunol 1993;92:95–104.

27. Gyorfi A, Fazekas A, Posch E, Irmes F, Rosivall L. Role of histamine in the development of neurogenic inflammation of rat oral mucosa. Agents Actions 1991;32:229–236.

28. Proud D, Macglashan DW, Newball HH et al. Immunoglobulin E-mediated release of kininogenase from purified human lung mast cells. Am Rev Respir Dis 1985;132:406–408.

29. Newball HH, Berninger RW, Talamo RC, Lichtenstein LM. Anaphylactic release of a basophil kallikrein-like activity. I. Purification and characterization. J Clin Invest 1979;64:457–465.
30. Holgate ST, Robinson C, Church M. In Middleton Jr E, Reed CE, Ellis E, Adkinson NF, Yunginger J, Busse W, eds. Mediators of immediate hypersensitivity. St Louis: Mosby-Year Book, 1993;267–303.
31. Lichtenstein LM, Gillespie E. The effects of the H_1- and H_2-antihistamines on allergic histamine release and its inhibition by histamine. J Pharmacol Exp Ther 1975;192:441–450.
32. Tauber R, Reimann HJ, Schmiedt E. Hat das Hastammeme pathophysiologische Bedeutung bei Nierentumoren? In Zioegler M. ed. Urologish-nephrologische Probleme 20 Tagung der Sudwestdeuschen Gesellschaft fur Urologie Konstanz Schmetzor, 1981:176.
33. Ring J, Behrendt H. H_1 and H_2-antagonists in allergic and pseudoallergic diseases. Clin Exp Allergy 1990;20:43–49.
34. Blandina J, Brunelleschi S, Fantozzi R, et al. The antianaphylactic action of histamine H_2-receptor agonists in the guinea-pig isolated heart. Br J Pharm 1987;90:459–466.
35. MacIntosh FC, Paton WDM. The liberation of histamine by certain organic bases. J Physiol 1949;109:190–219.
36. Kaliner M, Shelhamer JJ, Ottesen EA. Effects of infused histamine: correlation of plasma histamine levels and symptoms. J Allergy Clin Immunol 1982;69:283–289.
37. Woodward DF, Ledgard SE. Histamine-induced microvascular permeability increases in hamster skin: a response predominantly mediated by H_2-receptors. Agents Actions 1986;18:504–507.
38. Pearce FL. Biological effects of histamine: An overview. Agents Actions 1991;33:4–7.
39. Sitter H, Lorenz W, Doenicke A. The clinical and biological signs of histamine release during induction of anaesthesia and preparation of the surgical patient: A farewell party for the classical manifestations of anaphylaxis. Agents Actions 1992; Special Conference Issue:C219–229.
40. Hill SJ. Histamine receptors and interactions between second messenger transduction systems. Agents Actions 1991;33:145–159.
41. Moncada S, Martin J. Vasodilation—evolution of nitric oxide. Lancet 1993; 341:1511.
42. Palmer, RMJ, Ferrige AG, Moncada S. Nitric oxide release accounts for the biological activity of endothelium derived relaxing factor. Nature 1987; 327:524–526.
43. Leurs R, Brozius MM, Jansen W, et al. Histamine H_1-receptor-mediated cyclic GMP production in guinea-pig lung tissue is an L-arginine-dependent process. Biochem Pharmacol 1991;42:271–277.
44. Stark H, Bertaccini G. Histamine H_3-receptor activation inhibits acetylcholine release in the guinea pig myenteric plexus. Agents Actions 1991;33:167–169.
45. Ichinose M, Barnes PJ. Histamine H_3-receptors modulate antigen-induced

bronchoconstriction in guinea pigs. J Allergy Clin Immunol 1990;86: 491–495.

46. Borchard U, Hafner D, Hirth C. Electrophysiological actions of histamine and H_1–H_2-receptor antagonists in cardiac tissue. Agents Actions 1986;18: 186–190.

47. Felix SB, Baumann G, Niemczyk M, et al. Effects of histamine H_1- and H_2-receptor antagonists on cardiovascular function during systemic anaphylaxis in guinea pigs. Agents Actions 1991;32:245–252.

48. Thome U, Borchard U, Berger F, Hafner D. Electrophysiological characterization of histamine receptor subtypes in sheep cardiac Purkinje fibers. Agents Actions 1992; Special Conference Issue:C325–327.

49. Vleeming W, van Rooij H, Wemer J, Porsius AJ. Characterization and modulation of antigen-induced effects in isolated rat heart. J Cardiovasc Pharmacol 1991;18:556–565.

50. Lamparter B, Gross SS, Levi R. Nitric oxide and the cardiovascular actions of histamine. Agents Actions 1992; Special Conference Issue C187–C198.

51. Matsuyama K, Yasue H, Okumura K, et al. Effects of H_1-receptor stimulation on coronary arterial diameter and coronary hemodynamics in humans. Circulation 1990;81:65–71.

52. Parsons GH, Villablanca AC, Brock JM, et al. Bronchial vasodilation by histamine in sheep: characterization of receptor subtype. J Appl Physiol 1992;72:2090–2098.

53. Barker LA. Histamine and Antihistamines. In: Wingard LB, Brody TM, Larner J, Schwartz A, eds. Human Pharmacology, Molecular to Clinical. New York: Mosby, 1991;776–788.

54. Hofman J, Siergiejko Z, Bortkiewicz J, et al. Bronchial hyper-reactivity in atopic patients during H_2-receptor treatment. Agents Actions 1986;18: 239–241.

55. Barnes PJ. Histamine receptors in the lung. Agents Actions 1991;33: 103–121.

56. Kaliner M, Dyer J, Merlin S, et al. Increased urine histamine and histamine metabolites in contrast media reactions. Invest Radiol 1984;29:116.

57. Watkins J, Wild G. Problems of mediator measurement for a national advisory service to UK anaesthetists. Agents Actions 1990;30:247–249.

58. Barnett A, Kreutner W. Pharmacology of non-sedating H_1-antihistamines. Agents Actions 1991;33:181–197.

59. Simons, FER. Recent advances in H_1-receptor antagonist treatment. J Allergy Clin Immunol. 1990;86:995–999.

60. Sutherland, DC. Antihistamine agents: new options or just more drugs? Med J Aust 1989;151:158–162.

61. Rimmer SJ, Church MK. The pharmacology and mechanisms of action of histamine H_1-antagonists. Clin Exp Allergy 1990;20:3–17.

62. Nabe M, Agrawal DK, Sarmiento EU, Townley RG. Inhibitory effect of terfenadine on mediator release from human blood basophils and eosinophils. Clin Exp Allergy 1989;19:515–520.

63. Naclerio RM, Kagey-Sobotka A, Lichtenstein LM, et al. Terfenadine, an H_1-antihistamine, inhibits histamine release in vivo in the human. Am Rev Respir Dis 1990;142:167–171.

64. Charlesworth EN, Kagey-Sobotka A, Norman PS, Lichtenstein LM. Effect of cetirizine on mast cell-mediator release and cellular traffic during the cutaneous late-phase reaction. J Allergy Clin Immunol 1989;83:905–912.

65. Fadel R, Herpin-Richard N, Rihoux JP, Hennocq E. Inhibitory effect of cetirizine 2HCl on eosinophil migration in vivo. Clin Allergy 1987;17:373–379.

66. Church MK, Gradidge CF. Inhibition of histamine release from human lung in vitro by antihistamines and related drugs. Br J Pharmacol 1980;69:663–667.

67. Simons FER. The antiallergic effects of antihistamines (H_1-receptor antagonists). J Allergy Clin Immunol 1992;90:705–715.

68. Massey WA, Lichtenstein LM. The effects of antihistamines beyond H_1-antagonism in allergic inflammation. J Allergy Clin Immunol 1990;86:1019–1024.

69. Togias AG, Naclerio RM, Warner J, et al. Demonstration of inhibition of mediator release from human mast cells by azatadine base. JAMA 1985;255:225–229.

70. De Weck AL. Does the clinical experience with non-sedating H_1-antagonists justify a reassessment of antihistamines in allergy treatment? Clin Exp Allergy 1990;20:51–54.

71. Campbell AM, Bousquet J. Anti-allergic activity of H_1-blockers. Int Arch Allergy Immunol 1993;101:308–310.

72. Naclerio RM. Inhibition of mediator release during the early reaction to antigen. J Allergy Clin Immunol 1992;90:715–718.

73. Chand N, Diamantis W, Sofia RD. Antagonism of histamine and leukotrienes by azelastine in isolated guinea pig ileum. Agents Actions 1986;19:164–168.

74. Castro-Faria-Neto HC, Bozza PT, Silva AR, et al. Interference of azelastine with anaphylaxis induced by ovalbumin challenge in actively sensitized rats. Eur J Pharmacol 1992;213:183–188.

75. Twentyman OP, Ollier S, Holgate ST. The effect of H_1-receptor blockade on the development of early- and late-phase bronchoconstriction and increased bronchial responsiveness in allergen-induced asthma. J Allergy Clin Immunol 1993;91:1169–1178.

76. Marshall C, Lieberman P. Analysis of three pretreatment procedures to prevent anaphylactoid reactions to radiocontrast in previous reactors. J Allergy Clin Immunol 1989;83:245–254.

77. Lieberman P. Histamine and Antihistamine In: Rich RR, Fleisher TA eds. Clinical Immunology Yearbook. St Louis: Mosby, in press 1994.

78. Simons FER, Simons KJ. H_1-receptor antagonists: clinical pharmacology and use in allergic disease. Ped Clin North Am 1983;30:899–914.

79. Mota I, DaSilva WD. The anti-anaphylactic and histamine releasing proper-

ties of the antihistamines. Their effect on the mast cells. Br J Pharmacol 1960;15:396–404.

80. Wood-Baker R, Holgate ST. The comparative actions and adverse effect profile of single doses of H_1-receptor antihistamines in the airways and skin of subjects with asthma. J Allergy Clin Immunol 1993;91:1005–1014.

81. Kreutner W, Chapman RW, Gulbenkian A, Siegel MI. Antiallergic activity of loratadine, a nonsedating antihistamine. Allergy 1987;42:57–63.

82. Eiser N, Mills J, Snashall P, Guz A. The role of histamine receptors in asthma. Clin Sci 1981;60:363–370.

83. Cookson WOCM. Bronchodilator action of the antihistamine terfenadine. Br J Clin Pharm 1987;24:120–121.

84. Wasserfallen JB, Leuenberger P, Pecoud A. Effect of cetirizine, a new H_1-antihistamine, on the early and late allergic reactions in a bronchial provocation test with allergen. J Allergy Clin Immunol 1993;91:1189–1197.

85. Tryba M, Zenz M, Thole H. Therapeutic efficacy of combined H_1- + H_2-receptor antagonists on severe histamine-induced cardiovascular reactions in humans. Agents Actions 1992; Special Conference Issue:C238–241.

86. Ennis M, Lorenz W, Nehring E, Schneider C. In vitro and in vivo studies of radiographic contrast media-induced histamine release in pigs. Agents Actions 1991;33:26–29.

87. Zweiman G, Mishkin MM, Hildreth EA. An approach to the performance of contrast studies in contrast material-reactive persons. Ann Intern Med 1975;83:159–162.

88. Lieberman P. Anaphylactoid reactions to radiocontrast material. In Clinical Reviews in Allergy: Anesthesiol Allergy. 1991;9:319–338.

89. Lieberman P, Siele RL, Tredwell G. Radiocontrast reactions. Clin Rev Allergy 1986;4:229–245.

90. Patterson R, Pruzansky JJ, Dykewicz MS, Lawrence ID. Basophil-mast cell response syndromes: A unified clinical approach. Allergy Proc 1988; 9:611–620.

91. Greenberger PA. Life-threatening idiopathic anaphylaxis associated with hyperimmunoglobulinemia E. Am J Med 1984;76:553–556.

92. Ring J, Rothenberger KH, Clauss W. Prevention of anaphylactoid reactions after radiocontrast media infusion by combined H_1- and H_2-receptor antagonists: results of a prospective controlled trial. Int Arch Allergy Appl Immunol 1985;78:9–14.

93. Marshall GD, Lieberman PL. Comparison of three pretreatment protocols to prevent anaphylactoid reactions to radiocontrast media. Ann Allergy 1991;67:70–74.

94. Lieberman P. The use of antihistamines in the prevention and treatment of anaphylaxis and anaphylactoid reactions. J Allergy Clin Immunol 1990; 86(S):684–686.

95. Lieberman P. Anaphylactoid reactions to radiocontrast material. In: Reisman R, Lieberman P eds. Immunology and Allergy Clinics of North Amer-

ica—Anaphylaxis and Anaphylactoid Reactions. Philadelphia: Saunders, 1979;12:649–670.

96. Lorenz W, Doenicke A. Anaphylactoid reactions and histamine release by intravenous drugs used in surgery and anaesthesia. In: Watkins J, Ward AM, eds. Adverse response to intravenous drugs. London: Academic Press, 1978;83–112.

97. Apter AJ, Kaplan AA. An approach to immunologic reactions associated with plasma exchange. J Allergy Clin Immunol 1992;90:119–124.

98. Rohr AS, Pappano JE. Prophylaxis against fluorescein-induced anaphylactoid reactions. J Allergy Clin Immunol 1992;90:407–408.

99. Moscicki RA, Sockin SM, Corsello BF et al. Anaphylaxis during induction of general anesthesia: subsequent evaluation and management. J Allergy Clin Immunol 1990;86:325–332.

100. Philbin DM, Moss J, Akins CW, et al. The use of H_1- and H_2-histamine antagonists with morphine anesthesia: a double blind study. Anesthesiology 1981;55:292–296.

101. Williams D, Laska DA, Shetler TJ, McGrath JP, White SL, Hoover, DM. Vancomycin-induced release of histamine from rat peritoneal mast cells and a rat basophil cell line (RBL-1). Agents Actions 1991;32:217–223.

102. Weiler JM, Gellhaus MA, Carter JG, et al. A prospective study of the risk of an immediate adverse reaction to protamine sulfate during cardiopulmonary bypass surgery. J Allergy Clin Immunol 1990;85:713–719.

103. Moss J, Roizen M, Noroby E et al. Decreased incidence and mortality of anaphylaxis to chymopapain. Anesth Analog 1985;64:1197–1201.

104. Babe KS, Zebrowski M, Lieberman P, Kim R. Successful hypothermic cardiopulmonary bypass in a patient with primary cold urticaria. Ann Allergy 1993;70:90.

105. Madowitz J, Schweiger M. Severe anaphylactoid reactions to radiographic contrast media: Occurrence despite premedication with diphenhydramine and prednisone. JAMA 1979;241:2813.

106. Borish L, Matloff SM, Findlay SR. Radiographic contrast media-induced noncardiogenic pulmonary edema: Case report and review of the literature. J Allergy Clin Immunol 1984;74:104.

107. Kwiterovich KA, Maguire M, Murphy RP et al. Frequency of adverse reactions after fluorescein angiography. Ophthalmology 1991;98:1139–1142.

108. Greenberger PA. Idiopathic anaphylaxis. Immunology and Allergy Clinics of North America 1992;12:571–583.

109. Pasquariello CA, Lowe DA, Schwartz RE. Intraoperative anaphylaxis to latex. Pediatrics 1993;91:983–986.

110. Zeiss PA. Allergic diseases: Diagnosis and management. 4th ed. Philadelphia: JB Lippincott 1993:395.

111. Sheffer AL, Austen KF. Exercise-induced anaphylaxis. J Allergy Clin Immunol 1980;66:106–111.

112. Briner WW, Sheffer AL. Exercise-induced anaphylaxis. Med Sci Sports Exerc 1992;24:849–850.

113. Saryan JA, O'Loughlin JM. Anaphylaxis in children. Ped Ann 1992;21: 590–598.
114. Laroche D, Vergnaud M-C, Sillard, B et al. Biochemical markers of anaphylactoid reactions to drugs. Comparison of plasma histamine and tryptase. Anesthesiology 1991;75:945–949.
115. Withington DE, Elliot M, Man WK. Histamine release during paediatric cardiopulmonary bypass. Agents Actions 1991;33:200–202.
116. Marone G, Stellato C. Activation of human mast cells and basophils by general anaesthetic drugs. In: Assem E-SK, ed. Allergic Reactions to Anaesthetics. Clinical and Basic Aspects. Basel: Karger, 1992;30:54–73.
117. Tryba M, Zevounou F, Zenz M. Prevention of histamine-induced reactions during the induction of anaesthesia following premedication with H_1- + H_2-antagonists intramuscularly. Br J Anaesth 1986;58:478–482.
118. Lorenz W, Dick W, Junginger T, et al. Induction of anaesthesia and perioperative risk: influence of antihistamine H_1- + H_2-prophylaxis and volume substitution with Haemaccel-35 on cardiovascular and respiratory disturbances and histamine release. Theo Surg 1988;3:55–77.
119. Lorenz W, Doenicke A. H_1- + H_2-blockade: a prophylactic principle in anaesthesia and surgery against histamine-release responses of any degree of severity. N Eng Reg Allergy Proc Part I. 1985;6:37–57. Part II. 1985;6: 174–194.
120. Lorenz W, Duda D, Dick W, et al. Incidence and clinical importance of perioperative histamine release: randomised study of volume loading and antihistamines after induction of anaesthesia. Lancet 1994;343:933–940.
121. Dahl JB. Antihistamine prophylaxis and general anaesthesia. Lancet 1994; 343:929–930.
122. Weiss ME, Nyhan D, Peng Z, et al. Association of protamine IgE and IgG antibodies with life-threatening reactions to intravenous protamine. N Eng J Med 1989;320:886–892.
123. Ash AS, Schild HO. Receptors mediating some actions of histamine. Br J Pharmacol 1966;27:427–439.
124. Black JW, Duncan WAM, Durant GJ, Ganellin CR, Parsons EM. Definition and antagonism of histamine H_2-receptors. Nature 1972;236:385–390.
125. Dai S. Circulatory depression and ventricular arrhythmias induced by compound 48/80 in anaesthetized rats. Agents Actions 1991;34:316–323.
126. Lorenz W, Ennis M, Doenicke A, Dick W. Perioperative uses of histamine antagonists. J Clin Anesth 1990;2:345–360.
127. Lorenz W, Kubo K, Stinner B, et al. Studies on the effectiveness of H_1- + H_2-antagonist combinations in preventing life-threatening anaphylactoid reactions in anaesthesia and surgery: Problems with selecting the animal model from clinical data and with "equi-effective" dose. Agents Actions 1992; Special Conference Issue C231–237.
128. Dachman WD, Bedarida G, Blaschke TF, Hoffman BB. Histamine-induced venodilation in human beings involves both H_1- and H_2-receptor subtypes. J Allergy Clin Immunol 1994;93:606–614.

129. Vidovich RR, Heisleman DE, Hudock D. Treatment of urokinase-related anaphylactic reaction with intravenous famotidine. Ann Pharmacother 1992;26:782–783.
130. De Soto H, Turk M. Cimetidine in anaphylactic shock refractory to standard therapy. Anesth Analg 1989;69:260–269.
131. Yarbrough JA, Moffitt JE, Brown, DA, Stafford CT. Cimetidine in the treatment of refractory anaphylaxis. Ann Allergy 1989;63:235–238.

12

H₁-Receptor Antagonists in Children

F. Estelle R. Simons
The University of Manitoba, Winnipeg, Manitoba, Canada

I. INTRODUCTION

For decades, H_1-receptor antagonists have been commonly used in the treatment of pediatric allergic disorders, providing relief of sneezing, itching, and rhinorrhea in allergic rhinoconjunctivitis, and relief of itching in urticaria and atopic dermatitis (1). Older, potentially sedating first-generation H_1-antagonists are still widely used in infants and children, in part because of absence of regulatory approval of newer, less-sedating second-generation H_1-antagonists and lack of availability of pediatric dosage formulations of the newer medications in some countries (Table 1). Neither the older nor the newer H_1-antagonists have yet been optimally studied in young subjects. In this chapter, we will review the clinical pharmacology and use of H_1-antagonists in this age group, with special emphasis on recent studies and on the second-generation medications.

II. CLINICAL PHARMACOLOGY OF H₁-RECEPTOR ANTAGONISTS IN CHILDREN

The pharmacokinetics and pharmacodynamics of some of the first-generation H_1-receptor antagonists [chlorpheniramine, diphenhydramine, and

Table 1   Formulations and Dosages of Representative H_1-Receptor Antagonists

H_1-Receptor Antagonist	Formulation	Recommended Pediatric Dose
Chlorpheniramine maleate (Chlor-Trimeton®)	Tablets 4 mg, 8$^+$ mg, 12$^+$ mg Syrup 2.5 mg/5 mL Parenteral solution 10 mg/mL	0.35 mg/kg/24 h
Hydroxyzine hydrochloride (Atarax®)	Capsules 10, 25, 50 mg Syrup 10 mg/5 mL, parenteral 50 mg/mL	2 mg/kg/24 h
Diphenhydramine hydrochloride (Benadryl®)	Capsules 25 or 50 mg Elixir 12.5 mg/5 mL Syrup 6.25 mg/5 mL Parenteral solution 50 mg/mL	5 mg/kg/24 h
Terfenadine (Seldane®)[a]	Tablets 60 mg Suspension 30 mg/5 mL[b]	3–6 yrs: 15 mg bid 7–12 yrs: 30 mg bid
Astemizole (Hismanal®)[a]	Tablets 10 mg Suspension 10 mg/5 mL[b]	0.2 mg/kg/24 h
Loratadine (Claritin®)[a]	Tablets 10 mg Syrup 1 mg/mL[b]	2–12 yrs: 5 mg/day >12 yrs & >30 kg: 10 mg/day
Cetirizine (Reactine®)[a]	Tablets 10 mg	>12 yrs: 5–10 mg/day
Acrivastine (Semprex®)[a]	Tablets 8 mg[c]	>12 yrs: 8 mg tid
Ketotifen (Zaditen®)[b]	Tablets 1 mg; 2 mg[b] Syrup 1 mg/5 mL[b]	>3 yrs: 1 mg bid
Azelastine (Astelin®)[a]	Nasal solution 0.1% 0.137 mg/spray	Topical: 2 sprays/nostril/day or 2× daily
Levocabastine (Livostin®)	Microsuspension[d] Nasal spray 0.5 mg/mL or eye drops 0.5 mg/mL	Topical: 2 sprays/nostril 2–4× daily 1 drop in each eye 2–4× daily

[a] Not approved for use in children under age 12 years in the United States.
[b] Medication or formulation not available in the United States.
[c] Available as Semprex-D in combination with pseudoephedrine 60 mg.
[d] Available only as ophthalmic solution in the United States.
[+] Sustained-release.
b.i.d. = twice daily; t.i.d. = three times daily.

hydroxyzine (Fig. 1) Table 2a] and most of the second-generation H_1-antagonists [terfenadine, astemizole, loratadine, ketotifen, cetirizine (Fig. 2a), and ebastine] have been reasonably well-documented in children (Table 2b) (2–16), and cetirizine has recently been studied in infants, in whom a plasma elimination half-life of 2.8 h has been found (17). H_1-antagonists are well-absorbed after oral administration as liquid or solid formulations, with peak plasma concentrations usually being reached within 2 h. Terfenadine, astemizole, loratadine, and ebastine are metabolized by the hepatic microsomal mixed-function oxygenase systems. Plasma concentrations of these H_1-antagonists are low after single oral doses, which indicates considerable first-pass hepatic extraction. Active metabolites, terfenadine carboxylate, desmethylastemizole, descarboethoxyloratadine, and carebastine, respectively, are present in higher con-

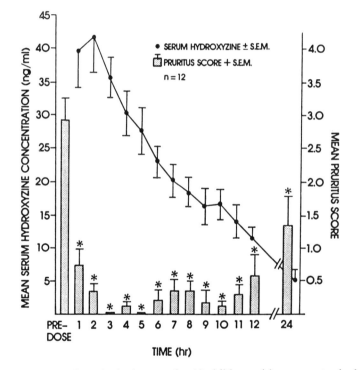

Figure 1 In a single-dose study, 12 children with severe atopic dermatitis, mean age 6.1 ± 4.6 years, and mean weight 22 ± 12 kg received hydroxyzine syrup 0.7 mg/kg by mouth. Mean serum hydroxyzine concentrations and mean pruritus scores are shown. The mean serum elimination half-life was 7.1 h. A single dose of hydroxyzine suppressed pruritus significantly for 24 h. (From Ref. 6.)

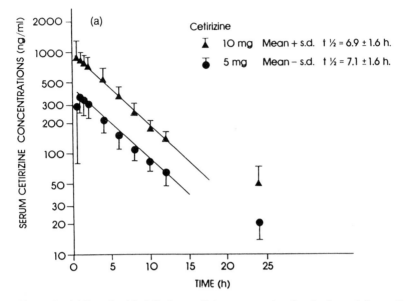

Figure 2 (a) In a double-blind, parallel-group study of a single oral dose of cetirizine, 5 mg, in ten children versus a single oral dose of cetirizine, 10 mg, in nine children, the serum elimination half-life of cetirizine was approximately 7 h. (From Ref. 10.)

centrations than the parent compound. Plasma elimination half-life values are variable, and the half-life values of active metabolites may be longer than those of the parent compound. In children, although the half-life values for some H_1-antagonists and their metabolites are shorter than they are in adults, once or twice daily dosing is possible for most medications.

Cetirizine, the active carboxylic acid metabolite of hydroxyzine, is excreted 40% unchanged in the urine within the first 24 h of a single dose (10). Compared to other H_1-antagonists, plasma concentrations are relatively high and the volume of distribution is relatively small. Acrivastine, and the topical H_1-antagonist levocabastine, which are not optimally studied in children to date, are also excreted mostly unchanged in the urine.

Potential medication interactions, including concomitant administration of H_1-antagonists with cytochrome P_{450} inhibitors, have not been studied in young subjects.

In children, as well as in adults, suppression of the histamine or antigen-induced wheal and flare is used as an objective test of the magnitude,

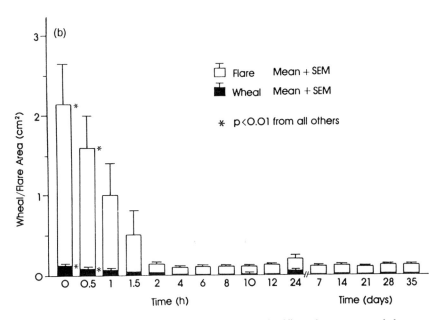

Figure 2   (b) A single dose of cetirizine, 5 mg, significantly suppressed the mean wheal and flare areas resulting from epicutaneous tests with histamine phosphate, 1 mg/ml, at various times from 1 to 24 h postdose. In the subsequent multiple-dose phase of the study, the wheal and flare suppression 12 h after the cetirizine dose on days 7, 14, 21, 28, and 35 did not differ from suppression 12 h after the first cetirizine dose. (From Ref. 10.)

onset, time to peak, and duration of peripheral H_1-blockade (4, 5, 7, 10, 12, 15, 16). The amount of wheal and flare suppression varies from one H_1-antagonist to another and with the dose administered. It probably correlates with H_1-antagonist effects in the airways but this has not been shown directly in children. The pharmacokinetics and pharmacodynamics of H_1-antagonists do not appear to change during chronic dosing in children (10) (Fig. 2b).

A single dose of chlorpheniramine, diphenhydramine, terfenadine, loratadine, cetirizine, or ebastine has a prompt onset of action (4, 5, 7, 10, 11, 12). A single dose of astemizole does not suppress the wheal and flare significantly (16). Offset of action of an H_1-antagonist, that is, the length of time required for the effect of a short course to "wear off," has not been studied in children. In adults, it is recommended that H_1-antagonists such as cetirizine be discontinued for 3–4 days, terfenadine and loratadine

Table 2a   Pharmacokinetics and Pharmacodynamics of First-Generation H_1-Receptor Antagonists in Children

Drug (metabolite)	Dosage (mg/kg)	No. of pts.	Mean Age (yr)	Cp_{max} (ng/mL)	t_{max} (h)	$t_{1/2}$ (h)	↓ Wheal/ Flare (h)	Ref.
Chlorpheniramine	0.12	11	11 ± 3.0	13.5 ± 3.5	2.5 ± 1.5	13.1 ± 6.6	1–24	4
Diphenhydramine	1.25	7	8.9 ± 1.7	81.8 ± 30.2	1.3 ± 0.5	5.4 ± 1.8	1–8/1–12	5
Hydroxyzine	0.7	12	6.1 ± 4.6	47.4 ± 17.3	2.0 ± 0.9	7.1 ± 2.3	n/a	6

Table 2b   Pharmacokinetics and Pharmacodynamics of Second-Generation H_1-Receptor Antagonists in Children

Drug (active metabolite)	Dosage	No. of pts.	Age (yr)	Cp_{max} (ng/mL)	t_{max} (h)	$t_{1/2}$ (h)	↓ Wheal/ Flare (h)	Ref.
Terfenadine (terfenadine carboxylic acid)	1–2 mg/ kg	13	7.5 ± 0.5	(242 ± 28)	(2.3 ± 0.2)	(2.0 ± 0.1[a])	1–8[b]	7
Astemizole (desmethyl-astemizole)	10 mg/ day × 10–12 wk	10 19	13.9 11.6	n/a	n/a	(10.8 days) (11.2 days)	n/a	8
Loratadine (descarbo-ethoxy-loratadine)	10 mg syrup	13	10.46	4.38 (3.79)	1.0 (1.69)	n/a (13.79)	1–12[c]	9, 15
Cetirizine	5 mg 10 mg	10 9	8.0 ± 0.6 8.0 ± 0.6	427.6 ± 144.2 978.4 ± 340.6	1.4 ± 1.1 0.8 ± 0.4	7.1 ± 1.6[d] 6.9 ± 1.6	1–24 0.5–24	10
Cetirizine	5 mg (solution)	8	2–4	560 ± 200 (est.)	1.44 ± 1.12	4.91 ± 0.6[e]	n/a	11
Ketotifen	1 mg bid	6	2–4	3.25	1.33	n/a	n/a	13, 14
Ebastine (carebastine)	5 mg 10 mg	10 10	7.3 7.8	(108.6 ± 11.8) (209.6 ± 24.2)	(2.8 ± 0.3) (3.4 ± 0.4)	(11.4 ± 0.7) (10.1 ± 1.1)	0.5–28 0.5–28	12

n/a = data not available. Cp_{max} (ng/mL) = maximum plasma concentration. t_{max} (h) = time of maximum plasma concentration. $t_{1/2}$ = plasma elimination half-life. ↓ wheal/flare (h) = suppression of wheal and flare vs baseline and vs placebo treatment ($p < 0.05$).
[a] Distribution phase.
[b] Duration of study limited to 8 h.
[c] Duration of study limited to 12 h.
[d] Urinary excretion of unchanged cetirizine = 40 ± 15% and 39 ± 14%.
[e] Urinary excretion of unchanged cetirizine = 37.8 ± 5.2%.

for 6–7 days, and astemizole for 4–6 weeks, before skin tests or bronchial challenge tests with histamine or allergen are performed (2).

A. Efficacy of H₁-Receptor Antagonists in Children

1. Allergic Rhinoconjunctivitis

H₁-antagonists are important first-line medications in the treatment of this common disorder. Histamine is the only chemical mediator of inflammation in allergic rhinitis for which a pharmacological antagonist is readily available. Intranasal challenge with histamine reproduces all the symptoms of allergic rhinitis: sneezing and pruritus, rhinorrhea, and nasal blockage. In children with allergic rhinitis challenged intranasally with antigens to which they have been naturally sensitized, H₁-antagonists prevent the sneezing, itching, and rhinorrhea from the early response to antigen, but are less effective for preventing or relieving the nasal blockage characteristic of the delayed response. In children with allergic conjunctivitis challenged with antigen, they are effective in preventing the ocular itching, tearing, and conjunctival erythema of the early response (18, 19) (Fig. 3).

H₁-antagonists are reasonably well studied in children aged 4–12 years with seasonal allergic rhinitis (20–40), although some of the studies can be criticized on the basis of lack of placebo control, or inadequate numbers of subjects. As determined using symptom scores, H₁-antagonists are more effective than placebo for relief of sneezing, pruritus, and rhinorrhea (Fig. 4) and have similar efficacy to each other (Fig. 5). In some countries, the H₁-antagonists levocabastine and azelastine are available for topical application to the nasal mucosa and the conjunctiva (18, 19, 39, 40) (Fig. 3; Table 1).

Additional dose-response efficacy studies, perennial allergic rhinitis studies, and comparative studies of H₁-antagonists versus intranasal cromolyn, nedocromil, and glucocorticoids are awaited in children.

In order to provide increased relief of nasal blockage, H₁-antagonists are sold in fixed-dose combinations with α-adrenergic agonists (decongestants) such as pseudoephedrine or phenylpropanolamine. First-generation H₁-antagonists are widely available in liquid formulations, often without prescription. They are presumed to be effective for use in infants and young children, although this has not been well-documented objectively in this age group. Some second-generation H₁-antagonist/decongestant liquid formulations are now available for study.

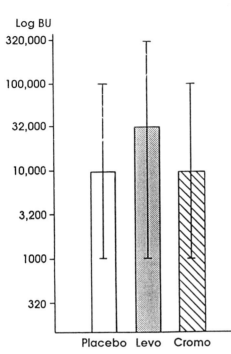

Figure 3 In a double-blind, cross-over study, 25 children with allergic rhinoconjunctivitis, aged 9–17 years, received 1 drop of levocabastine or cromolyn or placebo in the conjunctival sac of both eyes. After 15 min, a conjunctival provocation test was performed, starting with 320 BU pollen extract, increased every 10 min until a positive reaction occurred, or until 320,000 BU was reached. Levocabastine gave significantly better protection against eye redness, itching, and tearing than cromolyn or placebo. (From Ref. 19.)

2. Upper Respiratory Tract Infections

H_1-receptor antagonists are ubiquitously used for treatment of viral upper respiratory tract infections, but there is little scientific rationale to support this practice. Histamine concentrations are not increased in nasal secretions in subjects with symptomatic rhinovirus-induced "colds", in contrast to the increased levels of kinins, N-α-p-tosyl-L-arginine methyl ester (TAME)-esterase activity, and albumin which are found (41). In most studies of H_1-antagonists in children with "colds," the medications resulted in the same rate of improvement in respiratory symptoms as placebo

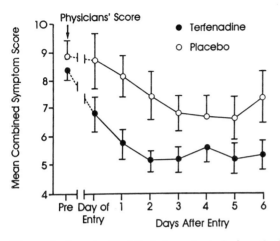

Figure 4 In a double-blind, 1-week study in 119 children aged 6–12 years with seasonal allergic rhinitis, control of nasal symptoms was observed in 85% of children taking terfenadine versus 60% taking placebo. Drowsiness was reported in 3.9% of children receiving terfenadine and 2.5% of children receiving placebo. (From Ref. 22.)

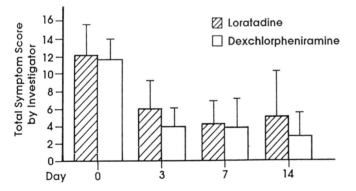

Figure 5 In a double-blind, 14-day, cross-over study in 40 children aged 4–12 years with allergic rhinitis, loratadine 0.11–0.24 mg/kg once daily was compared with dexchlorpheniramine 0.1–0.23 mg/kg every 4 h. There was no significant difference in the efficacy of the medications for relief of rhinitis symptoms. Drowsiness occurred only during dexchlorpheniramine treatment. (From Ref. 31.)

did (42–46); for example, in one recent study, no treatment, or placebo treatment, was as effective as H_1-antagonist/decongestant treatment (44).

3. Otitis Media

Acute otitis media and otitis media with effusion have high spontaneous remission rates. H_1-antagonists, often in combination with α-adrenergic agonist decongestants are frequently prescribed for infants and young children with otitis media, but there are no placebo-controlled, double-blind studies incorporating repeated objective assessment of tympanic membrane compliance to support a beneficial effect on eustachian tube function in these disorders (47–51). In one large, well-controlled, double-blind trial of a chlorpheniramine-pseudoephedrine combination versus placebo in nonallergic infants and children with otitis media with effusion, there was no difference in outcome between the two treatment regimens; in another study, the addition of a chlorpheniramine/pseudoephedrine combination to a 2-week course of amoxicillin provided no advantage in terms of efficacy, and increased the adverse effects (49).

In the context of these "negative" studies, it remains of interest that histamine concentrations are elevated in the middle ear effusions in otitis media (52, 53), and that the eustachian tube response to intranasal histamine and other chemical mediators of inflammation is increased in subjects with allergic rhinitis as compared to normal subjects .

4. Asthma

H_1-receptor antagonists do not play a major role in chronic asthma treatment, as histamine is only one of many chemical mediators of inflammation contributing to the pathophysiology of asthma. The early and late bronchoconstrictor responses produced by inhalation of allergen are associated with increased plasma concentrations of histamine. Increased circulating histamine has also been reported during spontaneous acute asthma episodes. In addition to production of H_1-blockade, terfenadine, astemizole, loratadine, azelastine, ketotifen, and other H_1-antagonists may have other effects such as mast-cell stabilization, platelet-activating factor or leukotriene antagonism, β_2-adrenergic receptor upregulation, or inhibition of eosinophil accumulation, which contribute to their modest efficacy in asthma.

Both first- and second-generation H_1-antagonists have been studied in asthma in childhood. Relatively high doses of H_1-antagonists seem to be required for H_1-blockade in the lung, in comparison to doses required for H_1-blockade in the nasal mucosa or skin. In some studies, H_1-antagonists such as chlorpheniramine, terfenadine, astemizole, loratadine, and ketotifen have been found to prevent histamine- and exercise-induced asthma (54–58), and relieve mild chronic asthma symptoms (59–65) (Fig. 6). In one study, cetirizine administered in relatively high doses provided significantly better asthma symptom control than placebo or cromolyn did, and children taking cetirizine used significantly less β_2-adrenergic agonist treatment (63) (Fig. 7). While the clinical importance of the antiasthma effect is small, previous concerns about adverse effects of H$_1$-antagonists in asthma, including specific concerns about potential drying of the secretions or bronchoconstriction (66) have not been substantiated.

Ketotifen is widely promulgated for use in children with asthma in some countries. When administered to infants with eczema and elevated total IgE levels, but no history of asthma, it has been reported to prevent or delay the development of asthma (67). Cetirizine is also being studied for its ability to prevent asthma in allergic infants.

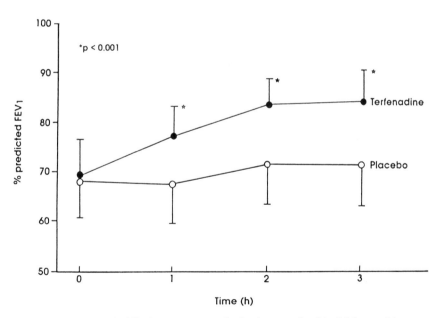

Figure 6 In a double-blind, cross-over, single-dose study, 20 children with exercise-induced bronchospasm received either terfenadine 60 mg or placebo. Terfenadine produced significant bronchodilation at 1, 2, and 3 h after the dose ($p < 0.001$). The increase in FEV$_1$ was 32% at 3 h. Terfenadine also reduced the maximum fall in peak expiratory flow after exercise by 10.5%. (From Ref. 58.)

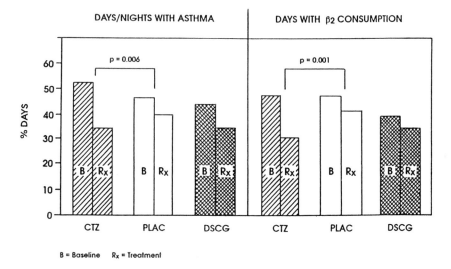

Figure 7 In a double-blind, cross-over study, 348 children and adolescents with mild or moderate asthma recorded symptoms and peak expiratory flows for 14 weeks. During a 2-week baseline period, only β_2-agonists by inhalation were permitted. For the next 12 weeks, cetirizine 5–15 mg twice daily, cromolyn, or placebo were given. During cetirizine treatment, the children required significantly less β_2-adrenergic agonist than they did while taking placebo, but despite this, asthma symptoms occurred on significantly fewer days and nights. The effect of cromolyn did not differ significantly from the effect of placebo. (From Ref. 63.)

5. Atopic Dermatitis

H_1-receptor antagonists relieve pruritus produced by histamine, an important pruritogen in atopic dermatitis. The evidence for the role of histamine in this disorder includes the following: the erythema and pruritus in atopic dermatitis resemble the effect of histamine introduced into the skin; mast-cell numbers and histamine concentrations are increased in the skin of subjects with atopic dermatitis; and subjects with atopic dermatitis may have elevated spontaneous release of histamine from basophils and elevated plasma histamine concentrations after antigen challenge (68).

Although H_1-antagonists are often used for suppression of pruritus in atopic dermatitis in infants and children, there are few studies of their efficacy in this disorder in this age group (6, 69–74) (Fig. 1). Many of these studies are not placebo-controlled, are of a short duration, and/or have been conducted in small numbers of subjects. The first-generation H₁-antagonists, especially hydroxyzine, are effective for relief of pruritus in atopic dermatitis. In one study that included children as young as 1 year of age, the antipruritic effect of a single dose of hydroxyzine was maintained for 24 h, when serum hydroxyzine concentrations had become low or negligible (6). In a 3-month, single-blind, cross-over study in young children, astemizole was significantly more effective than placebo, but less effective than hydroxyzine in reducing pruritus (71).

In other atopic dermatitis studies, terfenadine and loratadine were equally effective (72), ketotifen was superior to clemastine for relief of pruritus (73), and the effect of cetirizine was superior to placebo (74). Despite these studies, some investigators still believe that second-generation H₁-antagonists, lacking a sedative effect in recommended doses, should not replace the first-generation H₁-antagonists in the treatment of this disorder.

6. Urticaria, Mastocytosis, and Anaphylaxis

H₁-receptor antagonists are the most important medications available for relief of urticaria symptoms. Histamine, acting through its H₁-receptor, can mediate all the pathological features of urticaria: vasodilation, increased vascular permeability, whealing, flare, and sensory nerve stimulation leading to pruritus. When urticarial lesions are induced by heat, cold, or cholinergic stimuli, plasma histamine levels are transiently elevated in the veins draining the urticated area (68).

In many clinical trials in adults with urticaria, H₁-antagonists have proven to be significantly more effective than placebo in relieving itching and in reducing the number, size, and duration of urticarial lesions (2). Few satisfactory prospective, controlled, double-blind studies of H₁-antagonists have been performed in children with urticaria (75).

Plasma histamine concentrations are elevated in most children with pediatric mastocytosis syndrome. H₁-antagonists such as hydroxyzine are effective in the treatment of this disorder (76).

In anaphylaxis, H₁-antagonists may be helpful in controlling pruritus, urticaria, and other ancillary symptoms and signs, but they are not an adequate substitute for epinephrine and reliance on them may contribute to a fatal outcome (77).

B. Safety Issues

Few prospective studies of the potential adverse effects of H₁-receptor antagonists have been performed in young subjects (78–81), in contrast to the 60–70 prospective studies in adults described in recent reviews (82–85). Despite this, "recommended" doses of first-generation H₁-antagonists are available for infants and young children, and the liquid formulations of these older medications are advertised widely to physicians and to consumers for prescription and nonprescription use in this age group.

1. First-Generation H_1-Receptor Antagonists

Although most health care professionals and parents presume these medications to be safe for use in infants and children, this presumption may be incorrect. Diphenhydramine, promethazine, and hydroxyzine were formerly used to induce sleep electroencephalograms in this population and are still prescribed by some physicians for sedation, pain relief, and emesis prophylaxis in infants and children undergoing surgical procedures (86–94), although the safety of this practice has been questioned (93, 94).

The true incidence of adverse effects following manufacturers' recommended doses of first-generation H_1-antagonists in infants and children is unknown, and is documented chiefly in case reports (95–108). Adverse effects may occur following usual doses, as well as after overdose. Underreporting probably occurs, as the signs and symptoms of toxicity may be attributed to the illness for which the H_1-antagonist is being given. In infants and children with epidermal breakdown due to severe atopic dermatitis, varicella, or other pruritic skin disorders, toxic encephalopathy may occur after topical application of first-generation H_1-antagonists such as diphenhydramine or promethazine (109–113). Adolescents may intentionally misuse H_1-antagonists in search of a "high" (114).

First-generation H_1-antagonists cross the blood-brain barrier and have varying proclivity to cause fatigue, lassitude, somnolence, lethargy, and impairment of cognitive functioning and school performance (79–84). In infants and young children, they may also have paradoxical stimulatory effects on the central nervous system (CNS) and cause irritability, nervousness, hyperactivity, and convulsions (115). In addition to CNS adverse effects, they may produce blurred vision, dry mouth, and other anticholinergic effects. Some of them cause gastrointestinal symptoms. Cyproheptadine is also a 5-hydroxytryptamine (serotonin) antagonist and may cause appetite stimulation and weight gain (116). Trimeprazine or methdilazine may cause jaundice. Diphenhydramine and dimenhydrinate may cause fixed-drug eruptions (117). Very rarely, cytopenias occur. First-generation H_1-antagonists such as promethazine have been associated with apnea and sudden death in infants, although no cause and effect relationship has been established (118–120). The second-generation H_1-antagonist cetirizine does not increase the frequency of apnea episodes in infants (Personal Communication, Y. Brusquet).

2. Second-Generation H₁-Receptor Antagonists

The second-generation H_1-antagonists appear to be relatively free from CNS adverse effects in children, as evidenced by subjective information collected in thousands of children in allergic rhinoconjunctivitis studies; however, this advantage has been documented objectively with EEG or performance tests in only a few studies (78–81). In one study in which both central nervous system effects and peripheral H_1-blockade were assessed in children with allergic rhinitis, terfenadine proved to be safer and more effective than chlorpheniramine (Fig. 8) (79). In another study, the learning impairment observed in children with allergic rhinitis during the pollen season was exacerbated by administration of the first-generation H_1-antagonist diphenhydramine and partially ameliorated by administration of the second-generation H_1-antagonist loratadine (Fig. 9) (80).

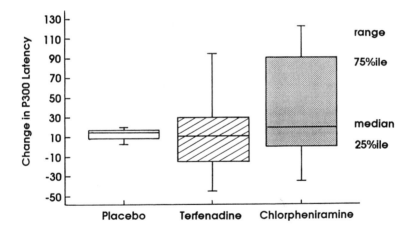

Figure 8 In 15 children with allergic rhinitis, mean age 8.5 ± 1.4 years, terfenadine 60 mg caused significantly less central nervous system dysfunction than chlorpheniramine 4 mg, as evidenced by lack of impairment of cognitive functioning (P300-event-related potential) and subjective symptom scores for somnolence (not shown). Tests were performed before and 2.5 h after a single dose. Terfenadine 60 mg also produced significantly greater peripheral blockade of H_1-receptors, as evidenced by suppression of the histamine-induced wheal and flare, than chlorpheniramine 4 mg or placebo (not shown). (From Ref. 79.)

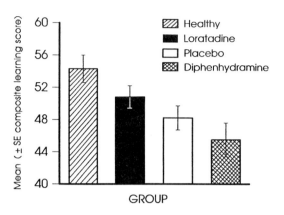

Figure 9 Seasonal allergic rhinitis adversely affected learning in children age 10–12 years who were given computer-assisted instruction in the form of a didactic simulation. Factual knowledge, conceptual knowledge, and knowledge application were tested. The adverse effect of the allergic rhinitis on learning was partially ameliorated by loratadine 10 mg, but exacerbated by diphenhydramine 25 mg. (From Ref. 80.)

3. Cardiovascular Effects

Astemizole and terfenadine have been reported to cause cardiac dysrhythmias after overdose or under other specific conditions such as hepatic dysfunction or concomitant administration with macrolide antibiotics such as erythromycin and clarithromycin or imidazole antifungals such as ketoconazole and itraconazole (83, 84, 121–124). Infants and children with preexisting cardiac problems such as long QT syndrome, or those taking medications that prolong the QTc interval may also be at increased risk. The mechanism involves excessive delay of repolarization that induces early after-depolarizations and delayed repolarization (124, 125), and produces marked QTc prolongation, bizarre T-wave changes, and a variety of dysrhythmias including polymorphic ventricular tachycardia described as torsade de pointes, literally "twisting of the points." Not all children at increased risk necessarily develop a dysrhythmia (Fig. 10) (121). Important warning symptoms include syncope at rest or with exercise, loss of consciousness, and palpitations. Additional prospective studies of the cardiovascular system adverse effects of H_1-antagonists are needed in young subjects, including pharmacoepidemiology studies, dose-response studies, and H₁-antagonist interaction studies with erythromycin and other commonly prescribed medications.

4. H₁-Receptor Antagonist Overdose

For decades, reports of severe toxic reactions and fatalities following overdose of first-generation H₁-antagonists in infants and children have appeared intermittently in the medical literature (95–108). All these medications, including diphenhydramine, dimenhydrinate, hydroxyzine, promethazine, cyproheptadine, doxylamine, tripelennamine, or pheniramine, are potentially lethal after overdose. In one 5-year pediatric study of accidental poisoning, the older H₁-antagonists were ranked as the second most common cause of hospital admission and death (126).

After overdose, infants and young children do not necessarily manifest lethargy, drowsiness, or coma, but may exhibit excitation, irritability, hyperactivity, visual hallucinations, and seizures, as well as anticholinergic effects such as dryness of the mucous membranes, fever, flushed fa-

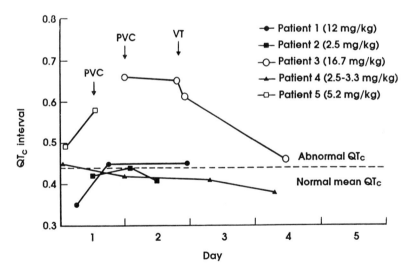

Figure 10 Six children, age 1.5 to 3.3 years, had an accidental astemizole overdose ranging from 2.5–16.7 mg/kg. Three of the children had no clinical symptoms; two were tired; and one collapsed and had 27 episodes of ventricular fibrillation. Electrocardiograms were obtained in five children, all of whom had a prolonged QTc interval. PVC = premature ventricular contractions; VT = episodes of ventricular tachycardia. (From Ref. 121.)

cies, pupillary dilation, urinary retention, and decreased gastrointestinal motility.

Hypotension secondary to α-adrenergic blockade, and sinus tachycardia secondary to the anticholinergic effects of the first-generation medications are reported. Occasionally, QTc prolongation and a variety of arrhythmias, including junctional rhythms, atrioventricular dissociation with a right bundle branch block or, rarely, torsade de pointes have been observed. Cardiorespiratory arrest and death may occur.

Treatment of infants and children who have had an overdose of a first-generation H_1-antagonist should include specific supportive measures such as use of anticonvulsants, or hemodialysis, if indicated. Centrally acting emetics such as ipecac are no longer recommended in poisonings and, in any case, are likely to be ineffective in infants or children who have received an overdose of H_1-antagonists such as diphenhydramine, dimenhydrinate, or promethazine, which have an antiemetic effect. Unless activated charcoal is administered very promptly after the overdose, it is useless in preventing H_1-antagonist absorption (127). There are no specific antidotes for H_1-antagonist poisoning.

There is little published clinical experience with children who have had an overdose of a second-generation H_1-antagonist (128). Very prompt administration of activated charcoal may prevent astemizole absorption (129). Most of the second-generation H_1-antagonists are not dialyzable (130, 131). After overdose of H_1-antagonists such as terfenadine or astemizole, continuous electrocardiographic monitoring should be performed for 24 h or until normalization occurs. This is important even if symptoms are absent and the electrocardiogram is normal at the time of presentation, as delayed torsade de pointes has been reported (121). If indicated, antiarrhythmic treatment should be instituted using cardioversion, pacing, and carefully selected antiarrhythmic medications. Magnesium sulfate and propranolol have been used successfully, but some antiarrhythmics such as quinidine, procainamide, and disopyramide (group 1A), and amiodarone and sotalol (group III), are contraindicated, as they may cause further QT_c prolongation and increase bradycardia.

5. H_1-Receptor Antagonist Use During Pregnancy and Lactation

H_1-antagonists cross the placenta. Some first-generation H_1-antagonists are teratogenic in animals, and have been suspected, although never proven, to cause fetal anomalies in humans. In the Collaborative Perinatal Project, an epidemiological investigation of possible teratogenic effects of drugs in pregnancy, 50,282 mother/child pairs were studied. Mean number of pharmaceutical products used was 3.8/mother. No overall evidence incriminating H_1-antagonists as teratogenic agents could be found. Of the H_1-antagonists, brompheniramine had a higher risk than chlorpheniramine, pheniramine, and diphenhydramine; and tripelennamine had a lower risk (132, 133).

Withdrawal symptoms such as tremulousness and irritability occurring several days after parturition have occurred in infants whose mothers received large therapeutic doses of first-generation H_1-antagonists such as hydroxyzine or diphenhydramine immediately before delivery (134, 135).

While there is no evidence that second-generation H_1-antagonists cause fetal anomalies in humans, and no evidence that withdrawal symptoms occur in infants whose mothers received second-generation H_1-antagonists in late pregnancy or during labor, many of these medications are classified as Pregnancy Category C in the United States and should be used during pregnancy *only* if the expected benefits to the mother exceed the unknown risks to the fetus.

H_1-antagonists are excreted in breast milk (136–141), for example, 0.03% of the total administered dose of loratadine is excreted via this route during 48 h (140). Drowsiness or irritability have been reported in nursing infants whose mothers have ingested older H_1-antagonists (137, 138).

III. SUMMARY

In children, H_1-antagonists are useful in allergic rhinoconjunctivitis treatment. Proof of their efficacy in this disorder has been obtained in limited numbers of pediatric studies, corroborated by many studies in adults. H_1-antagonists are important medications for relief of pruritus in urticaria and atopic dermatitis, although they have been inadequately studied in young subjects with these disorders. H_1-antagonists are not harmful to children with chronic asthma, but in the doses recommended for allergic rhinoconjunctivitis, they have a very minimal antiasthma effect. There is little scientific rationale for the use of H_1-antagonists in upper respiratory tract infections or otitis media.

First-generation H_1-antagonists are presumed to be safe for use in infants and young children. While they have undoubtedly been administered to millions of subjects in this age group without incident, they may cause CNS dysfunction more commonly than generally realized. Their benefit/risk ratio should be critically assessed before they are recommended for use in any infant or young child. Second-generation H_1-antagonists are relatively free from CNS adverse effects; however, in some countries, these newer H_1-antagonists are still not approved for use in children under age 12 years. Additional objective studies of H_1-antagonists are required in infants and children with regard to both efficacy and safety.

REFERENCES

1. Vaughan VC. Allergic disorders. In: Nelson WE, ed. Textbook of Pediatrics. Philadelphia: W. B. Saunders Company, 1964;1453–1481.
2. Simons FER, Simons KJ. The pharmacology and use of H_1-receptor antagonist drugs. N Engl J Med 1994;330:1663–1670.
3. Thompson JA, Bloedow DC, Leffert FH. Pharmacokinetics of intravenous chlorpheniramine in children. J Pharm Sci 1981;70:1284–1286.
4. Simons FER, Luciuk GH, Simons KJ. Pharmacokinetics and efficacy of chlorpheniramine in children. J Allergy Clin Immunol 1982;69:376–381.
5. Simons KJ, Watson WTA, Martin TJ, Chen XY, Simons FER. Diphenhydramine: pharmacokinetics and pharmacodynamics in elderly adults, young adults, and children. J Clin Pharmacol 1990;30:665–671.
6. Simons FER, Simons KJ, Becker AB, Haydey RP. Pharmacokinetics and antipruritic effects of hydroxyzine in children with atopic dermatitis. J Pediatr 1984;104:123–127.
7. Simons FER, Watson WTA, Simons KJ. The pharmacokinetics and pharmacodynamics of terfenadine in children. J Allergy Clin Immunol 1987;80:884–890.
8. Möller C, Andlin-Sobocki P, Blychert L-O. Pharmacokinetics of astemizole in children. Rhinology 1992;13:21–25.
9. Lin CC, Radwanski E, Affrime MB, Cayen MN. Pharmacokinetics of loratadine in pediatric subjects. Am J Therapeutics 1995;2:504–508.
10. Watson WTA, Simons KJ, Chen XY, Simons FER. Cetirizine: a pharmacokinetic and pharmacodynamic evaluation in children with seasonal allergic rhinitis. J Allergy Clin Immunol 1989;84:457–464.
11. Desager JP, Dab I, Horsmans Y, Harvengt C. A pharmacokinetic evaluation of the second-generation H_1-receptor antagonist cetirizine in very young children. Clin Pharmacol Ther 1993;53:431–435.
12. Simons FER, Watson WTA, Simons KJ. Pharmacokinetics and pharmacodynamics of ebastine in children. J Pediatr 1993;122:641–646.
13. Kennedy GR. Metabolism and pharmacokinetics of ketotifen in children. Res Clin Forums 1982;4:17–20.
14. Schmidt-Redemann B, Brenneisen P, Schmidt-Redemann W, Gonda S. The determination of pharmacokinetic parameters of ketotifen in steady state in young children. Int J Clin Pharmacol Ther Toxicol 1986;24:496–498.
15. Simons FER, Lukowski JL, Becker AB, Simons KJ. Comparison of the effects of single doses of the new H_1-receptor antagonists loratadine and terfenadine versus placebo in children. J Pediatr 1991;118:298–300.

16. Simons FER, Watson WTA, Becker AB, Simons KJ. Histamine blockade after astemizole in children: a single-dose, placebo-controlled study. Pediatr Allergy Immunol 1994;5:214–217.

17. Spicak V. The pharmacokinetics of cetirizine in infants. J Allergy Clin Immunol 1995;95:197.

18. Kjellman N-IM, Andersson B. Terfenadine reduces skin and conjunctival reactivity in grass pollen allergic children. Clin Allergy 1986;16:441–449.

19. Rimas M, Kjellman N-IM, Blychert L-O, Björkstén B. Topical levocabastine protects better than sodium cromoglycate and placebo in conjunctival provocation tests. Allergy 1990;45:18–21.

20. Lockhart JDF, Maneksha S. Children with allergies. Terfenadine suspension versus placebo. The Practitioner 1983;227:1313–1315.

21. Molkhou P, Beaumont D. Efficacy and tolerance of terfenadine suspension in children with allergic rhinitis. A double-blind trial versus placebo. Acta Therap 1985;11:99–107.

22. Guill MF, Buckley RH, Rocha Jr W, et al. Multicenter, double-blind, placebo-controlled trial of terfenadine suspension in the treatment of fall-allergic rhinitis in children. J Allergy Clin Immunol 1986;78:4–9.

23. Grillage MG, Harcup JW, Mayhew SR, Huddlestone L. Astemizole suspension in the maintenance treatment of paediatric hay fever: a comparison with terfenadine suspension. Pharmatherapeutica 1986;4:642–647.

24. Hedley K, Main PGN, Tristram SJ. Astemizole suspension for maintenance therapy of hay fever in children aged 6 to 12 years. Pharmatherapeutica 1984;3:18–26.

25. Naspitz CK, Solé D, Wandalsen NF, Prado E, Rosario Filho N, Sano F. Comparative multicentric double-blind study between dexchlorpheniramine and astemizole in the treatment of children with perennial allergic rhinitis. Am J Rhin 1990;4:109–114.

26. Bronsky EA, Berkowitz R, Dockhorn RJ, et al. Pediatric dose-ranging trial of astemizole suspension in seasonal allergic rhinitis. J Allergy Clin Immunol 1992;89:157.

27. Wood SF. Clinical experience with non-sedating antihistamines in paediatric allergic rhinitis. Rhinology 1992;13:27–37.

28. Meltzer EO, Ellis EF, Rosen JP, et al. A comparison of loratadine, chlorpheniramine and placebo suspensions in children with seasonal allergic rhinitis. J Allergy Clin Immunol 1988;81:177.

29. Novembre E, Bernardini R, Marano E, Iudice A, Vergallo G, Vierucci A. Comparative efficacy and tolerability of terfenadine versus astemizole in the treatment of seasonal allergic rhinoconjunctivitis in children. Curr Ther Res 1990;47:765–771.

30. Siegel S, Ellis E, Kemp J, et al. Efficacy and safety of loratadine in the treatment of allergic rhinitis in children. Allergy 1988;43:5.

31. Boner AL, Miglioranzi P, Richelli C, Marchesi E, Andreoli A. Efficacy and safety of loratadine suspension in the treatment of children with allergic rhinitis. Allergy 1989;44:437–441.

32. Boner AL, Richelli C, Castellani C, Marchesi E, Andreoli A. Comparison of the effects of loratadine and astemizole in the treatment of children with seasonal allergic rhinoconjunctivitis. Allergy 1992;47:98–102.

33. Lutsky BN, Klöse P, Melon J, et al. A comparative study of the efficacy and safety of loratadine syrup and terfenadine suspension in the treatment of 3- to 6-year old children with seasonal allergic rhinitis. Clin Ther 1993; 15:855–865.

34. Baelde Y, Dupont P. Cetirizine in children with chronic allergic rhinitis. A multicentre double-blind study of two doses of cetirizine and placebo. Drug Invest 1992;4:466–472.

35. Masi M, Candiani R, van de Venne H. A placebo-controlled trial of cetirizine in seasonal allergic rhinoconjunctivitis in children aged 6 to 12 years. Ped Allergy Immunol 1993;4:47–52.

36. Allegra L, Paupe J, Wieseman HG, Baelde Y. Cetirizine for seasonal allergic rhinitis in children aged 2–6 years. A double-blind comparison with placebo. Ped Allergy Immunol 1993;4:157–161.

37. Storms WW, Garcia JD, Tobey RE, et al. Once-a-day ebastine is effective therapy for seasonal allergic rhinitis in children. J Allergy Clin Immunol 1993;91:197.

38. Ostrom N, Welch M, Morris R, et al. Evaluation of ebastine, a new non-sedating antihistamine, in children with seasonal allergic rhinitis. J Allergy Clin Immunol 1994;93:163.

39. Falconieri P, Monteleone AM, Mancuso T, Arcese G, Businco L. Double-blind placebo-controlled study with levocabastine eye drops in children with allergic conjunctivitis. J Allergy Clin Immunol 1994;93:272.

40. Watson WTA, Roberts JR, Knight A, et al. Levocabastine-D nasal spray in the management of pediatric patients with seasonal allergic rhinitis. J Allergy Clin Immunol 1994;93:296.

41. Proud D, Naclerio RM, Gwaltney JM, Hendley JO. Kinins are generated in nasal secretions during natural rhinovirus colds. J Infect Dis 1990;161: 120–123.

42. Jaffé G, Grimshaw JJ. Randomized single-blind trial in general practice comparing the efficacy and palatability of two cough linctus preparations, 'Pholcolix' and 'Actifed' Compound, in children with acute cough. Curr Med Res Opin 1983;8:594–599.

43. Weippl G. Therapeutic approaches to the common cold in children. Clin Ther 1984;6:475–482.

44. Hutton N, Wilson MH, Mellits ED, et al. Effectiveness of an antihistamine-decongestant combination for young children with the common cold: A randomized, controlled clinical trial. J Pediatr 1991;118:125–130.

45. Smith MBH, Feldman W. Over-the-counter cold medications. A critical review of clinical trials between 1950 and 1991. JAMA 1993;269:2258–2263.

46. Kogan MD, Pappas G, Yu SM, Kotelchuck M. Over-the-counter medication use among US preschool-age children. JAMA 1994;272:1025–1030.

47. Schnore SK, Sangster JF, Gerace TM, Bass MJ. Are antihistamine-decongestants of value in the treatment of acute otitis media in children? J Fam Pract 1986;22:39–43.
48. Cantekin EI, Mandel EM, Bluestone CD, et al. Lack of efficacy of a decongestant-antihistamine combination for otitis media with effusion ("secretory" otitis media) in children. N Engl J Med 1983;308:297–301.
49. Mandel EM, Rockette HE, Bluestone CD, Paradise JL, Nozza RJ. Efficacy of amoxicillin with and without decongestant-antihistamine for otitis media with effusion in children. N Engl J Med 1987;316:432–437.
50. Cantekin EI, McGuire TW, Griffith TL. Antimicrobial therapy for otitis media with effusion ("secretory" otitis media). JAMA 1991;266:3309–3317.
51. The Otitis Media Guideline Panel. Managing otitis media with effusion in young children. Pediatrics 1994;94:766–773.
52. Skoner DP, Doyle WJ, Tanner EP, Fireman P. Histamine in middle ear fluids. Ann Allergy 1985;54:350.
53. Chonmaitree T, Patel JA, Lett-Brown MA, et al. Viruses and bacteria enhance histamine production in middle ear fluids of children with acute otitis media. J Infect Dis 1994;169:1265–1270.
54. Woenne R, Kattan M, Orange RP, Levison H. Bronchial hyperreactivity to histamine and methacholine in asthmatic children after inhalation of SCH 1000 and chlorpheniramine maleate. J Allergy Clin Immunol 1978;62:119–124.
55. Klein G, Urbanek R, Matthys H. Long-term study of the protective effect of ketotifen in children with allergic bronchial asthma. The value of a provocation test in assessment of treatment. Respiration 1981;41:128–132.
56. Graff-Lonnevig V, Hedlin G. The effect of ketotifen on bronchial hyperreactivity in childhood asthma. J Allergy Clin Immunol 1985;76:59–63.
57. Backer V, Bach-Mortensen N, Becker U, et al. The effect of astemizole on bronchial hyperresponsiveness and exercise-induced asthma in children. Allergy 1989;44:209–213.
58. MacFarlane PI, Heaf DP. Selective histamine blockade in childhood asthma; the effect of terfenadine on resting bronchial tone and exercise induced bronchoconstriction. Respir Med 1989;83:19–24.
59. Groggins RC, Milner AD, Stokes GM. The bronchodilator effects of chlorpheniramine in childhood asthma. Br J Dis Chest 1979;73:297–301.
60. Graff-Lonnevig V, Kusoffsky E. Comparison of the clinical effect of ketotifen and DSCG in pollen-induced childhood asthma. Allergy 1980;35:341–348.
61. Lewiston NJ, Johnson S, Sloan E. Effect of antihistamine on pulmonary function of children with asthma. J Pediatr 1982;101:458–460.
62. Simons FER, Luciuk GH, Becker AB, Gillespie CA. Ketotifen: a new drug for prophylaxis of asthma in children. Ann Allergy 1982;48:145–150.
63. Kjellman N-IM. Is there a place for antihistamines in the treatment of perennial asthma? Ped Allergy Immunol 1993;4:38–43.
64. Menardo JL, Clavel R, Couturier P, et al. Evaluation of prophylactic treat-

ment with loratadine (L) vs sodium cromoglycate (SC) in allergic children with mild to moderate bronchial asthma. Allergy 1993;48:31.

65. Rosario NA, Kantor O Jr. Acute bronchodilator effect of loratadine in asthmatic children. Allergy 1993;48:36.

66. Schuller DE. Adverse effects of brompheniramine on pulmonary function in a subset of asthmatic children. J Allergy Clin Immunol 1983;72:175–179.

67. Iikura Y, Naspitz CK, Mikawa H, et al. Prevention of asthma by ketotifen in infants with atopic dermatitis. Ann Allergy 1992;68:233–236.

68. Wahlgren C-F. Pathophysiology of itching in urticaria and atopic dermatitis. Allergy 1992;47:65–75.

69. Klein GL, Galant SP. A comparison of the antipruritic efficacy of hydroxyzine and cyproheptadine in children with atopic dermatitis. Ann Allergy 1980;44:142–145.

70. Fiocchi A, Riva E, Borella E, Arensi D, Giovannini M. Ketotifen treatment of atopic dermatitis in childhood. Curr Ther Res 1985;37:1113–1123.

71. Roberts JR, Simons FER, Simons KJ, Gillespie CA. The antipruritic effect of placebo versus hydroxyzine versus astemizole in children with severe atopic dermatitis. J Allergy Clin Immunol 1988;81:210.

72. Lutsky BN, Schuller JL, Cerio R, et al. Comparative study of the efficacy and safety of loratadine syrup and terfenadine suspension in the treatment of chronic allergic skin diseases in a pediatric population. Arzneim-Forsch/Drug Res 1993;43:1196–1199.

73. Yoshida H, Niimura M, Ueda H, Imaura S, Yamamoto S, Kukita A. Clinical evaluation of ketotifen syrup on atopic dermatitis: a comparative multicenter double-blind study of ketotifen and clemastine. Ann Allergy 1989;62:507–512.

74. La Rosa M, Ranno C, Musarra I, Guglielmo F, Corrias A, Bellanti JA. Double-blind study of cetirizine in atopic eczema in children. Ann Allergy 1994;73:117–122.

75. Önes Ü. Astemizole treatment in childhood chronic urticaria. NER Allergy Proc 1988;9:355.

76. Kettelhut BV, Metcalfe DD. Pediatric mastocytosis. J Invest Dermatol 1991;96:15S–18S.

77. Yunginger JW, Sweeney KG, Sturner WQ, et al. Fatal food-induced anaphylaxis. JAMA 1988;260:1450–1452.

78. Feldman W, Shanon A, Leiken L, Ham-pong A, Peterson R. Central nervous system side-effects of antihistamines in schoolchildren. Rhinology 1992;13:13–19.

79. Simons FER, Reggin JD, Roberts JR, Simons KJ. Benefit/risk ratio of the antihistamines (H_1-receptor antagonists) terfenadine and chlorpheniramine in children. J Pediatr 1994;124:979–983.

80. Vuurman EFPM, van Veggel LMA, Uiterwijk MMC, Leutner D, O'Hanlon JF. Seasonal allergic rhinitis and antihistamine effects on children's learning. Ann Allergy 1993;71:121–126.

81. Simons FER, Fraser TG, Reggin JD, Roberts JR, Simons KJ. Adverse

central nervous system effects of older antihistamines in children. Pediatr Allergy Immunol (in press).

82. Meltzer EO. Comparative safety of H₁-antihistamines. Ann Allergy 1991; 67:625–633.

83. Simons FER. The therapeutic index of newer H₁-receptor antagonists. Clin Exp Allergy 1994;24:707–723.

84. Simons FER. H₁-receptor antagonists. Comparative tolerability and safety. Drug Safety 1994;10:350–380.

85. Rombaut NEI, Hindmarch I. Psychometric aspects of antihistamines: a review. Human Psychopharmacol Clin Exp 1994;9:157–169.

86. Russo RM, Gururaj VJ, Allen JE. The effectiveness of diphenhydramine HCl in pediatric sleep disorders. J Clin Pharmacol 1976;16:284–288.

87. Korein J, Fish B, Shapiro T, Gerner EW, Levidow L. EEG and behavioral effects of drug therapy in children. Chlorpromazine and diphenhydramine. Arch Gen Psychiatry 1971;24:552–563.

88. Kapetansky D, Warren R, Hawtof D. Cleft lip repair using intramuscular hydroxyzine sedation and local anesthesia. Cleft Palate-Craniofacial J 1992; 29:481–484.

89. Lacouture PG, Gaudreault P, Lovejoy Jr. FH. Chronic pain of childhood: a pharmacologic approach. Pediatr Clin North Am 1984;31:1133–1151.

90. Hawk W, Crockett RK, Ochsenschlager DW, Klein BL. Conscious sedation of the pediatric patient for suturing: A survey. Ped Emerg Care 1990;6: 84–88.

91. Blanc VF, Ruest P, Jacob J-L, Tang A. Antiemetic prophylaxis with promethazine or droperidol in paediatric outpatient strabismus surgery. Can J Anaesth 1991;38:54–60.

92. O'Brien JF, Falk JL, Carey BE, Malone LC. Rectal thiopental compared with intramuscular meperidine, promethazine, and chlorpromazine for pediatric sedation. Ann Emerg Med 1991;20:644–647.

93. Snodgrass WR, Dodge WF. Lytic/"DPT" cocktail: time for rational and safe alternatives. Pediatr Clin North Am 1989;36:1285–1291.

94. Nahata MC, Clotz MA, Krogg EA. Adverse effects of meperidine, promethazine, and chlorpromazine for sedation in pediatric patients. Clin Pediatr 1985;24:558–560.

95. Wyngaarden JB, Seevers MH. The toxic effects of antihistaminic drugs. JAMA 1951;145:277–282.

96. Lavenstein BL, Cantor FK. Acute dystonia. An unusual reaction to diphenhydramine. JAMA 1976;236:291.

97. Berger M, White J, Travis LB, et al. Toxic psychosis due to cyproheptadine in a child on hemodialysis: a case report. Clin Nephrol 1977;7:43–44.

98. Lewith GT, Davidson F. Dystonic reactions to Dimetapp elixir. J R Coll Gen Pract 1981;31:241.

99. DeGrandi T, Simon JE. Promethazine-induced dystonic reaction. Ped Emerg Care 1987;3:91–92.

100. Dollberg S, Hurvitz H, Kerem E, Navon P, Branski D. Hallucinations and

hyperthermia after promethazine ingestion. Acta Paediatr Scand 1989;78: 131–132.

101. Huxtable RF, Landwirth J. Diphenhydramine poisoning treated by exchange transfusion. Am J Dis Child 1963;106:496–500.

102. Magera BE, Betlach CJ, Sweatt AP, Derrick CW Jr. Hydroxyzine intoxication in a 13-month-old child. Pediatrics 1981;67:280–283.

103. Krenzelok EP, Anderson GM, Mirick M. Massive diphenhydramine overdose resulting in death. Ann Emerg Med 1982;11:212–213.

104. Kumar VV, Devi KR. Acute cyproheptadine poisoning. Indian J Pediatr 1989;56:521–523.

105. Baehr GR, Romano M, Young JM. An unusual case of cyproheptadine (Periactin) overdose in an adolescent female. Ped Emerg Care 1986;2: 183–185.

106. Richmond M, Seger D. Central anticholinergic syndrome in a child: a case report. J Emerg Med 1986;3:453–456.

107. Köppel C, Ibe K, Tenczer J. Clinical symptomatology of diphenhydramine overdose: an evaluation of 136 cases in 1982 to 1985. J Toxicol Clin Toxicol 1987;25:53–70.

108. Goetz CM, Lopez G, Dean BS, Krenzelok EP. Accidental childhood death from diphenhydramine overdosage. Am J Emerg Med 1990;8:321–322.

109. Shawn DH, McGuigan MA. Poisoning from dermal absorption of promethazine. Can Med Assoc J 1984;130:1460–1461.

110. Filloux F. Toxic encephalopathy caused by topically applied diphenhydramine. J Pediatr 1986;108:1018–1020.

111. Tomlinson G, Helfaer M, Wiedermann BL. Diphenhydramine toxicity mimicking varicella encephalitis. Pediatr Infect Dis J 1987;6:220–221.

112. Woodward GA, Baldassano RN. Topical diphenhydramine toxicity in a five year old with varicella. Ped Emerg Care 1988;4:18–20.

113. Reilly JF Jr, Weisse ME. Topically induced diphenhydramine toxicity. J Emerg Med 1990;8:59–61.

114. Gott PH. Cyclizine toxicity—intentional drug abuse of a proprietary antihistamine. N Engl J Med 1968;279:596.

115. Yokoyama H, Iinuma K, Yanai K, Watanabe T, Sakurai E, Onodera K. Proconvulsant effect of ketotifen, a histamine H_1-antagonist, confirmed by the use of d-chlorpheniramine with monitoring electroencephalography. Meth Find Exp Clin Pharmacol 1993;15:183–188.

116. Arisaka O, Shimura N, Nakayama Y, Yabuta K. Cyproheptadine and growth. Am J Dis Child 1988;142:914–915.

117. Coskey RJ. Contact dermatitis caused by diphenhydramine hydrochloride. J Am Acad Dermatol 1983;8:204–206.

118. Kahn A, Blum D. Phenothiazines and sudden infant death syndrome. Pediatrics 1982;70:75–78.

119. Hickson GB, Altemeier WA, Clayton EW. Should promethazine in liquid form be available without prescription? Pediatrics 1990;86:221–225.

120. Sandyk R. Phenothiazine-induced sleep apneas and the opioid system. Pediatrics 1986;77:261.

121. Hoppu K, Tikanoja T, Tapanainen P, Remes M, Saarenpää-Heikkilä O, Kouvalainen K. Accidental astemizole overdose in young children. Lancet 1991;338:538–540.

122. Tobin JR, Doyle TP, Ackerman AD, Brenner JI. Astemizole-induced cardiac conduction disturbances in a child. JAMA 1991;266:2737–2740.

123. Wiley II JF, Gelber ML, Henretig FM, Wiley CC, Sandhu S, Loiselle J. Cardiotoxic effects of astemizole overdose in children. J Pediatr 1992;120: 799–802.

124. Woosley RL, Chen Y, Freiman JP, Gillis RA. Mechanism of the cardiotoxic actions of terfenadine. JAMA 1993;269:1532–1536.

125. Rampe D, Wible B, Brown AM, Dage RC. Effects of terfenadine and its metabolites on a delayed rectifier K⁺ channel cloned from human heart. Mol Pharmacol 1993;44:1240–1245.

126. Pearn J, Nixon J, Ansford A, Corcoran A. Accidental poisoning in childhood: five year urban population study with 15 year analysis of fatality. Br Med J 1984;288:44–46.

127. Guay DRP, Meatherall RC, Macaulay PA, Yeung C. Activated charcoal adsorption of diphenhydramine. Int J Clin Pharmacol Ther Toxicol 1984; 22:395–400.

128. Le Blaye I, Donatini B, Hall M, Krupp P. Acute ketotifen overdosage. A review of present clinical experience. Drug Safety 1992;7:387–392.

129. Laine K, Kivistö KT, Neuvonen PJ. The effect of activated charcoal on the absorption and elimination of astemizole. Human Exp Toxicol 1994;13: 502–505.

130. Awni WM, Yeh J, Halstenson CE, Opsahl JA, Chung M, Matzke GR. Effect of haemodialysis on the pharmacokinetics of cetirizine. Eur J Clin Pharmacol 1990;38:67–69.

131. Zazgornik J, Scholz N, Heykants J, Vanden Bussche G. Plasma concentrations of astemizole in patients with terminal renal insufficiency, before, during and after hemodialysis. Int J Clin Pharmacol Ther Toxicol 1986;24: 246–248.

132. Aselton P, Jick H, Milunsky A, Hunter JR, Stergachis A. First-trimester drug use and congenital disorders. Obstet Gynecol 1985;65:451–455.

133. Schatz M, Hoffman CP, Zeiger RS, Falkoff R, Macy E, Mellon M. The course and management of asthma and allergic diseases during pregnancy. In: Middleton E Jr, Reed CE, Ellis EF, Adkinson NF Jr, Yunginger JW, Busse WW, eds. Allergy Principles and Practice. St. Louis: Mosby-Year Book, Inc., 1993;1301–1342.

134. Parkin DE. Probable Benadryl withdrawal manifestations in a newborn infant. J Pediatr 1974;85:580.

135. Prenner BM. Neonatal withdrawal syndrome associated with hydroxyzine hydrochloride. Am J Dis Child 1977;131:529–530.

136. Kauffman RE, Banner W Jr, Berlin CM Jr, et al. The transfer of drugs and other chemicals into human milk. Pediatrics 1994;93:137–150.

137. Ito S, Blajchman A, Stephenson M, Eliopoulos C, Koren G. Prospective follow-up of adverse reactions in breast-fed infants exposed to maternal medication. Am J Obstet Gynecol 1993;168:1393–1399.

138. Kok THHG, Taitz LS, Bennett MJ, Holt DW. Drowsiness due to clemastine transmitted in breast milk. Lancet 1982;1:914–915.

139. Findlay JWA, Butz RF, Sailstad JM, Warren JT, Welch RM. Pseudoephedrine and triprolidine in plasma and breast milk of nursing mothers. Br J Clin Pharmacol 1984;18:901–906.

140. Hilbert J, Radwanski E, Affrime MB, Perentesis G, Symchowicz S, Zampaglione N. Excretion of loratadine in human breast milk. J Clin Pharmacol 1988;28:234–239.

141. Lucas BD Jr, Purdy CY, Scarim SK, et al. Terfenadine pharmacokinetics in breast milk in lactating women. Clin Pharmacol Ther 1995;57:398–402.

13

Adverse Effects of H₁-Receptor Antagonists in the Central Nervous System

Eli O. Meltzer and Michael J. Welch
University of California, San Diego, California

I. INTRODUCTION

The use of H_1-antihistamines has increased steadily over the years. In the United States over 30 million people are treated with an antihistamine each year (1). In 1988, single-entity prescription and over-the-counter (OTC) antihistamine sales exceeded $500 million. This amount approaches nearly $2 billion when antihistamine-decongestant combinations are included in the data (2).

In addition to prescription pharmacological agents, approximately 300,000 OTC medications are currently available. Of the highest selling nonprescription medications launched after 1975, 6 of the top 10 in 1990 (Dimetapp®, Benadryl®, Actifed®, Drixoral®, Comtrex®, Chlor-Trimeton®) were single-entity antihistamines or antihistamine-decongestant combinations (3). Patients who choose these drugs are often inadequately informed about their potential adverse effects. This problem is compounded by the fact that some adults are functionally illiterate and cannot understand the labeling on the OTC medication. Furthermore, the print on the labels is often quite small, also placing the consumer at a disadvantage.

Whether medications are prescribed by the physician or purchased directly over the counter, it is important that the physician and the patient know their potential risks. Physicians are committed always to obey the cardinal law of medicine, "primum non nocere"—first of all, do no harm. Therefore, the safety profile of agents used as commonly as antihistamines needs to be clearly delineated. The purpose of this chapter is to examine the adverse effects of H_1-receptor antagonists on the central nervous system (CNS). The tests used to assess alertness and performance will be described and the effects of the various antihistamines will be discussed.

II. CENTRAL NERVOUS SYSTEM EFFECTS OF ANTIHISTAMINES

A medication's ability to cross the blood-brain barrier is the critical determinant of its effect on central nervous system function. Entry of chemicals into the brain is regulated by an ill-defined mechanism that resides in the endothelial lining of the cerebral capillaries. Antihistamines with a small lipophilic molecular structure that circulate unbound to protein enter the brain relatively easily. Many newer H_1-antagonists (e.g., terfenadine, astemizole) are large lipophilic molecules with a charged side chain. They are bound extensively to protein and, consequently, have difficulty crossing the blood-brain barrier (4). Radioactive tracer tests in animals confirm the absence of terfenadine, astemizole, and their metabolites in the CNS (5).

Four types of adverse nervous system effects from antihistamines have been reported: stimulatory, neuropsychiatric, peripheral, and depressive reactions. Stimulatory effects include dyskinesia with muscle spasms of the facial muscles, tongue, neck and hands (6); activation of epileptogenic foci (7); euphoria, hyperreflexia, hypertension, and headaches, and anticholinergic reactions like insomnia, irritability, nervousness, tachycardia, and tremor (8). Neuropsychiatric reactions have produced anxiety, confusion, depression, hallucinations, and psychosis (9). Adverse peripheral neurological effects include paresthesias and paralysis and, with cholinergic blockade, dilated pupils, impairment of accommodation, blurred vision, and diplopia (10).

The most common antihistamine CNS adverse effects are depressive or suppressive. The experience of sedation or drowsiness from H_1-receptor antagonists approved in the U.S. prior to 1985 (first-generation agents) occurs in 10 to 25% of users (11–14). The specific symptoms include drowsiness, fatigue, lassitude, sedation, dizziness, and weakness. These adverse effects are frequently dose-related and correlate with the serum H_1-antagonist concentration (15). Among the first-generation antihistamines, ethanolamines and phenothiazines have marked sedative effects (16). Ethylenediamines cause moderate sedation, and alkylamines typically cause mild sedation. The newer, second-generation H_1-receptor antagonists generally have fewer central nervous system effects and, in the recommended doses, most appear to cause no more sedation than placebo does.

The mechanism by which the various H_1-blocking drugs produce their CNS depressant effects is uncertain. It is assumed to be inhibition of histamine N-methyl-transferase or blockage of one of the central receptors (histaminergic, serotoninergic, cholinergic, or alpha-adrenergic). Older antihistamines interact with these receptors in the brain. The newer, less sedating agents appear to have both decreased penetration of the blood-brain barrier and greater affinity and specificity for peripheral H_1-receptors. They are therefore less likely to cause these central effects (17).

A number of factors can further influence an individual's susceptibility to the adverse effects of antihistamines. These include the age of the patient, circadian rhythm, quality of nocturnal sleep, disease state, and concomitant drugs which might further compromise the central nervous system (18).

III. ASSESSMENT OF THE CENTRAL NERVOUS SYSTEM EFFECTS OF H₁-ANTAGONISTS

Numerous researchers have evaluated the effects of H_1-antihistamines on the CNS using various methods (Table 1). The decrease in central nervous system function from antihistamines is complex and cannot be reflected in one measurement. There have been attempts to distinguish between a drug's tendency to induce sleepiness and its potential to induce cognitive and psychomotor performance impairment. Sleepiness or drowsiness is clinically defined as a state of reduced mental alertness (19). It can be rated subjectively by methods listed in Table 1 or measured objectively using electrophysiological tests. Drug-induced impairment consists of a transient but inescapable state of decreased cognitive or psychomotor performance during which an individual cannot rally to overcome the impairment (20). It is measured objectively by task parameters. Some of these performance functions and task tests are listed in Table 2. The disappearance of drug-induced impairment is related more to biological drug clearance from the central nervous system than to patient motivation (21).

The term sedation has been used to include both drowsiness and impairment. It can be defined as a global reduction of alertness as well as intellectual and motor performances (22).

Sleepiness/alertness level provides a basic capacity for performance. Performance is usually poorest at the trough of alertness. Performance is then modified by factors such as motivation and task familiarity. Various

Table 1 Methods to Assess Sedation Effects of Antihistamines

DROWSINESS TESTS
Subjective Evaluations
    Diary cards to record presence or absence of daytime sleepiness
    Diary cards with visual analog scale or numerical scales to record the magni-
        tude of daytime sleepiness
    Stanford Scale for Sleepiness
    Self-rating of mood using Profile-of-Moods questionnaire
    Positive Affect-Negative Affect scales
Objective Evaluation
    Multiple sleep latency test with electroencephalographic potentials
COGNITIVE AND PSYCHOMOTOR PERFORMANCE IMPAIRMENT
  TESTS
    P3-evoked electroencephalographic potential latency
    Blinking speed with electrooculography
    Visual function
      Dynamic visual acuity
      Critical flicker fusion
    Simple reaction time
    Choice reaction time
    Vigilance tests
    Simulated assembly line tests
    Word-color test
    Digit-symbol substitution
    Office clerical battery
    Computer-simulated driving test
    Computer-simulated flying test
    Actual driving test
      Coarse steering frequency
      Obstacle test
      Weaving index test
    Computer-simulated scholastic test

researchers have reported an inconsistent correlation between the symptoms and measures of drowsiness and antihistamine-induced mental impairment (23–26). Therefore, both should be assessed to determine the potential for adverse central nervous system effects of H_1-antihistamines. Also, because the patient's subjective perception does not always agree with objective findings, quantifiable evaluations should be used (26).

Tolerance to antihistamine-induced drowsiness has been reported (19, 27). However, no trend toward tolerance to the subjective symptoms of

Table 2 Central Nervous System: Psychomotor
Function Tests

Function	Test
Stimulus detection	Auditory vigilance
Perception	Letter cancellation
Recognition	Digit-symbol substitution
Processing	Mental arithmetic
Integration	Critical flicker fusion
Memory	Digit span
Learning	Verbal learning
Ballistic	Finger tapping
Gross motor	Stabilometer
Fine motor	Hand steadiness
Coordination	Peg board

Adapted from Ref. 28.

drowsiness was seen in other studies (26). Tolerance to antihistamine-induced impairment also seems to occur to a variable degree. In some performance evaluations tolerance did not occur with short-term use (19, 26), while in others, the acute performance decrement was no longer noted within a few days of antihistamine administration (19, 27). The full clinical significance of these findings has yet to be resolved.

Neuropsychologists have developed tests to measure specific mental skills and processes (28). A quantitative evaluation of sedation can be performed by testing these functions separately or considering them globally. What follows are examples of the CNS assessments listed in Table 1 involving the various first- and second-generation antihistamines.

A. Drowsiness Tests

1. Subjective Tests/Self-Rating

Subjective tests include self-reports of what is alternatively called sleepiness, drowsiness, somnolence, or fatigue. The ratings are either in absolute terms, citing presence or absence, or on a numerically scored scale (26, 29, 30). Results of studies using these tests have shown that terfenadine 60 mg twice daily, astemizole 10 mg daily, loratadine 10 mg daily and cetirizine 5 mg daily have an incidence of sedation not significantly different from placebo. All the first-generation antihistamines in recommended doses cause sedation. This symptom is not experienced by every individual with single doses (31, 32). In certain patients it persists without the development of tolerance.

The Stanford Scale for Sleepiness (Table 3) is a seven-point list that has been used to further delineate this central nervous system effect (22).

Subjective mood rating scores of apathy, depression, irritability, and anxiety have also been used. Such a test is the Positive Affect-Negative Affect Scales (PANAS) (33). The PANAS provides a positive affect score that reflects the degree to which an individual feels alert, active, and enthusiastic. High PA is a state of full concentration, high energy, and pleasurable engagement. Low PA is a dimension of lethargy and sadness. The negative affect score is a state of subjective distress and unpleasurable engagement. High negative affect implies anger, fear, loneliness, disgust, and guilt while low negative affect is characterized by calmness and serenity (34).

2. Sleep Latency

Drowsiness can be objectively measured through the use of electrophysiological potentials such as the multiple sleep latency test. This test measures the time needed to induce EEG signs of stage 1 sleep in individuals given repeated opportunities to fall asleep. In a double-blind cross-over study by Roehrs et al. (35), subjects were almost 50% sleepier (defined by sleep latency) after taking diphenhydramine (50 mg three times daily) than after terfenadine (60 mg twice daily) or placebo (Fig. 1). These findings with terfenadine were confirmed by Murri et al. (36), who evaluated daytime sleepiness in healthy men given 120 mg or placebo once daily in the morning for three consecutive days. All mean stage-1 sleep latencies throughout the study failed to show any significant difference between terfenadine and placebo. In another study of drowsiness, Seidel et al. (37) compared a first-generation antihistamine, hydroxyzine, with a second-generation antihistamine, cetirizine, and placebo. Sleep latency tests showed that the 25-mg hydroxyzine dose group was significantly sleepier

Table 3 Stanford Scale for Sleepiness

1. Feeling active, vital, alert, wide awake.
2. Functioning at high level, but not at peak, able to concentrate.
3. Relaxed awake but not fully alert, responsive.
4. A little foggy.
5. Foggy, beginning to lose track, difficulty in staying awake.
6. Sleepy, prefer to lay down, "woozy."
7. Almost in reverie, cannot stay awake, sleep onset appears imminent.

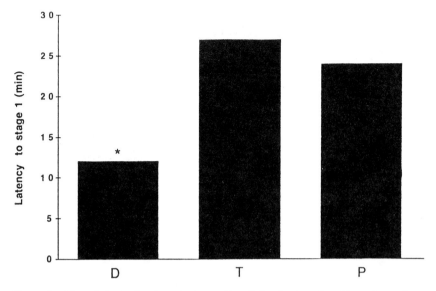

Figure 1   Mean latency to stage 1 sleep after diphenhydramine 50 mg three times daily (D), terfenadine 60 mg twice daily (T), and placebo (P). * = different from placebo ($p < 0.05$). (Adapted from Ref. 35.)

than the placebo group at 2 and 4 h after receiving medication. Sleepiness among subjects given cetirizine 5, 10, or 20 mg was not significantly different from that noted in subjects who received placebo. For clinical measurement purposes, sleep latencies below 5 min represent pathological sleepiness (38). Hydroxyzine resulted in pathological sleepiness throughout most of the daytime testing period. Despite the abnormal multiple sleep latency test EEG findings in the hydroxyzine group, patients' self-ratings of sleepiness and fatigue did not differ significantly among the various antihistamine treatment groups. This finding suggests that clinical reports that depend on self-awareness may have minimal reliability in safety assessments of this adverse effect, a phenomenon that has been previously noted (39, 40). Just as drinkers are often unable to recognize their alcohol-related dysfunction, subjects and patients are often unable to recognize antihistamine-induced sedation.

In a study comparing the first-generation H₁-receptor antagonist diphenhydramine (50 mg three times daily) and the second-generation agent cetirizine (10 mg daily) using the multiple sleep latency test, an acute first-day increase in sleepiness was found with diphenhydramine. However, when the medications were continued for 3 days and the tests repeated on the third day, no differences were observed, apparently because of development of tolerance to the sedative effect (27). A similar development of tolerance has been noted with subchronic administration of triprolidine (5 mg twice daily) (19).

The central effects of a newly developed second-generation, long-acting antihistamine, loratadine (10 mg and 40 mg) were also compared to the standard H_1-antihistamine diphenhydramine (50 mg three times daily) using the sleep latency test (41). Mean latency to sleep was reduced significantly with diphenhydramine compared to placebo, whereas neither loratadine dose reduced sleep latency. In this study, the sleep latency test was much more sensitive in assessing sedation than the cognitive performance testing used (see below).

B. Cognitive and Psychomotor Performance Impairment Tests

1. Evoked Response Latency

The P300 (or P3), a positive auditory EEG response, is an objective and sensitive measure of sustained attention and cerebral processing speed (42, 43). Its latency depends on the amount of time required for evaluation of the stimulus. The latency can be prolonged if the stimulus discrimination is made more difficult. Medications that cause cognitive impairment can also prolong P300 latency.

The tonal oddball test is frequently used to record the P300. In this test, the subject listens to a series of regularly occurring tones at one pitch with an occasional oddball tone of a different pitch. The subject's task is to count the oddball tone, which occurs infrequently and randomly. If the subject attends to the stimuli and performs the task, a P300 will be elicited by the oddball tone.

A single-dose study by Meador et al. (44) of a common first-generation antihistamine, chlorpheniramine (4 mg), compared with a standard therapeutic dose of terfenadine (60 mg) and placebo, used the P300 auditory-evoked potential as an objective measure of cerebral processing speed. This double-blind, randomized, three-way cross-over study with premedication and 3 h postmedication recordings involved 24 healthy volunteers. All subjects performed the task of counting the actual number of oddball tones with a less than 5% error rate in all treatment states. The mean pretreatment/posttreatment differences in the P300 latencies (ms) were longer with chlorpheniramine in more subjects than with placebo ($p < 0.01$), whereas such a difference was not seen with terfenadine. A subjective assessment of drowsiness also was recorded, and subjects reported sedation significantly more frequently with chlorpheniramine than with terfenadine ($p < 0.05$) but not significantly more often with placebo versus terfenadine ($p = 0.23$). Unlike the Seidel study (37), the objective EEG findings of significantly more cognitive slowing correlated well with subjective feelings of increased drowsiness in those treated with the first-generation antihistamine chlorpheniramine. Simons et al. (45) studied 15 children with well-controlled allergic rhinitis in a single-dose, double-blind fashion and also found that chlorpheniramine, but not terfenadine, significantly increased the P300 latency.

2. Visual Function

The dynamic visual acuity test shows four rings on a screen with gaps at the oblique meridians. The subject must determine the locations of the gaps as the images sweep by. Another visual function test, the critical flicker fusion test, requires the subject to define the frequency (Hz) at which the flickers become fused. A first-generation antihistamine, triprolidine, impaired subjects' performance in both of these tests, whereas the second-generation antihistamines terfenadine and astemizole had no adverse effect (46).

3. Visual-Motor Coordination

Adaptive tracking is a visual-motor coordination performance measure. With this test, Nicholson (47) demonstrated that the safety of various antihistamines differed due to their CNS effects. When subjects given antihistamines were asked to guide a simulated vehicle along a road with a joystick, the investigators found that chlorpheniramine 4 mg, produced modest impairment shortly after ingestion. The subjects who received promethazine 10 mg, or clemastine 1 mg, were even more profoundly sedated, although the onset of this effect was delayed. Terfenadine, 60 mg, did not appear to change performance in adaptive tracking (Fig. 2). As with the results of Seidel's (37) study of self-awareness (see above), objective and subjective findings differed. In these tests, Nicholson's subjects reported little subjective sense of impaired performance when the medications were exerting maximum effect (48). In other studies with diphenhydramine (31) and chlorpheniramine (32) the opposite discrepancy was noted: subjective feelings of fatigue worsened, even though there was no objective evidence of impaired performance. Thus, no consistent relationship appears to exist between objective performance and subjective ratings of performance by patients treated with antihistamines.

4. Reaction Time

Goetz et al. (23) used simple and choice reaction time tasks to measure performance impairment after oral hydroxyzine given at bedtime. The patient operated a bidirectional toggle switch with left, neutral, and right

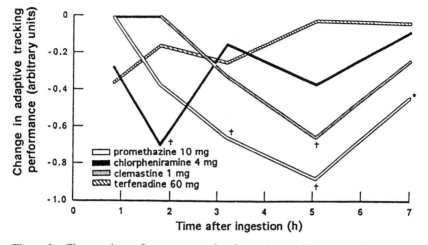

Figure 2 Changes in performance on visual-motor coordination testing after various single-dose antihistamines. $*p < 0.01$; $†p < 0.001$ when compared to terfenadine. (Adapted from Ref. 47.)

positions. A light was positioned left and right of the toggle. When the light illuminated, the subject moved the toggle toward the light, and a reaction time was measured. If the subject knew which of the two test lights would illuminate, this was the *simple* reaction time. *Choice* reaction time was measured if the subject did not know which of the lights would illuminate. Results showed that nighttime hydroxyzine can cause subjective symptoms of drowsiness the next morning without objective signs of performance impairment based on simple and choice reaction time measurements. These findings again suggest that a discrepancy can occur between objective and subjective assessments of sedation.

In a study comparing hydroxyzine (25 mg) with cetirizine (5, 10, and 20 mg) and placebo, Seidel et al. (37) assessed performance as reflected by the subjects' simple reaction time with a vigilance test. The subjects were instructed to press a button when a point of light appeared on a video screen at random intervals rather than at its previously regular, somewhat tedious, frequency. The reaction times were significantly slower in hydroxyzine-treated subjects than in placebo controls. These subjects were also unaware that they were performing less well.

Volkerts et al. (19) used choice reaction time tasks that place high demands on both cognitive and psychomotor functions. In one task, the subject had to decide whether a presented character was a letter or a digit.

In a second, the subject decided if two letters, displayed simultaneously, belonged to the same category (i.e., vowels or consonants). The third was a memory scanning task wherein the subject had to determine whether a presented digit was part of a previously displayed list of five digits. The study found that triprolidine impaired performance, but surprisingly, so did terfenadine (60 mg twice daily). The effect of cetirizine (10 mg daily), a second-generation antihistamine, was no different from placebo. The impairment with terfenadine, unlike triprolidine, was not associated with any performance problems on a real highway driving test (see below), or with any reduction in sleep latency (see above).

Gengo and colleagues (49) studied reaction time using a driving simulator. In response to either a traffic threat or a traffic sign, the subject needed to react by either braking or turning the steering wheel. In this cross-over study, 15 male volunteers received either diphenhydramine (50 mg), cetirizine (5, 10, or 20 mg), or placebo. The effects of all doses of cetirizine were indistinguishable from placebo at all time points in contrast to the greater effect produced by diphenhydramine in prolonging reaction time compared to placebo ($p < 0.001$). Testing at specific times revealed that diphenhydramine produced a greater prolongation of reaction time than did placebo ($p < 0.02$) 2 h after dose administration.

5. Recognition

The symbol-substitution test forces the subject to write the appropriate letterlike symbol found in a digit key under the digits one to nine, which appear in random sequence in printed rows. Scores are based on the number of correct written symbols. Gengo et al. (49) showed that at 2 h following medication administration, diphenhydramine produced significantly poorer performances than did placebo or any cetirizine dose (5 mg, 10 mg, or 20 mg). At 6 and 8 h postdose, there was a small, but significant, decrement in performance following the 20-mg dose of cetirizine compared with the other treatments (Fig. 3). Roth et al. (41) used symbol substitution as a performance measure when studying the CNS effects of loratadine at low and high dose (10 mg and 40 mg) and, unlike diphenhydramine, loratadine had no effect on this test.

The Stroop word/color test, another recognition test, measures delays in naming word/color incongruities. With this test, Gengo and Gabos (50) documented that the CNS-depressant effects on performance of a first-generation antihistamine like hydroxyzine are related to the time course of the drug-blood levels and its ability to cross the blood-brain barrier. Although cetirizine, the metabolite of hydroxyzine, produced the same degree and duration of skin wheal suppression—a peripheral antihistamine effect—as hydroxyzine in this study, cetirizine did not impair perfor-

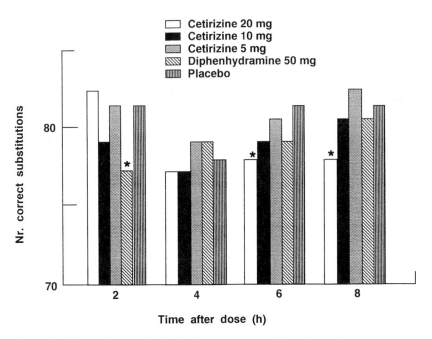

Figure 3   Mean digit symbol substitution scores following each treatment. ∗ = $p < 0.05$ when compared to placebo. (Adapted from Ref. 49.)

mance as evaluated by the Stroop word/color test, most likely because it does not penetrate into the CNS to the same degree as hydroxyzine.

6. *Driving Tests*

In an epidemiological study of Canadian drivers killed in automobile accidents attributed to their own error, Warren et al. (51) determined that such drivers were 1.5 times more likely to have been using a first-generation antihistamine than were drivers who were not responsible for the accidents in which they were killed.

Actual car driving performance was compared among subjects given chlorpheniramine, terfenadine, or placebo (52). The incidence and degree of sedation and performance decrement were determined from changes in EEG, eye blinking, and steering operation. Drowsiness is known to be associated with increased numbers of alpha waves, slow eye blinking, and coarse steering maneuver of greater than 10 degrees. The subjects were asked to drive for 2.5 h, beginning 2 h after the study medication was administered. In the chlorpheniramine group, only four of the ten subjects completed the full driving time; the remainder were judged to be incapable of continuing. In the terfenadine and placebo segments of this cross-over study, all ten subjects completed the 2.5 h of driving. In all three groups, the density of alpha waves increased as driving time increased. EEG changes occurred earlier in the chlorpheniramine group, who had significantly more alpha energy than did the other groups. There was no difference between the terfenadine and placebo group. The frequency of slow blinking also increased with driving time; chlorpheniramine produced significantly more slow blinking than did either terfenadine or placebo; no difference was detected between terfenadine and placebo recipients. Coarse steering maneuvers of greater than 10 degrees occurred significantly more often after chlorpheniramine treatment than placebo, but again no difference was noted between terfenadine and placebo. After the driving test was completed, self-rated sleepiness was greater than baseline in almost all cases, but the mean score reported by the chlorpheniramine group was more than twice that for subjects assigned to terfenadine or placebo treatments.

Betts and colleagues (53) compared triprolidine and terfenadine in an obstacle test of driving performance. A significant increase in the time required to complete the turns between barriers in the course, in the number of barriers hit, and in the number of mistakes made during the drive was reported after the triprolidine treatment versus terfenadine and placebo. In addition, the subjects noted that the triprolidine caused them to feel significantly more drowsy, withdrawn, and clumsy.

The mental processes involved in driving include attention, short-term memory, recognition, sensorimotor integration, decision-making, and judgment (21). Attention and concentration are particularly important for routine driving. Generally, these processes are best tested by monotonous and boring tasks. Antihistamines frequently affect such tasks (48).

O'Hanlon (54) devised a specially instrumented automobile to evaluate "real world" driving performance by measuring lane weaving on the open road and in traffic. They studied triprolidine 10 mg and two second-generation antihistamines, terfenadine (60 mg) and loratadine (10 mg). The weaving tendency produced by triprolidine was comparable with that seen in drivers with blood alcohol concentrations of 0.05%. Although this effect of triprolidine persisted for up to 4 h, most drivers reported sedative effects and impaired performance at 1 to 2 h after drug ingestion, but not at 3 and 4 h. This suggests that the drivers had adapted to the sensation of sedation, but not to the objective adverse effects of sedation on driving. In these doses, terfenadine and loratadine had no undesirable effect on the weaving index (Fig. 4).

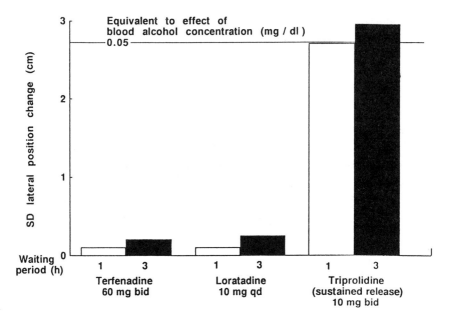

Figure 4 Effect of single-dose antihistamines on driving performance using the weaving index. qd = daily; bid = twice daily. (Adapted from Ref. 54.)

Ramaekers et al. (55), utilizing the same driving test, compared the effects of single doses of cetirizine 10 mg, loratadine 10 mg, placebo, and alcohol. The effect of cetirizine on driving performance resembled that of alcohol, each causing subjects to weave excessively. Loratadine had no effect on driving performance. Volkerts et al. (19) evaluated the possible sedative effects on driving performance of cetirizine 10 mg daily, terfenadine 60 mg twice daily and 120 mg daily, and triprolidine 5 mg twice daily. Twenty-seven healthy men who had driven at least 10,000 km/year for 5 years were administered drug or placebo over a 4-day driving trial. An instrumented vehicle recorded driver ability to maintain a constant 90 km/h speed and a steady lateral position (absence of weaving) in normal highway traffic. Possible effect of drug on performance was evaluated on day 1 (acute effect) and day 4 (subchronic effect). Neither cetirizine nor terfenadine was found to affect the subjects' perceived quality of driving, deviation of speed, or lateral weaving on either the first or fourth day. Triprolidine impaired all of these on day 1. On day 4, the triprolidine effects were variably diminished (see Sec. IV). It was concluded in this study that newer antihistamines such as cetirizine 10 mg daily, and terfenadine 60 mg twice daily or 120 mg daily, in contrast to a first-generation antihistamine such as triprolidine, are safe for patients who drive.

7. Piloting Tests

Offenloch and Zochner (56) studied the effect of terfenadine during an instrument flight simulator procedure in 10 pilots who were training to obtain their instrument license. Pilots were rated on their ability to maintain compass course, proper altitude, and speed. No difference was noted in performance between pilots on or off terfenadine. A placebo-controlled study of loratadine given to experienced Brazilian Airline captains and Brazilian Air Force pilots similarly found no effect on performance in a sophisticated flight simulator test (57).

8. Clerical Tests

Adelsberg and D'Amico-Beadon (58) tested the effects of loratadine 10 mg, diphenhydramine 50 mg, and placebo on skills required for successful functioning in an office environment. Four groups of subjects from a retail business were recruited: clerical/secretarial, data entry, accounting, and managerial personnel. A set of 12 tests of performance skills were developed, including the Clerical Abilities Battery, which evaluated various office skills such as reading, filing, arithmetic and accounting, visual attention, and both visual and auditory memory. Loratadine did not alter the performance in any of the 12 tests whereas diphenhydramine impaired the subjects in 8. This sedating effect was significant at 90 min following diphenhydramine but not at 4 h after drug administration.

9. Tests in Children

There are few objective studies of the adverse effects of antihistamines in children. Simons et al. (45) measured the CNS effects of terfenadine (60 mg), chlorpheniramine (4 mg), and placebo in children aged 6 to 12 years before and 2 to 2.5 h after single dosing. A visual analog score was used for somnolence, and P300-event-related potentials were used as a measure of cognitive processing. As seen in adults, terfenadine produced significantly less impairment of cognitive function and subjective complaints of somnolence than chlorpheniramine. The degree of peripheral H₁-blockade was also measured using epicutaneous histamine tests, and terfenadine was found to produce greater wheal and flare suppression than chlorpheniramine, making the relative benefit/risk ratio much more favorable for terfenadine than chlorpheniramine. The authors make a plea that given the widespread use of antihistamines in children, additional studies of this type should be performed in this age group.

IV. TOLERANCE TO SEDATION FROM ANTIHISTAMINES

Tolerance to antihistamine-induced drowsiness has been reported by some, but not all, investigators (26). Volkerts et al. (19) treated volunteers with triprolidine 5 mg twice daily or placebo for four consecutive days and conducted testing on the first and fourth treatment day. Subjects underwent self-assessment ratings, objective sleep latency tests, cognitive performance tests, and actual driving tests. Triprolidine significantly reduced scores on all of these tests on the first day. By the fourth day, both the subjective (self-rating) and objective (sleep latency) measures of drowsiness were no longer statistically different between the triprolidine and placebo groups. The speed performance driving test was also not different between the groups. In contrast to this evidence of the development of tolerance, some of the cognitive studies and the lateral weaving driving test continued to show statistical, although diminished, differences between groups, implying perhaps partial tolerance. Walsh et al. (27) treated subjects with diphenhydramine (50 mg three times daily) for 3 days and measured multiple sleep latency, a simulated assembly-line task, and sleepiness ratings by visual analog system on day 1 and day 3. Diphenhydramine produced marked impairment and drowsiness only on the first day of treatment; no difference between diphenhydramine and placebo was noted on the third day, suggesting that tolerance occurred. Manning et al. (59) using diphenhydramine, and Bye et al. (60) using triprolidine, also reported tolerance to the CNS effects of these two first-generation antihistamines.

V. DRUG INTERACTION

In alcohol-induced impairment, blood concentrations of 0.07 g/dl, although rarely causing drowsiness, produce a fourfold increase in the relative risk of a fatal accident (61). O'Hanlon (54) demonstrated by the instrumented car method that the noncentrally acting antihistamines terfenadine and loratadine did not potentiate the effect of alcohol on driving performance.

Bhatti and Hindmarch (62) utilized a simulated driving task and measured brake reaction time in response to critical stimuli in patients given terfenadine with and without alcohol. Healthy volunteers received acute doses of terfenadine (60 mg, 120 mg, 240 mg) or placebo followed by a "social dose" of alcohol. Terfenadine, at the 240-mg dose, but not the 60- and 120-mg dose, was found to significantly impair performance alone and following alcohol. This study points out the need for establishing behavioral effects of antihistamines over a range of doses.

Other studies have looked at astemizole, loratadine, and ebastine and their potential interactive effects on performance when taken in conjunction with alcohol. As with terfenadine at standard doses, these second-generation antihistamines did not potentiate the depressive effects of alcohol on cognitive and driving performance tests (55, 63, 64).

Moser and co-workers (31) studied the safety of antihistamine-tranquilizer combinations by a series of tests. Diazepam, a widely used antianxiety agent, was added to diphenhydramine, terfenadine, or placebo. Subjective feelings of well-being were unaffected in the terfenadine and placebo groups. In contrast, the subjects receiving the diazepam/diphenhydramine combination reported significant complaints of being tired, dull, or ill. This combination also produced significant decreases in performance on several psychomotor tests, including the Vienna determination apparatus, the Vienna reaction apparatus, the ball cylinder test, and the D-2 test. None of these tests was adversely affected by diazepam alone or by diazepam in combination with terfenadine. The effect of diazepam with and without antihistamine treatment was also studied by Mattila et al. (65). The sedative effect of diazepam as documented by subjective drowsiness scores and impairment of performance was not enhanced by the antihistamine, ebastine. In this report, the effects of ebastine alone appeared no different from those of placebo.

VI. SEDATIVE EFFECTS OF ALLERGIC DISEASE

It should be noted that research on the sedative effects of antihistamines is often conducted in individuals not experiencing allergic disease symptoms. The results from these studies may be misleading since recent data suggest it is possible that the allergic disease process by itself can cause performance and learning impairment. The CNS effects of allergic rhinitis have been studied using a quality of life questionnaire, the SF-36. This instrument assesses nine health dimensions, many of which reflect central nervous system functioning. One study reported a significant impairment in eight of the nine dimensions in patients with allergic rhinitis by comparison with healthy subjects (66). A follow-up study found treatment of allergic rhinitis with the antihistamine cetirizine (10 mg q.d.) actually improved eight of the nine health concepts including physical functioning, energy-fatigue, general perception of health, social functioning, physical limitations, social limitations, mental health, and pain (67).

Marshall and Colon (33) studied depressed individuals with allergic rhinitis in and out of their "allergy season" and demonstrated effects of the season on mood and cognitive functions. Atopic and control subjects were given tests to assess mood, psychomotor speed, and cognitive functioning. Atopic subjects exhibited declines in verbal learning, slower decision-making, and psychomotor speed on both simple and choice reaction times, and lower positive affect during their "allergy season" in comparison to out of their "allergy season."

Vuurman et al. (68) found similar adverse effects of seasonal allergic rhinitis alone on learning in children. Children suffering from seasonal allergic rhinitis and matched controls were instructed on the use of a computer game and tested 2 weeks later on their knowledge of the game. Only the atopic children received different treatments just before their instruction: diphenhydramine (first-generation), loratadine (second-generation), or placebo. Both the placebo and diphenhydramine groups learned significantly less than the nonallergic untreated controls. The loratadine group's learning performance was superior to either of the other atopic groups but still inferior to the normal group (Fig. 5). These findings suggest that allergic symptoms alone reduce learning ability in children and that this effect is partially counteracted by treatment with loratadine and aggravated by diphenhydramine. Taken together, the above studies point out the importance of controlling for the presence or absence of allergic symp-

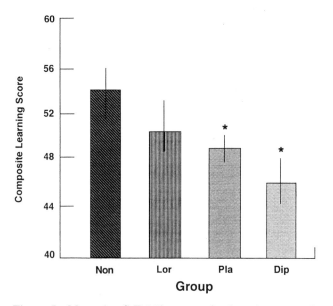

Figure 5 Mean ($\pm$ S.E.M.) composite learning score for different treatment groups. $* = p < 0.05$ compared to nonallergic. (Adapted from Ref. 68.)

toms when conducting research on the effects of new antihistamines on mental function.

VII. SUMMARY

An extensive body of research exists regarding the sedative effects of H_1-antihistamines. There is great interest in this area due to the well-known adverse CNS effects associated with first-generation antihistamines, and the advent of numerous second-generation agents claimed to have nonsedative properties. In assessing the CNS effects of antihistamines, there are a number of questions that should be addressed (Table 4).

The tendency for an antihistamine to alter CNS function can be arbitrarily separated into two components: drowsiness and performance impairment. Although these two manifestations of sedation are closely related, studies have clearly shown that drowsiness can exist in the absence of performance impairment and, even more commonly, cognitive and psychomotor function can be diminished without sleepiness being noted. Since either or both may be present without an individual's awareness, *symptoms should not be relied on to predict decrements in alertness or performance.*

Because the CNS effects of antihistamines are complex and cannot be reflected in one measurement, a variety of assessments are required. These may range from the subjective (e.g., self-rating of drowsiness) to the objective (e.g., sleep latency), and from the simple (e.g., critical flicker fusion) to the complex (e.g., actual driving). When these tests are applied to the evaluation of the H_1-antihistamines currently available, it is clear there is a real distinction between the older first-generation antihistamines and the newer second-generation ones. At the recommended doses, the second-generation antihistamines are clearly less sedating in more patients than their predecessors. In fact, nearly all studies to date of the second-generation antihistamines terfenadine, astemizole, and loratadine have shown, at the recommended doses, sedative properties no greater than those seen with placebo. However, the nonsedating feature of this new class of antihistamines is not absolute. A report has suggested that if the dose of terfenadine is increased enough, sedation can occur (62). This is also true of loratadine (41) and cetirizine (29). Other second-generation antihistamines (ebastine, azelastine, acrivastine) are presently being evaluated utilizing the various probes of CNS function outlined in this chapter.

The potential development of tolerance to the sedative effects of first-generation antihistamines over time raises an interesting question: if tolerance is a real phenomenon and can be relied upon to occur, do the classic

Table 4 Questions to Assess CNS Effects of Antihistamines

Does the drug cause sleepiness and/or impair performance?
  Varies from individual to individual
  Varies from drug to drug: more with some, especially
    with first-generation antihistamines
Are the effects present at therapeutic doses?
  Varies from drug to drug: more with some, especially with first-generation anti-
    histamines, such as ethanolamines, phenothiazines
Are the effects dependent on dose or concentration?
  Dose-dependent with many first- and second-generation antihistamines
  Blood concentration dependent with many
What are the durations of the effects?
  Varies from drug to drug based on the pharmacokinetics and pharmacodynamics
Do both subjective perceptions and objective measures need to be assessed?
  Subjective and objective assessments do not always correlate, so objective,
    quantifiable evaluations should be used
Do both the effects on sleepiness and performance impairment need to be as-
  sessed?
  As these do not always correlate, both should be assessed
  As these are complex functions, multiple assessments should be made
Are the effects with continuous dosing continuous, cumulative, or reduced?
  Varies from drug to drug
    Cumulative effects do not appear to occur unless the individual has physiolog-
      ical abnormalities which affect the pharmacokinetics of the drug
    Continuous effects appear to occur some of the time
    Reduction or tolerance to the CNS effects appears to occur; however, not
      necessarily in all patients in all parameters or to a complete degree
  Varies from individual to individual
Are the effects compounded by other drugs?
  Drugs that also affect the CNS add to the effects
Are certain populations more at risk than others to these effects?
  Varies with age of individual
  Varies with pharmacokinetics and pharmacodynamics including contributions
    of abnormalities of gastrointestinal, hepatic, and renal systems
  Varies with quality of nocturnal sleep
  Varies with disease state

Modified from Ref. 21.

antihistamines still have a role to play in the pharmacotherapy of allergic disease? As long as the patient remains on steady-state dosing, and toler-ance develops, use of a first-generation antihistamine could be justified. The studies on tolerance are limited in number, and more work needs to be done to determine if all the different tests used to assess drowsiness and impairment demonstrate the same degree of tolerance in all individuals over multiple days of use. *It appears that tolerance to both the drowsiness and multiple performance effects does not universally, regularly, or completely occur.* Therefore, it would be safer to select a second-generation antihistamine proven not to cause sedation from the initial dose rather than a first-generation antihistamine that can cause sedation that may or may not disappear over time.

As all second-generation antihistamines are found to be relatively nonsedating, their risk/benefit ratios will be determined more by their non-CNS adverse effects (e.g., potential to cause cardiac arrhythmias), and by their degree of H$_1$-antagonism, antiallergic and anti-inflammatory properties, onset of action, duration of action, and overall potency. It will continue to be important to recognize the differences as well as the similarities within this valuable group of medications.

REFERENCES

1. Hess A. National probability sample of sufferers of upper respiratory allergy. Presented at the Forty-first Annual Meeting of the American Academy of Allergy and Immunology. New York, March, 1985.
2. Meltzer EO. Antihistamine- and decongestant-induced performance decrements. J Occup Med 1990;32:327–34.
3. Medical Advertising News, Engel Communications Inc., Plainview, New York, August 1991.
4. Simons FER, Simons KJ. Antihistamines. In: Allergy: Principles and Practice (Middleton E, Reed CE, Ellis EF, Adkinson NF, Yunginger JW, Busse WW, eds.), 4th ed. St. Louis: CV Mosby, 1993;856–92.
5. Lesson GA, Chan KY, Knapp WC, et al. Metabolic disposition of terfenadine in laboratory animals. Arzneimittelforschung 1982;32:1173–8.
6. Thach BT, Chase TN, Bosma JF. Oral facial dyskinesia associated with prolonged use of antihistaminic decongestants. N Engl J Med 1975;293:486–7.
7. King G, Weeks SD. Pyribenzamine activation of the electroencephalogram. Electroencephalogr Clin Neurophysiol 1965;18:503–7.
8. Wyngaarden JB, Seevers MH. The toxic effects of antihistaminic drugs. JAMA 1951;145:277–82.
9. Koppel C, Ibe K, Tenczer J. Clinical symptomatology of diphenhydramine overdose: an evaluation of 136 cases in 1982 to 1985. J Toxicol Clin Toxicol 1987;25:53–70.
10. Douglas WW. Histamine and 5-hydroxytryptamine and their antagonists. In: Gilman AG, Goodman LS, Rall TW, Murad F, eds. The Pharmacological Basis of Therapeutics. 7th ed. New York: Macmillan, 1985:605–638.
11. Simons FER, Simons KJ. Second-generation H$_1$-receptor antagonists. Ann Allergy 1991;66:5–21.

12. Kemp JP, Buckley CE, Gershwin ME, et al. Multicenter, double-blind, placebo-controlled trial of terfenadine in seasonal allergic rhinitis and conjunctivitis. Ann Allergy 1985;54:502–9.

13. Friedman HM. Loratadine: a potent, nonsedating and long-acting H_1 antagonist. Am J Rhinol 1986;1:95–9.

14. Meltzer EO, Storms WW, Pierson WE, et al. Efficacy of azelastine in perennial allergic rhinitis: clinical and rhinomanometric evaluation. J Allergy Clin Immunol 1988;82:447–55.

15. Simons FER, Simons KJ. H_1 receptor antagonist treatment of chronic rhinitis. J Allergy Clin Immunol 1988;81:975–80.

16. United States Pharmacopeia. Drug information for the health care professional, antihistamines (systemic), 13th ed. Rockville, MD, United States Pharmacopeial Convention, 1993;1:320.

17. Timmerman H. Factors involved in the incidence of the central nervous system effects of H_1 blockers. In: Church MK, Rihoux JP, Lewiston NY, eds. Therapeutic Index of Antihistamines. Hogrefe and Huber Publishers, 1992:19–31.

18. Roth T. Antihistamines and daytime sleepiness. J Respir Dis (Monograph Ser) 1987;7:32–5.

19. Volkerts ER, van Willigenburg APP, van Laar MW, et al. Does cetirizine belong to the new generation of antihistamines? An investigation into its acute and subchronic effects on highway driving, psychometric test performance and daytime sleepiness. Human Psychopharmacol 1992;7:227–238.

20. Gengo FM, Gabos C. Antihistamines, drowsiness, and psychomotor impairment: CNS effects of cetirizine. Ann Allergy 1987;59II:53–8.

21. Gengo FM, Manning C. A review of the effects of antihistamines on mental processes related to automobile driving. J Allergy Clin Immunol 1990;186:1034–9.

22. Passalacqua G, Scordamaglia A, Ruffoni S, et al. Sedation from H_1 antagonists: evaluation methods and experimental results. Allergol Immunopathol 1993;21:79–83.

23. Goetz DW, Jacobson JM, Apaliski SJ, et al. Objective antihistamine side effects are mitigated by evening dosing of hydroxyzine. Ann Allergy 1991;67:448–54.

24. Kulshrestha VK, Gupta PP, Turner P, Wadsworth J. Some clinical pharmacological studies with terfenadine, a new antihistamine drug. Brit J Clin Pharmacol 1978;6:25–29.

25. Seppala T, Nuotto E, Korttila K. Single and repeated dose comparison of three antihistamines and phenylpropanolamine: psychomotor performance and subjective appraisals of sleep. Br J Clin Pharmacol 1981;12:179–88.

26. Goetz DW, Jacobson JM, Murnane JE, et al. Prolongation of simple and choice reaction times in a double-blind comparison of twice-daily hydroxyzine versus terfenadine. J Allergy Clin Immunol 1989;84:316–22.

27. Walsh JK, Muehlbach MJ, Humm T, et al. Sleepiness and performance during three days' use of cetirizine or diphenhydramine. J Allergy Clin Immunol 1994;93:235.

28. Hindmarch I. Psychomotor function and psychoactive drugs. Br J Clin Pharmacol 1980;10:189–209.
29. Falliers CJ, Brandon ML, Buchman E, et al. Double-blind comparison of cetirizine and placebo in the treatment of seasonal rhinitis. Ann Allergy 1991; 66:257–62.
30. Del Carpio J, Kabbash L, Turenne Y, et al. Efficacy and safety of loratadine (10 mg once daily), terfenadine (60 mg twice daily), and placebo in the treatment of seasonal allergic rhinitis. J Allergy Clin Immunol 1989;84:741–6.
31. Moser L, Huther KJ, Koch-Weser J, et al. Effects of terfenadine and diphenhydramine alone or in combination with diazepam or alcohol on psychomotor performance and subjective feelings. Eur J Clin Pharmacol 1978;14:417–23.
32. Kulshrestha VK, Gupta PP, Turner P, et al. Some clinical pharmacological studies with terfenadine, a new antihistaminic drug. Br J Clin Pharmacol 1978;6:25–9.
33. Marshall PS, Colon EA. Effects of allergy season on mood and cognitive function. Ann Allergy 1993;71:251–8.
34. Watson D, Clark LA, Tellegen A. Development and validation of brief measures of positive and negative affect: The PANAS scales. J Personality Social Psychol 1988;54:1063–70.
35. Roehrs TA, Tietz EI, Zorick FJ, et al. Daytime sleepiness and antihistamines. Sleep 1984;7:137–41.
36. Murri L, Massetani R, Krause M, et al. Evaluation of antihistamine-related daytime sleepiness. Allergy 1992;47:532–4.
37. Seidel WF, Cohen S, Bliwise NG, et al. Cetirizine effects on objective measures of daytime sleepiness and performance. Ann Allergy 1987;59:58–62.
38. Richardson GS, Carskadon MA, Flagg W, et al. Excessive daytime sleepiness in man: multiple sleep latency measurement in narcoleptic and control subjects. Electroencephalogr Clin Neurophysiol 1978;45:621–7.
39. Vollmer R, Matejcek M, Greenwood C, et al. Correlation between EEG changes indicative of sedation and subjective responses. Neuropsychobiology 1983;10:249–53.
40. Nicholson AN. The significance of impaired performance. In: Burley D, Silverstone T, eds. Medicines and road traffic safety. Int Clin Psychopharmacol 1988;3(suppl 1):117–27.
41. Roth T, Roehrs T, Koshorek G, Sicklesteel J, Zorick F. Sedative effects of antihistamines. J Allergy Clin Immunol 1987;80:94–8.
42. Sutton S, Braren M, Zubin J, et al. Evoked potential correlates of stimulus uncertainty. Science 1965;150:1187–8.
43. Pritchard WS. Psychophysiology of P300. Psychol Bull 1981;89:506–40.
44. Meador KJ, Loring DW, Thompson EE, et al. Differential cognitive effects of terfenadine and chlorpheniramine. J Allergy Clin Immunol 1989;84:322–5.
45. Simons FER, Reggin JD, Roberts JR, Simons KJ. Benefit/risk ratio of the antihistamines (H₁-receptor antagonists) terfenadine and chlorpheniramine in children. J Pediatrics 1994;124:979–83.
46. Nicholson AN, Smith PA, Spencer MB. Antihistamines and visual function:

studies on dynamic acuity and the pupillary response to light. Br J Clin Pharmacol 1982;14:683–90.

47. Nicholson AN. Central effects of H_1- and H_2-antihistamines. Aviat Space Environ Med 1985;56:293–8.

48. Clarke LH, Nicholson AN. Performance studies with antihistamines. Br J Clin Pharmacol 1978;6:31–5.

49. Gengo FM, Gabos C, Mechtler L. Quantitative effects of cetirizine and di-phenhydramine on mental performance measured using an automobile driving simulator. Ann Allergy 1990;64:520–6.

50. Gengo FM, Gabos C. Antihistamines, drowsiness and psychomotor impairment: central nervous system effect of cetirizine. Ann Allergy 1987;59, Part II):53–7.

51. Warren R, Simpson H, Hilchie J, et al. Drugs detected in fatally injured drivers in the province of Ontario. In: Goldberg L, ed. Alcohol, Drugs, and Traffic Safety. Vol 1. Stockholm: Almquist and Wiksell, 1981:203–17.

52. Aso T, Sakai Y. Effects of terfenadine on actual driving performance. Jpn J Clin Pharmacol Ther 1988;19:681–8.

53. Betts T, Markman D, Debenham S, et al. Effects of two antihistamine drugs on actual driving performance. Br Med J 1984;288:281–2.

54. O'Hanlon JF. Antihistamines and driving performance: the Netherlands. J Respir Dis 1988;9(suppl 7A):S12–S17.

55. Ramaekers JG, Uiterwijk MMC, O'Hanlon JF. Effects of loratadine and cetirizine on actual driving and psychometric test performance, and EEG during driving. Eur J Clin Pharmacol 1992;42:363–9.

56. Offenloch K, Zahner G. Rated performance, cardiovascular and quantitative EEG parameters during simulated instrument flight under the effect of terfenadine. Arzneim.-Forsch./Drug Res. 1992;42(I):864–8.

57. Neves-Pinto RM, Moreira Lima G, da Mota Teixeira R. A double-blind study of the effects of loratadine versus placebo on the performance of pilots. Am J Rhin 1992;6:23–27.

58. Adelsberg BR, D'Amico-Beadon A. The effects of loratadine, diphenhydra-mine and placebo on worker productivity: results of a double blind trial. J Allergy Clin Immunol 1990;85:296.

59. Manning C, Scandale L, Manning EJ, Gengo FM. Central nervous system effects of meclizine and dimenhydrinate. Evidence of acute tolerance to antihistamines. J Clin Pharmacol 1992;32:996–1002.

60. Bye CE, Claridge R, Peck AW, et al. Evidence for tolerance to the central nervous system effects of the histamine antagonist, triprolidine, in man. Eur J Clin Pharmacol 1977;12:181–6.

61. Mann P. Arrive alive: how to keep drunk and pot-high drivers off the highway. 1st ed. New York: Woodmore Press, 1983:16.

62. Bhatti JZ, Hindmarch I. The effects of terfenadine with and without alcohol on an aspect of car driving performance. Clin Exp Allergy 1989;19:609–11.

63. Hindmarch I, Bhatti JZ. Psychomotor effects of astemizole and chlorphenira-mine, alone and in combination with alcohol. Int Clin Psychopharmacol 1987;2:117–9.

64. Mattila MJ, Kuitunen T, Pletan Y. Lack of pharmacodynamic and pharmacokinetic interactions of the antihistamine ebastine with ethanol in healthy subjects. Eur J Clin Pharmacol 1992;43:179–84.

65. Mattila MJ, Aranko K, Kuitunen T. Diazepam effects on the performance of healthy subjects are not enhanced by treatment with the antihistamine ebastine. Br J Clin Pharmacol 1993;35:272–77.

66. Bousquet J, Bullinger M, Fayol C, et al. Assessment of quality of life in perennial allergic rhinitis using the SF-36 questionnaire. (SF-36) J Allergy Clin Immunol 1994, 93:271.

67. Burtin B, Fayol C, Marquis P, et al. Cetirizine improves quality of life assessed by the SF-36 questionnaire (SF-36) in perennial allergic rhinitis. J Allergy Clin Immunol 1994;93:163.

68. Vuurman EFPM, van Veggel LMA, Uiterwijk MMC, et al. Seasonal allergic rhinitis and antihistamine effects on children's learning. Ann Allergy 1993; 71:121–6.

14

Adverse Effects of H_1-Receptor Antagonists in the Cardiovascular System

Peter Honig and James N. Baraniuk
Georgetown University, Washington, D.C.

I. INTRODUCTION

Selective H_1-antagonists have long been known to possess cardioactive properties (1). In fact, as early as 1946, it was known that some of the early antihistamine analogs had "quinidinelike" electrophysiological actions on isolated rabbit auricular tissue (2). The recent identification of antihistamine-associated cardiac arrhythmias has precipitated a reevaluation of the risk-benefit ratios of these agents by both international drug regulatory bodies and health care practitioners (3).

Terfenadine and astemizole use have been associated with the development of a characteristic ventricular arrhythmia, torsade de pointes, in a very small percentage of patients (Table 1). Other antihistamines, including older, over-the-counter antihistamines, may be capable of inducing supraventricular and other arrhythmias (4, 5). It has become clear that there are several possible mechanisms of antihistamine-precipitated arrhythmia induction. By exploring the electrical and ion channel events of the cardiac cycle, the potential influences of different classes of antihistamines and their metabolites on these ion channels, the metabolism of antihistamines and alterations of metabolism that may affect cardiac actions, and the epidemiology of antihistamine-related arrhythmias, an appreciation of the difficulties involved in determining the strength of this association can be reached.

Table 1 U.S. Cases of Torsade de Pointes Associated with Terfenadine Use
Reported to the Food and Drug Administration Through April 1, 1992

Age/Sex	Daily dose (mg)	Duration of use (days)	Identified risk factors	Maximum QT (ms)
47/F	120	7	Macrolide, history of prolonged QT	Not reported
79/M	120	5	Cirrhosis	580
16/M	3360	1	Intentional overdose	560 QT_c
55/F	2160	?	Intentional overdose	Not reported
45/F	900	1	Intentional overdose, hypokalemia	600
80/F	180	35	Ketoconazole, ischemic heart disease	600
31/F	600	900	Ketoconazole	Not reported
45/F	120	2	Hypokalemia, history of ethanol abuse	520
53/M	120	7	Ketoconazole, history of prolonged QT	600
39/F	120	12	Ketoconazole	655 QT_c
61/F	120	14	Ethanol abuse	653 QT_c
57/M	120	3	Hypokalemia, ethanol abuse	600
64/M	120	10	Ketoconazole, hypokalemia, ischemic heart disease	600 QT_c
66/M	120	10	Ketoconazole, ischemic heart disease	600
62/M	120	5	Cirrhosis	620
72/F	120	?	Not reported	600
59/F	120	7	Cirrhosis	500
69/F	?	?	None reported	Not reported
65/F	120	365	Hypokalemia, congestive heart failure	580
69/F	120	3	Ketoconazole	430
22/F	180	60	Ketoconazole	660 QT_c
47/F	120	540	Itraconazole	Not reported
74/F	?	?	Not reported	Not reported
18/M	240	>21	None reported	Not reported
38/F	120	5	Macrolide antibiotic	Not reported

ms = milliseconds.
Adapted from Ref. 25, with permission.

The purpose of this review is to provide a historical perspective as
well as an overview of the current understanding of the pathogenesis and
etiology of antihistamine-induced dysrhythmias in an effort to allow practitioners to maximize the therapeutic indices of these chemically diverse
and clinically valuable drugs.

II. HISTAMINE AND THE HEART

The cardioactivity of histamine may be mediated through electrophysiological, vascular, and direct mechanical effects on the myocardium (6). H_1-, H_2-, and the more recently identified H_3-receptors have been localized in the heart (7). H_1-receptors are abundant in guinea pig heart nodal tissue and cardiac vessels but are also heterogeneously distributed in the myocardium (8). Photoaffinity labeling and electrophoresis studies with ^{125}I-iodoazidophenpyrane have suggested the existence of at least two isoforms of the H_1-receptor (8). The functional and clinical implications of a unique cardiac isoform remain unknown. Histamine-induced increase in action potential duration (APD) and rate of depolarization (V_{max}) in isolated rabbit left atrial tissue are completely reversed by the selective H_1-antagonist dimetindene (9). Stimulation of the H_1-receptor prolongs atrioventricular conduction and also induces coronary artery vasoconstriction (6).

The molecular biology of histamine receptors is an active field of investigation. H_1-receptors are coupled to phospholipase C, while the H_2-receptor is connected to adenyl cyclase (10). H_1-receptor messenger RNA is present in bovine lung and other organs, but has not been detected in the heart, suggesting that H_2-, H_3-, or possibly other uncharacterized histamine H_1-receptors are responsible for cardiac effects (11). These mRNA studies require confirmation in humans (12).

The H_2-receptors in atrial and ventricular myocardium mediate positive chronotropic and inotropic responses and vasodilatory responses in the coronary arteries (13). The selective H_2-agonists dimaprit and impromidine yield marked decreases in APD90 of guinea pig ventricular tissue. These effects are completely blocked by cimetidine but not by the selective H_1-antagonist dimetindene (13).

The role of the cardiac H_3-receptor is less well understood. H_3-receptors on bronchial parasympathetic ganglion cells may act as "autoreceptors" to inhibit cholinergic neurotransmission as shown in guinea pig airways (14). H_3-receptors appear to act as autoreceptors on sympathetic nerves and can inhibit sympathetic tone that may potentiate vasodilation (15). Stimulation of the H_3-receptor in isolated guinea pig atria inhibits the sympathetic contractile response to electrical stimulation, suggesting that H_3-receptors modulate adrenergic neurotransmission in the heart (7).

Other potential effects of antihistamines in the cardiovascular system are suggested from the actions of histamine in vivo, for example, by the cutaneous histamine wheal and flare reaction (16). The itch and flare are due to sensory nerve stimulation with the recruitment of axon responses that release neuropeptides and increase superficial cutaneous blood flow.

Histamine acts upon endothelial cell H_1-receptors to cause local vascular permeability and edema (wheal) (16). Antihistamines and tricyclic antidepressants with antihistaminic properties will block histamine wheal and flare responses and reduce early phase mast-cell-mediated reactions.

Histamine H_1-receptors on endothelial cells in human nasal mucosa (17) probably contribute to the vascular permeability and plasma extravasation of allergic rhinitis. Stimulation of human nasal H_1-receptors leads to vascular permeability (measured by albumin exudation), nasal congestion, and decreased airway resistance, but has no net effect on nasal blood flow (18). Histamine stimulates H_1-receptors on nociceptive sensory nerves and recruits systemic reflexes such as sneeze, and local parasympathetic reflexes that mediate glandular secretion (19).

In anaphylaxis, H_1-, and H_2-receptors on peripheral blood vessels may induce profound dilation of resistance vessels, decrease blood pressure, and permit plasma extravasation (20). As described, histamine causes direct coronary artery vasoconstriction via an H_1-receptor mechanism, cardiac positive inotropy by an H_2-receptor mechanism, and positive chronotropy by combined H_1- and H_2-mechanisms (21). Despite these actions, however, antihistamines do not markedly alter the cardiopulmonary effects of anaphylaxis (22). The addition of an H_2-antagonist may be beneficial in treating anaphylactic and anaphylactoid (for example, radiocontrast material allergy) conditions.

III. THE ELECTROCARDIOGRAPHIC CYCLE AND ARRHYTHMOGENESIS

The electrophysiological and arrhythmogenic activity of second-generation antihistamines can be better understood through an appreciation of the predominant ion currents and resulting depolarizations and repolarizations that constitute the cardiac cycle (Fig. 1). The surface ECG is a summary vector representation of this electrical activity and is composed of the P-R followed by QRS complex, flat S-T interval, and T wave (23). The cardiac cycle is initiated in the sinoatrial (SA) node of the left atrium and is conducted through the atrioventricular (AV) node via Purkinje fibers of the His bundle and left and right bundle branches to the septum and ventricles. The time required for atrial depolarization and conduction through the Purkinje fibers is the PR interval. The wave of depolarization passes first into the septum (Q wave) and then to the entire ventricle (R and S waves). Conduction of this wave of depolarization from Purkinje fibers through the ventricular myosyncytium involves a rapid influx of extracellular sodium, followed by calcium, into the myocytes. The instan-

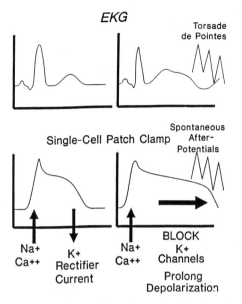

Figure 1 The surface electrocardiogram shows the P, PR, QRS, ST, and T waves. Single ventricular myocyte patch clamp experiments illustrate depolarization (influx of Na⁺ and Ca²⁺ into cell) and repolarization. Phase 3 repolarization is indicated by the action of outward K^+ rectifier current (I_K). Blockade of the K^+ channel(s) responsible for the I_K prolongs depolarization and, when coupled with other factors such as pauses in cardiac rhythm, hypokalemia, or hypomagnesemia, can permit the generation of spontaneous afterpotentials. On the surface EKG, this electrical activity appears as a wide QRS, multifocal ventricular tachycardia with a wandering electrical axis, or torsade de pointes.

taneous depolarization can be measured by patch clamp techniques in individual ventricular myocytes in vitro. This rapid depolarization is termed Phase 0. The maximal rate of rise of the action potential upstroke (V_{max}) is a major determinant of conduction velocity and duration of the QRS complex. This is predominantly mediated by the fast inward sodium current (I_{Na}).

This initial depolarization is followed by the period of peak depolarization (Phase 1) that is due to fluxes through sodium and calcium ion channels. The depolarization is maintained at a constant level (Phase 2) before repolarization (Phase 3) to the resting potential (Phase 4).

Integration of the three-dimensional depolarization and repolarization vectors onto the surface electrocardiogram generates the characteristic Q wave of initial depolarization (Phase 0 and 1), R-S wave of subsequent depolarization (Phase 1), flat ST-segment of continued depolarization (Phase 2), T wave of repolarization (Phase 3), before the return of ventricular myocytes to their resting potential during the T-Q interval (Phase 4). The measured QT interval incorporates ventricular myocyte depolarization, but repolarization accounts for the majority of the time. The length of this interval varies with heart rate: the slower the heart rate, the longer the RR interval, and the longer the QT interval. Several mathematical corrections for this variability exist although the long-standing Bazett's correction ($QT_c = QT/(R-R)^{0.5}$) remains in widespread clinical use (24). Each of these phases is due to the integrated effects of activation and inactivation of a score of ion channels and the fluxes of sodium, calcium, and potassium ions in and out of the cell.

The key ion flux regulating the repolarization of Phase 3 is the "potassium rectifier current" (I_K) (25). Blocking the potassium rectifier current leads to prolongation of Phase 3 and an increased QT_c on the surface electrocardiogram. In large population studies, prolongation of the QT_c to a value greater that 440 ms has been associated with an increased risk of sudden cardiac death (26, 27). The electrophysiological mechanism at the basis of ventricular dysrhythmias secondary to prolonged cardiac repolarization is not completely understood; however, several hypotheses exist. A leading hypothesis contends that heterogeneous generation of "early after-depolarization" potentials leads to a dispersion of refractoriness (28, 29). Such regional discrepancies in action potential duration may favor the development of ventricular reentrant arrhythmias. Other investigators disagree and maintain that the ventricular arrhythmias are a result of multifocal, triggered automaticity (30).

Two types of polymorphic wide QRS tachycardia (23) are recognized depending upon whether the QT_c is normal or increased.

1. Polymorphic Ventricular Tachycardia with a Normal QT_c

Polymorphic ventricular tachycardia with a normal QT_c occurs in Wolff-Parkinson-White syndrome, myocardial infarction, and ischemia, and responds to Class IA antiarrhythmics such as quinidine and procainamide. There is no alteration of T waves, QT_c interval, or pause dependency for the onset of the arrhythmia.

2. Polymorphic Ventricular Tachycardia with an Increased QT_c

Polymorphic ventricular tachycardia with an increase in QT_c interval is subdivided into *pause-dependent torsade de pointes* and *autonomic-dependent polymorphic ventricular tachycardia with prolonged QT_c interval* (23, 28, 29).

a. Pause-Dependent Torsade de Pointes   Pause-dependent torsade de pointes was first described by Desertenne (31) and occurs after pauses when the QT_c is prolonged (generally over 440 ms). A cardiac pause, or lengthening of the R-R interval following a previous normal beat, may occur after a premature ventricular or atrial beat, heart block with nonconduction of an atrial beat, or during other bradycardias. When a pause occurs, there is an absence of ventricular depolarization and repolarization for one beat. This, in turn, causes changes in electrical mechanisms so that the period of ventricular myocyte repolarization (Phase 3) that will follow the subsequent QRS is lengthened. Thus, after a pause, the QT_c interval is lengthened. The electrical mechanism for this pause-induced alteration is poorly understood. During the prolonged Phase 3 repolarization, spontaneous after-potentials can occur in the partially repolarized ventricular myocyte. Cell-to-cell transmission of these impulses can generate the wide QRS, polymorphic ventricular tachycardia recognized on the surface EKG. Approximately 2 to 4% of patients with prolonged QT_c intervals (over 440 ms) may progress to torsade de pointes (23, 27).

Torsade de pointes is causally related to drugs or electrolyte imbalances (23, 27). It can be induced by Class IA antiarrhythmics such as quinidine and procainamide. Quinidine-induced torsade de pointes usually occurs in the first 3 days of therapy, and is more likely in the setting of hypokalemia (including diuretic use), hypomagnesemia, and heart disease (32). Procainamide may be a particular problem in patients who are slow acetylators and thus have a reduced procainamide clearance. Psychotropic agents and certain antibiotics are potassium channel antagonists that have been associated with torsade de pointes. Up to 30% of patients on pentamidine, an anti-*Pneumocystis* antibiotic with Mg^{2+} chelating properties, may develop prolonged QT_c and have the potential to progress to torsade de pointes (23). Other antibiotics with this propensity are erythromycin and trimethoprim-sulfamethoxazole (23, 33, 34). A list of drugs and conditions known to affect cardiac repolarization and to be associated with torsade de pointes is shown in Table 2 (35).

Treatment of pause-dependent torsade de pointes involves (23, 35):

1. Discontinuation of drug(s) that may be inducing the arrhythmia
2. Correction of electrolyte imbalances
3. Electrical overdrive pacing of the heart to increase heart rate and decrease the QT_c
4. Administration of isoproterenol, if necessary, to increase heart rate
5. Intravenous replacement of magnesium, if indicated.

Hypomagnesemia, hypokalemia, and hypocalcemia may potentiate torsade de pointes by altering the magnitudes of the cardiac depolarization

Table 2 Drugs and Toxins Associated with Torsade de Pointes

Antiarrhythmic Agents	Vasodilators	Miscellaneous
Group Ia	Prenylamine	Terfenadine
Quinidine	Lidoflazine[a]	Astemizole
Procainamide	Fenoxidil[a]	Tirodilene
Disopyramide	Bepridil	Corticosteroids
Ajmaline[a]	Psychotropics	Diuretics (e.g.,
Group Ib	Thioridazine	furosemide)
Lidocaine[a]	Chlorpromazine[a]	Suxamethonium[a]
Mexilitene[a]	Amitriptyline[a] (and	Isoproterenol[a]
Aprindine	other tricyclic	Amantadine[a]
Group Ic	antidepressants)	Chloral hydrate[a]
Encainide[a]	Doxepin[a]	Vasopressin[a]
Group III	Maprotilene[a]	Atropine[a]
N-acetylprocainamide	Antibiotics	Halofantine
Amiodarone	Pentamidine	Cisapride
Sotalol	(parenteral and	Toxins
Group IV	inhaled)	Arsenic[a]
Nifedipine[a]	Erythromycin (p.o.	Organophosphate
	and parenteral)	insecticides
	Trimethoprim-	Liquid protein
	sulfamethoxazole	diets
	Quinine, Chloroquine[a]	

[a] Occasional or unconfirmed incidence of torsades de pointes.
p.o. = by mouth
Adapted from Ref. 35.

and repolarization currents. CNS disorders, liquid protein diets, and vasopressin have also been associated with development of torsade de pointes, but the mechanisms are unclear.

  b. *Autonomic-Dependent Torsade de Pointes* Autonomic-dependent torsade de pointes with increased QT_c interval occurs in congenital syndromes and in sporadic noncongenital cases (23). In the Jervell and Lange-Neilsen syndrome, inheritance is autosomal recessive and associated with deafness. The subjects usually die in childhood. In the Romano-Ward syndrome, there is normal hearing and the inheritance is autosomal dominant. Sporadic cases of torsade de pointes in these patients may be precipitated by stress, exercise, loud noises, or other stimuli that activate the sympathetic nervous system and cause an outpouring of norepinephrine (from sympathetic neurons) and epinephrine (from the adrenal glands) that can act upon cardiac β-receptors. Sympathetic tone regulates the rate of repolarization. Isoproterenol increases the QT_c interval, perhaps because of β-receptor-mediated potassium influx and relative serum hypokalemia. The treatment of choice for autonomic torsade de pointes is propranolol (β-blockade). If the arrhythmia is unremitting, left cervical ganglionectomy (sympathectomy) may be performed. Note that isoproterenol therapy is detrimental in autonomic-dependent torsade de pointes, but is beneficial in the pause-dependent form. Therefore, it is essential to make the correct diagnosis and determine the cause of the arrhythmia when treating torsade de pointes.

IV. HISTORICAL PERSPECTIVE: THE HYDROXYZINE STORY

As we have seen, understanding the electrophysiological basis for normal and abnormal rhythms allows for a more enlightened evaluation and interpretation of the cardiac activity of the antihistamines. However, researchers and clinicians have known that some antihistamines were cardioactive without the benefit of such sophisticated understanding. Hydroxyzine is a p-chlorobenzhydryl piperazine derivative that was developed as an antihistamine over 40 years ago. With the knowledge that other H_1-antagonists prolonged the atrial refractory periods, Hutcheon et al. investigated the cardiovascular activity of hydroxyzine in 1956 (1, 36). The effect of hydroxyzine in preventing experimentally induced ventricular tachycardia, its effect on isolated papillary muscle as well as its effects on coronary and peripheral vasculature were explored in animal models. Intravenous hydroxyzine caused a transient hypotensive effect that was not prevented by pretreatment with atropine; however, it increased coronary blood flow 40 to 126%. More interestingly, in feline models hydroxyzine (5 mg/kg) reduced the incidence of harman-methosulfate-induced ventricular fibrillation by 40% compared to controls. The total duration of arrhythmias was also significantly less in the hydroxyzine-treated animals (mean duration 4.6 min vs 29.1 for controls). Procainamide used as an active control (5 mg/kg) did not arrest these experimental arrhythmias. Because vagal stimulation was thought to be an important mediator of ventricular arrhythmias, hydroxyzine was assessed for its vagal blocking activity and found to have 0.2% the activity of atropine. The research revealed that hydroxyzine reduced epinephrine-induced automaticity of papillary muscle preparations. It was also observed that hydroxyzine decreased the contractile force by 50% and raised the threshold of excitability by 49%

in feline left atrial preparations. The conclusion of the authors was that hydroxyzine showed promise as an antifibrillatory agent but they did not speculate on the pharmacological or electrophysiological mechanisms of this activity. Subsequently, Burrell et al. investigated the use of hydroxyzine in a variety of acute arrhythmias in human subjects (37). The trial was uncontrolled and hydroxyzine was given "orally, intramuscularly or intravenously (25–100 mg), depending on the exigency of the clinical situation." The authors' conclusions were overly optimistic ("excellent results in 30/50 patients"), but there was suggestion of an effect on ventricular and atrial automaticity.

In dogs, Birdsong and Pate investigated the role of hydroxyzine in preventing ventricular fibrillation induced by deep hypothermia (38). This was a blind, placebo, and active control study using quinidine. Although hydroxyzine offered no survival protection over placebo, the authors noted that hydroxyzine resulted in similar ECG patterns to quinidine including "prolonged QT intervals with low voltage, marked prolongation, and depression of the ST segment and large, frequently diphasic, T waves."

Interest in hydroxyzine as a potential antiarrhythmic persisted throughout the 1960s. Abaza et al. published their anecdotal investigations of hydroxyzine as a case series of 10 patients with atrial and supraventricular arrhythmias (39). Large doses (1–6 g) were used with good results in that 9 of 10 patients returned to sinus rhythm.

Finally, the resurgence in popularity of hydroxyzine as a potential anxiolytic prompted Hollister to publish the results of a study performed 17 years earlier (40). He investigated hydroxyzine as a psychotherapeutic agent at doses up to 300 mg daily over 9 weeks in 27 elderly psychotic patients. Serial ECG changes were mild except for prolongation of the QT intervals and morphological changes in T waves in 9 subjects.

In retrospect, it appears that hydroxyzine has cardiac electrophysiological effects that may or may not have clinical relevance at usual doses. The findings of Hollister and others suggest that hydroxyzine possesses potassium channel blocking activity. Published reports confirm that hydroxyzine overdose is associated with cardiac effects (41). However, adverse drug reaction databases contain a paucity of reports on hydroxyzine induced dysrhythmias for a variety of reasons that include the age of the drug, and lack of widespread awareness of its cardioactivity by health care practitioners. What can be learned from this example is that antihistamines have long been known to possess cardiac electrophysiological effects that may have therapeutic or safety ramifications. Therefore, it should not be surprising that new antihistamines (including those under development) may be proarrhythmic under selected circumstances.

V. THE TERFENADINE STORY: PHARMACOEPIDEMIOLOGY AND HYPOTHESIS GENERATION

Although effects of antihistamines on the cardiovascular system have been described since their introduction, it is only recently that an association between antihistamines and cardiac death has been demonstrated. In June 1990, the Pulmonary-Allergy Drug Advisory Committee of the U.S. Food and Drug Administration reviewed 25 case reports of ventricular arrhythmias, syncope, and cardiac arrest in persons taking the second-generation, nonsedating H₁-antagonist terfenadine at either the recommended 60 mg b.i.d. dose or in overdose (Table 1). This number has now increased to over 200 reported cases (42–58). In August 1990, the FDA required that a "Dear Doctor" letter be sent to all practicing physicians warning of the risks of concomitant terfenadine use with other drugs identified from the 25 original cases (59). Astemizole-related torsade de pointes has also been reported (60–64).

The original 25 cases of terfenadine-associated cardiac events could be divided into two general groups depending upon the presence of factors that could affect either terfenadine metabolism or cardiac depolarization/repolarization (45). Detailed analysis of the reports suggested that metabolic factors may have been involved. Identified factors that could possibly interfere with hepatic terfenadine metabolism included liver disease (six cases) and coadministration of ketoconazole (nine cases), macrolide antibiotics (two cases), and itraconazole (one case). Concomitant additive or synergistic pharmacodynamic factors included hypokalemia or use of diuretics (five cases), congenital prolonged Q-T syndrome (two cases), and other underlying atherosclerotic/ischemic cardiac disease (four cases). These initial 25 cases suggested that drug metabolism abnormalities, often associated with the use of other drugs, and underlying cardiac repolarization abnormalities contributed to terfenadine-induced arrhythmias.

A similar pattern has more recently been suggested for astemizole and the FDA has required appropriate relabeling of this product and the mailing of a "Dear Doctor" letter cautioning against exceeding the recommended dose and the concomitant use of imidazole antifungal agents and macrolide antibiotics (3).

VI. TERFENADINE AND ASTEMIZOLE METABOLISM

The documentation of a case of a 39-year-old female who had recurrent syncope and torsade de pointes ventricular arrhythmia after taking terfenadine concomitantly with ketoconazole opened the door for a more complete understanding of terfenadine metabolism. The woman had easily quantifiable concentrations of unmetabolized terfenadine which were associated with abnormal cardiac repolarization and a prolonged QT interval (43). In the majority of normal individuals, terfenadine can be considered a prodrug that is rapidly and almost completely metabolized by a cytochrome P_{450} enzyme into an active acid metabolite (TAM) (Fig. 2) (65). The cytochrome P_{450} system in the liver and other organs has the general property of metabolizing hydrophobic chemicals into more hydrophilic forms by hydroxylation, demethylation, or related chemical oxidative processes (66, 67). These enzymes can be divided into two general groups that are involved in either steroidogenesis (68) or xenobiotic (foreign chemical) metabolism (Table 3). Xenobiotic metabolism is essential for the activation or inactivation of many drugs and food components, and the neutralization of ingested plant and fungal alkaloids and other toxins. Each of

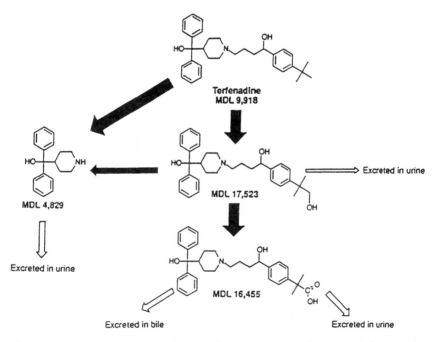

Figure 2 Structure of terfenadine and its major metabolites. MDL 16,455, the "terfenadine acid metabolite" is the predominant H_1-antagonist in vivo. Reproduced with permission from Ref. 85.

Table 3   The Cytochrome P$_{450}$ System

Steroidogenesis	Xenobiotic Metabolism
CYP4	CYP1A1, A2
CYP7	CYP2A6, A7, B6, B7, C8, C9, C17, C18, C19, D6, E1, F1
CYP11	
CYP17	
CYP19	CYP3A3, A4, A5, A7
CYP21	CYP4B1
CYP27	

the many subtypes of P$_{450}$ isoenzymes has its own spectrum of chemical substrates. Each enzyme can be regulated at the transcriptional and protein level, and many partial and total inhibitors for each specific isoenzyme have been identified.

Terfenadine and astemizole are at least partially metabolized by P$_{450}$ CYP3A4 (69). This isoenzyme also metabolizes erythromycin, cyclosporine, nifedipine, and testosterone, and activates aflatoxin B$_1$ and other procarcinogens (Table 4) (66, 67). There is variable expression of P$_{450}$ CYP3A4 in the human population, with this isoenzyme accounting for 10 to 60% of the total hepatic P$_{450}$ activity (70). Although no true metabolic polymorphism is likely to exist, the nearly tenfold variation in CYP3A4 expression suggests that some people may be poor terfenadine and astemizole metabolizers as has been previously well described for nifedipine (71). CYP3A4 is expressed in the epithelial layer of the gut as well as in hepatocytes, although the role of intestinal P450 in the biotransformation of terfenadine is unknown. Related enzymes that may also be active on terfenadine include CYP3A3, CYP3A5, and CYP3A7. CYP3A5 is ex-

Table 4   P$_{450}$CYP3A4

Metabolizes terfenadine, astemizole, erythromycin, cyclosporin, nifedipine, testosterone; activates aflatoxin B$_1$ and other procarcinogens.
Variable expression in human population: 3A4 = 10 to 60% of total P450 activity.
P$_{450}$CYP3A4 can be inhibited by erythromycin and ketoconazole.
Ingestion of ketoconazole or erythromycin can increase serum terfenadine concentrations and prolong QT$_c$ interval (Ref. 75).

pressed in only 10 to 20% of human livers. CYP3A7 is present only in fetal human liver (72).

Significant in vitro inhibitors of CYP3A4 include erythromycin, ketoconazole, cimetidine, ranitidine, and naringenin (the predominant flavonoid in grapefruit juice and pulp (66–68, 73, 74). This enzyme can be induced by glucocorticoids, barbiturates, and rifampicin. Honig et al. showed that administration of erythromycin or ketoconazole with terfenadine led to increased serum concentrations of terfenadine (see below) (75–81). Recent investigations suggest that erythromycin and ketoconazole may also increase plasma concentrations of astemizole or metabolites (61). These results indicate that interference with cytochrome P_{450} CYP3A4 can lead to increased levels of terfenadine and astemizole that may be of clinical significance.

Terfenadine and astemizole have intrinsic potassium channel antagonist properties (see below). As previously described, blocking the outward K^+ rectifier current (I_K) prolongs repolarization. This may facilitate the generation of spontaneous after-depolarizations and refractory dispersion. A pause-dependent lengthening of the QT_c interval plus antagonism of the outward K^+ rectifier current by terfenadine or astemizole could trigger an episode of torsade de pointes under certain circumstances.

1. Terfenadine in Animal Models

The effects of terfenadine on cardiac intervals have been studied in individual feline cardiac myocytes grown in culture using the patch clamp technique. In these experiments, the potassium rectifier current I_K was blocked by terfenadine with an EC_{50} of 150 nanomolar (nM) (82). Quinidine, an antiarrhythmic agent known to block potassium channels and to cause torsade de pointes, had an EC_{50} of 180 nM. In this model system, the acid metabolite of terfenadine had no potassium channel antagonist activity up to a concentration of 5 μM.

This model has also been used to investigate the electrophysiological properties of astemizole which was shown to be a potent blocker of the delayed rectifier potassium current (83).

QT_c interval prolongation by terfenadine has also been evaluated in guinea pig hearts in vitro (84). D-, 1- and racemic terfenadine increased QT_c interval, action potential duration, and effective refractory period by 8%.

2. Terfenadine in Human Models

Terfenadine and its metabolites were studied on rapidly activating delayed rectifier potassium channels cloned from human heart (85). Terfenadine, but not its metabolites, blocks at least one type of human K^+ channel at clinically relevant concentrations (Table 5).

Table 5 Effect of Terfenadine and Metabolites on the Activity of the Cloned
K$^+$ Channel *fHK*, the Presumed Mediator of the Outward K$^+$ Rectifier Current

| Drug | Plasma concentration | | IC$_{50}$*fHK* nanomolar |
	Therapeutic	Overdose or ketoconazole	
Terfenadine	Undetectable 43 nM "accumulators"	63 to 172 nM	367
MDL 17,523 (intermediate)	<10 nM	?	18,000
MDL 16,455 (acid metabolite)	2,800 nM	4,200 nM	214,000
MDL 16,829 (acid metabolite)	200 nM	?	414,000

From Ref. 85.

The effect of terfenadine on the QT$_c$ interval has been examined in humans. In a large population study, normal subjects taking terfenadine had a mean increase of QT$_c$ of 8 ms above normal (86). Subjects with cardiovascular disease taking terfenadine showed an increase of 10 ms in QT$_c$. Honig et al. have demonstrated a wide population variability in terfenadine biotransformation (87). Of 150 normal, healthy subjects administered terfenadine at steady state, 11 (5 males/6 females) had quantifiable concentrations of unmetabolized terfenadine (range 5.2–14.4 ng/ml). Further studies are under way to determine if such poor metabolizers are more susceptible to metabolic inhibition, and if so, at increased risk for adverse cardiac events.

Sanders (88) performed cross-over studies with 16 subjects and compared the effects of 60, 180, or 300 mg terfenadine, 50 mg diphenhydramine, and placebo on the QT$_c$. Six hours after ingesting 300 mg of terfenadine, QT$_c$ was increased by 23 ms. Diphenhydramine decreased the heart rate and so increased the QT$_c$.

Terfenadine pharmacokinetics have also been investigated in patients with marked hepatic insufficiency. In 16 patients with hepatic disease, 4 had easily quantifiable concentrations of unmetabolized terfenadine after one week of therapy (89). This risk factor for cardiac arrhythmia was identified by the FDA Pulmonary-Allergy Drugs Advisory Committee, and its importance was underscored by the description of a 57-year-old female with hepatic cirrhosis who was treated with terfenadine for 4 days

(240 mg/day) before experiencing QT_c prolongation and sustained ventricular arrhythmias (55).

3. Drug Interactions

a. Terfenadine–Macrolide Antibiotic Interactions Honig et al. examined the interaction of terfenadine and erythromycin (77, 79). Nine subjects who were extensive metabolizers of CYP2D6 (debrisoquin hydroxylator phenotype) took 60 mg of terfenadine twice daily for 7 days. After 1 week, no terfenadine was measurable in the blood of any subject (Table 6). Concentrations of the terfenadine acid metabolite were 490 ± 38 nM. The mean QT_c interval was increased by 18 ± 10 ms (not significant). In the second week, the nine subjects continued to take 60 mg of terfenadine twice daily plus erythromycin stearate 500 mg three times a day. At the end of 7 days, three of the nine subjects had detectable terfenadine levels in serum (mean, 43 ± 16 nM), while the other six had no detectable terfenadine. The mean terfenadine acid metabolite concentration was 1014 ± 110 nM. The three subjects with detectable terfenadine levels were called "accumulators," and developed significant prolongation of their QT_c interval. Increases were 64 ± 12 ms compared to 39 ± 11 ms for the entire group ($p < 0.05$). ST-U wave changes occurred in one of the three "accumulators." Erythromycin itself is known to affect the QT_c, and this was demonstrated in a separate part of the study in which the mean QT_c increased from 389 ± 11 to 399 ± 7 ms (not significant). This

Table 6 Interaction of Terfenadine and Erythromycin

Nine subjects (debrisoquin hydroxylator phenotype)	Week 1 (terfenadine 60 mg p.o. b.i.d. × 7 days)	Week 2 (terfenadine 60 mg p.o. b.i.d. × 7 days, erythromycin 500 mg p.o. t.i.d. × 7 days)
[Terfenadine] "accumulators"	Not detected	6/9: not detected 3/9: 43 ± 16 nM
[TAM]	490 ± 38 nM	1014 ± 110 nM
Increase in QT_c	18 ± 10 ms (ns)	39 ± 11 ms (ns)
"accumulators" (3/9)	33 ± 21 ms (ns)	64 ± 12 ms ($p < 0.05$) ST-U changes in 1/3

Erythromycin alone increases QT_c from 389 ± 11 to 399 ± 7 ms.
Terfenadine + erythromycin = 6% of U.S. terfenadine prescriptions in 1992.
p.o. = by mouth; b.i.d. = twice daily; t.i.d. = three times daily; ms = milliseconds; ns = not significant.

interaction is of clinical importance because, at the time, 6% of terfenadine prescriptions were written for patients also taking erythromycin.

In a study with similar design (79), terfenadine was detected in the blood of four out of six volunteers who had taken clarithromycin plus terfenadine for 1 week. In contrast, azithromycin did not increase terfenadine levels. This suggests that erythromycin and clarithromycin, but not azithromycin, can alter terfenadine metabolism.

b. Terfenadine–Ketoconazole Interaction Ketoconazole targets a fungal P_{450} enzyme (65, 66). Interactions of terfenadine and ketoconazole were studied in six normal volunteers (76, 80) who ingested 60 mg of terfenadine twice daily for 7 days. Terfenadine was detected in the blood of 1 out of 6 subjects in a concentration of 15 nM (Table 7). The QT_c interval was increased from 408 ± 8 ms to 416 ± 6 ms for the entire group (not significant). In the second week, 60 mg of terfenadine twice daily plus 200 mg of ketoconazole twice daily were taken concomitantly for 7 days. At the end of the second week, terfenadine was detected in the blood of 6 out of 6 subjects, in concentrations ranging from 11 to 172 nM. The QT_c intervals were increased to 490 ± 16 ms ($p < 0.01$). This is of clinical significance since increases of QT_c greater than 440 ms are associated with a 2.3 times higher risk for sudden death in some patient subsets. Ketoconazole should not be prescribed with terfenadine or astemizole.

c. Terfenadine–Itraconazole Interaction Itraconazole is a recently approved oral systemic antifungal agent with an enhanced spectrum of activity over ketoconazole (90). Itraconazole inhibition of terfenadine metabolism resulting in cardiac arrhythmia has been reported (91, 92). The effect of steady-state levels of itraconazole on the pharmacokinetics and

Table 7 Interaction of Terfenadine and Ketoconazole

Six volunteers	Terfenadine (60 mg p.o. b.i.d. × 7 days)	Terfenadine (60 mg p.o. b.i.d. × 7 days, ketoconazole 200 mg p.o. b.i.d. × 7 days)
Terfenadine	Detected in 1/6 (15 nM)	Detected in 6/6 range: 11 to 172 nM
QT_c 408 ± 8 ms	416 ± 6 ms (ns)	490 ± 16 ms ($p < 0.01$)

Terfenadine + ketoconazole = 0.2% of U.S. terfenadine prescriptions in 1992.
$QT_c > 440$ ms = 2.3 × higher risk for sudden death.
Ketoconazole targets fungal P_{450} enzyme.
b.i.d. = twice daily; p.o. = by mouth; ms = milliseconds; ns = not significant.

electrocardiographic pharmacodynamics of a single dose of terfenadine, was examined in six healthy human volunteers (93). A single 120-mg dose of terfenadine was administered and a 12-h pharmacokinetic and ECG profile obtained. The subjects were then started on itraconazole (200 mg daily) for 1 week at which time a repeat dose of terfenadine was administered accompanied by serial blood sampling and electrocardiograms. All six had easily quantifiable terfenadine levels (Cmax 10–22 ng/ml) which were associated with an increase in QT_c interval (Fig. 3). The design of this single-dose interaction study produced a different pattern in the before/after concentrations of the acid metabolite than seen in the previously cited studies. One explanation could be the fact that the acid metabolite may be further oxidatively metabolized to an inactive second metabolite (69). The steady-state interaction studies allow for inhibition of this postulated second process and accumulation of the acid metabolite.

Fluconazole, at recommended doses, does not appear to affect terfenadine metabolism and blood levels (94).

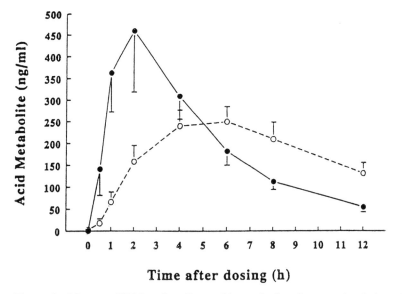

Time after dosing (h)

Figure 3 Mean ± SEM terfenadine acid metabolite-time graph of six subjects receiving single-dose terfenadine alone (solid line) and with concomitant itraconazole (broken line). Reproduced with permission from Ref. 93.

d. Terfenadine–Grapefruit Juice Interaction   Concomitant administration of grapefruit juice has been shown to inhibit the metabolism of other highly cleared P_{450} 3A4 substrates such as felodipine and nisoldipine. The effects of grapefruit juice 240 mL every 12 h on terfenadine pharmacokinetics and ECG pharmacodynamics were studied in six healthy volunteers in a manner analogous to the above studies (81). Grapefruit juice given concomitantly with terfenadine caused significant increases in the mean AUC of the acid metabolite (921 to 1428 ng h/ml) and detectable concentrations of unmetabolized terfenadine in all six subjects. This was associated with a statistically significant 14-ms mean increase in the QT_c interval.

e. Terfenadine–H₂-Antagonist Interactions   Human in vivo steady-state terfenadine interaction studies have been conducted with cimetidine (600 mg twice daily) and ranitidine (150 mg twice daily) without demonstration of a pharmacokinetic or pharmacodynamic interaction (95). Cimetidine is a well-known inhibitor of certain P_{450}-mediated oxidative reactions and has been shown to inhibit *in vitro* microsomal terfenadine biotransformation at high concentrations (personal communication, L.R. Cantilena), but does not apparently affect terfenadine pharmacokinetics at clinically relevant doses.

4. Summary of Metabolism and Cardiac Effects

These data suggest that inhibition or compromise of terfenadine and astemizole metabolism resulting in the accumulation of the cardiotoxic parent compounds may be the mechanism leading to the reported arrhythmias (Table 8). Terfenadine is normally totally metabolized on first-pass through the liver and cannot be detected in the blood. However, very high doses, as in overdose, could lead to increased absorption and appearance of unmetabolized drug in the blood. Increased blood levels could also be the result of reduced terfenadine metabolism as could occur in subjects with low P_{450} 3A4 expression (70) or in the presence of P_{450} 3A4 inhibitors such as ketoconazole, itraconazole, other azoles, erythromycin, or other drugs. Liver disorders such as hepatitis, cirrhosis, and alcoholic hepatitis may also be responsible for decreased P_{450} 3A4 activity or vascular shunting that could reduce terfenadine first-pass biotransformation. Since P_{450} 3A4 is expressed in the gut epithelium, it is theoretically possible that diarrhea or other enteric diseases could also affect metabolism of this drug. The end result of decreased metabolism would be an increase in terfenadine blood levels from undetectable to detectable, and a delay in the appearance of terfenadine acid metabolite in the blood.

Astemizole may have similar patterns of metabolism and drug interac-

Table 8 Proposed Theory for Cardiotoxicity

Overdose→	Decreased conversion → to metabolites	Intrinsic K^+ channel antagonism in heart
Terfenadine Astemizole "Prodrugs"	Variable P_{450} 3A4 expression in liver and intestinal wall since P_{450} 3A4 contributes 10 to 60% of total P_{450} activity in different individuals.	Terfenadine and astemizole appear to block K^+ rectifier channel responsible for repolarization and to induce pause-dependent torsade de pointes
	P_{450} 3A4 inhibitors: ketoconazole itraconazole erythromycin clarithromycin others Cirrhosis, EtOH	Potential additional risk factors: cardiac conditions congenital prolongation of QT coronary artery disease high-grade AV block
	Terfenadine appears in the blood when not metabolized (Metabolites are usually active antihistamines in vivo)	Potential synergistic interactions with other K^+ channel blockers (see Table 2): Hypokalemia Hypomagnesemia Hypocalcemia Hypothyroidism

tion but, except for case reports of astemizole-induced torsade de pointes, published information is considerably less extensive than for terfenadine.

If significant concentrations of unmetabolized terfenadine or astemizole were present in the blood, then these drugs could act by their intrinsic potassium channel antagonist properties to induce altered cardiac repolarization and, possibly, torsade de pointes. Factors that could enhance this effect on cardiac myocytes include cardiac abnormalities, synergistic reactions with other potassium channel blockers, and other electrolyte abnormalities. Cardiac conditions associated with increased frequency of wide QRS tachycardias include congenital prolonged QT syndrome, coronary artery disease, ischemic heart disease, and high-grade AV block. The risk of terfenadine and astemizole use in atherosclerotic heart disease is unclear. Four of the initial 25 cases of terfenadine-associated arrhythmias were related to cardiac disease. The nature of the underlying defect(s) is not fully detailed, although it may be due to acquired or congenital conduction abnormalities. There have been published reports of torsade de pointes in patients who apparently had been taking astemizole at normal doses without identifiable risk factors (96, 97).

It is theoretically possible that other drugs that promote pause-dependent torsade de pointes may potentiate the pharmacodynamic cardiac effects of terfenadine and astemizole. These may include other potassium channel antagonists, antiarrhythmic agents including quinidine and procainamide, tricyclic antidepressants, antipsychotic drugs, intravenous pentamidine, intravenous erythromycin, trimethoprim/sulfamethoxazole, cisapride, and oral hypoglycemic agents. Hypokalemia, hypomagnesemia, and hypocalcemia associated with diuretic use or other causes may also affect ventricular repolarization. Hypothyroidism is another condition associated with repolarization abnormalities and hypokalemia, but as yet no cases of terfenadine or astemizole-related torsade de pointes have been reported in subjects with altered thyroid metabolism.

VII. OTHER ANTIHISTAMINES AND CARDIAC TOXICITY

As we have previously seen, hydroxyzine has been known to possess cardiotropic properties. In preliminary studies, cetirizine, the carboxylated metabolite of hydroxyzine (98) did not prolong Qt_c in healthy male volunteers (99). Promethazine has been evaluated as an antiarrhythmic in guinea pig ventricular myocytes and found to block inward sodium current (I_{Na}) in a manner similar to the Class I antiarrhythmic drugs such as quinidine and procainamide (100). Cyproheptadine decreases plateau levels and action potential duration in rabbit purkinje and ventricular myocardial cells in a dose-dependent manner (101). Cyproheptadine may affect slow inward calcium currents (I_{Ca}) as well as the sodium-mediated fast inward current (I_{Na}) (102). Chlorpheniramine is a chiral compound and both enantiomers increase action potential duration and the rate of membrane depolarization. Interestingly, in specialized atrial conducting tissue, the l-isomer has a more profound effect on action potential duration (103). Azelastine is another selective H_1-antagonist that exhibits fast sodium channel activity and a prolonged inhibitory effect on contraction (104). To date, there are no published studies investigating the clinical electrocardiographic changes associated with these drugs administered with concomitant metabolic inhibitors. Other antihistamines have been associated with cardiac arrhythmias (4, 5) possibly by an anticholinergic mechanism.

Loratadine has been studied in doses up to four times the clinically recommended dose and taken concomitantly with ketoconazole (105). There were no demonstrable effects on the surface electrocardiogram in either of these studies. Loratadine had no effects on electrocardiographic parameters in guinea pigs in doses up to 100 mg/kg (106).

VIII. LABELING AND USE

Terfenadine and astemizole are now specifically contraindicated in the presence of hepatic compromise or if used concomitantly with ketoconazole, itraconazole, troleandomycin, or erythromycin (107, 108). Furthermore, product labeling cautions against their use in patients taking drugs that are known to affect cardiac repolarization (e.g., Class IA antiarrhythmics, sotalol, bepridil, etc.).

Patient education is of paramount importance for these nonsedating antihistamines because the lack of side effects allows patients to unwisely titrate dosing to higher levels in order to achieve a greater or more rapid effect. The problem with this practice is well illustrated by the case of a patient who experienced ventricular fibrillation after "self-adjusting" her astemizole dose due to lack of perceived efficacy (109).

Finally, since the pharmacokinetic variability of $P_{450}3A4$ substrates is being increasingly appreciated, health care practitioners and patients should always be leery of exceeding the recommended dose of any medication eliminated by this isoenzyme system. Clinical phenotyping of cytochrome P_{450} CYP3A4 levels is currently a research tool. At present, there is no routine screening test to identify patients at risk for development of torsade de pointes.

IX. SUMMARY

The response to the 25 cases of cardiotoxicity associated with terfenadine use is instructive to our prescribing practices and concerns about product safety. Numbers of prescriptions written for terfenadine dropped after the 1990 FDA actions (110, 111) but worldwide it remains the most commonly prescribed antihistamine.

An important consequence of the decrease in second-generation antihistamine use has been an increase in the use of more sedating, first-generation, over-the-counter antihistamines. It is estimated that 30 million Americans bought over-the-counter antihistamines in 1988. These first-generation drugs cross the blood–brain barrier and may antagonize serotonin, acetylcholine, and α-adrenergic receptors in addition to histamine H_1-receptors. Subjective sedation occurs in 10 to 25% of subjects using these antihistamines (112–116). The effects of these drugs on psychomotor reflexes, driving, and interactions with alcohol and tranquilizers have been studied extensively. The first-generation agents cause greater performance decrements than the newer, nonsedating antihistamines (112). Oral administration of triprolidine, diphenhydramine, hydroxyzine, azatadine, clemastine, cyclizine, chlorpheniramine, brompheniramine and promethazine has been associated with general physical and mental sedation characterized by sleepiness and drowsiness, lethargy, dullness, lowered ability to concentrate, and feelings of incompetence, boredom and being self-centered (reviewed in Ref. 117). These neurocognitive effects must be considered when choosing an appropriate antihistamine. In addition, the identification and study of the infrequent but serious cardiac events once again underscores the importance of individualizing pharmacotherapy. A cavalier attitude toward prescribing should never be tolerated, and the patient's underlying medical conditions as well as current drug regimen should be determined before any new drug is added. Thankfully, most of the pharmacological agents in current use have wide therapeutic indices and are "forgiving" with regard to severe untoward consequences. When properly used, terfenadine and astemizole can be valuable tools in the treatment of histamine-mediated diseases; however, the potentially devastating side effects should never be considered trivial. The lessons learned from the terfenadine and astemizole experiences will be used by the United States Food and Drug Administration and other drug regulatory bodies. The critical importance of elucidating the metabolism of a new drug and using in vitro techniques to evaluate the potential for adverse clinical events is becoming standard procedure in the drug development and approval processes (117).

REFERENCES

1. Wyngaarden JB, Seevers MH. The toxic effects of antihistaminic drugs. JAMA 1951;145:277–282.
2. Loew ER. Pharmacology of antihistamine compounds. Physiol Rev 1947; 27:542–573.
3. Anonymous. FDA Med. Bull. 1992;22:2–3.
4. Clark RF, Vance MV. Massive diphenhydramine poisoning resulting in a wide-complex tachycardia: successful treatment with sodium bicarbonate. Ann Emerg Med 1992;21:318–321.
5. Hestand HE, Teske DW. Diphenhydramine hydrochloride intoxication. J Pediatr 1977;90:1017–1018.
6. Bristow MR, Ginsburg R, Harrison DC. Histamine and the human heart: the other receptor system. Am J Cardiol 1982;49:249–250.
7. Luo XX, Tan YH, Sheng BH. Histamine H₃-receptors inhibit sympathetic neurotransmission in guinea pig myocardium. Eur J Pharmacol 1991;204: 311–314.
8. Ruat M, Bouthenet ML, Schwartz JC, Ganellin CR. Histamine H₁-receptor

in heart: unique electrophoretic mobility and autoradiographic localization. J Neurochem 1990;55:379–385.

9. Bouchard U, Hafner D. Electrophysiological characteristics of histamine receptor subtypes in mammalian heart preparations. Naunyn Schmiedebergs Arch Pharmacol 1986;334:294–302.

10. Timmerman H. Histamine agonists and antagonists. Acta Otolaryngol (Stockh) 1991;suppl 479:5–11.

11. Yamashita M, Fukui H, Sugama K, Horio Y, Ito S, Mizuguchi H, Wada H. Expression cloning of a cDNA encoding the bovine histamine H_1-receptor. Proc Natl Acad Sci USA 1991;88:11515–11519.

12. Chowdhury BA, Kaliner MA, Fraser CM. Cloning of a gene encoding the human H_1-histamine receptor. J Allergy Clin Immunol 1994;93:215.

13. Kecskemeti V. Cardiac electrophysiological effects of histamine, H_1- and H_2-agonists. Agents Actions 1993;38:C292–C294.

14. Ichinose M, Stretton CD, Schwartz JC, Barnes PJ. Histamine H_3-receptors inhibit cholinergic neurotransmission in guinea pig airways. Br J Pharmacol 1989;7:13–5.

15. Ishikawa, S, Sperelakis N. A novel class (H_3) of histamine receptors on perivascular nerve terminals. Nature 1987;327:158–60.

16. Baraniuk JN, Kowalski M, Kaliner M. Neuropeptides in the skin. In: Skin Immune System (SIS) (Bos JB, ed.) Baton Rouge, LA; CRC Press. 1990: 307–326.

17. Okayama M, Baraniuk JN, Hausfeld JN, Merida M, Kaliner MA. Characterization and autoradiographic localization of histamine H_1-receptors in human nasal turbinates. J Allergy Clin Immunol 1992;89:1144–1150.

18. Raphael GD, Baraniuk JN, Kaliner MA. How and why the nose runs. J Allergy Clin Immunol 1991; 87:457–467.

19. Raphael GD, Meredith SD, Baraniuk JN, Druce HM, Banks SM, Kaliner MA. The pathophysiology of rhinitis. II. Assessment of the sources of protein in histamine-induced nasal secretions. Am Rev Respir Dis 1989;139: 791–800.

20. Flynn SB, Owen DAA. Histamine receptors in peripheral vascular beds in the cat. Br J Pharmacol 1975;55:181–8.

21. Kang YH, Wei HM, Fischer H, Merrill JF. Histamine-induced changes in coronary circulation and myocardial oxygen consumption: influences of histamine receptor antagonists. FASEB J 1987;1:483–490.

22. Silverman HJ, Taylor WR, Smith PL, Kagey-Sobotka A, Permutt S, Lichtenstein LM, Bleecker ER. Effects of antihistamines on the cardiopulmonary changes due to canine anaphylaxis. J Appl Physiol 1988;64:210–217.

23. Taskforce of the working group on arrhythmias of the European Society of Cardiology. The Sicilian gambit. Circulation 1991;84:1831–1851.

24. Bazett HC. An analysis of the time relationships of electrocardiograms. Heart 1920;7:353–370.

25. Woosley RL. Chen Y, Freiman JP, Gillis RA. Mechanisms of the cardiotoxic actions of terfenadine. JAMA 1993;269:1532–1536.

26. Algra A, Tijsssen JGP, Roelandt JRTC, Pool J, Lubsen J. QTc prolongation measured by standard 12-lead electrocardiography is an independent risk factor for sudden death due to cardiac arrest. Circulation 1991;83: 1888–1894.

27. Goldberg RJ, Bengtson J, Chen Z, Anderson KM, Locati E, Levy D. Duration of the QT interval and total cardiovascular mortality in healthy persons. The Framingham heart study experience. Am J Cardiol 1991;67: 55–58.

28. January CT, Riddle JM. Early after-depolarizations: mechanism of induction and block. Circ Res 1989;64:977–990.

29. Cranefield PF, Aronson RS. Torsades de pointes and early after-depolarizations. Cardiovasc Drugs Ther 1991;5:531–537.

30. Surawicz B. Electrophysiologic substrate of torsade de pointes: dispersion of repolarization or early after-depolarizations. J Am Coll Cardiol 1989;14: 172–184.

31. Dessertenne F. La tachycardie ventriculaire à deux foyes opposés variables. Arch Mal Coeur 1966;59:263–272.

32. Roden DM, Woosley RL, Primm RK. Incidence and clinical features of the quinidine-associated long QT syndrome: implications for patient care. Am Heart J 1986;111:1088–1093.

33. Freedman RA, Anderson KP, Green LS, Mason JW. Effect of erythromycin on ventricular arrhythmias and ventricular repolarization in idiopathic long QT syndrome. Am J Cardiol 1987;59:168–9.

34. Nattel S, Ranger S, Talajic M, Lemery R, Roy D. Erythromycin-induced long QT syndrome: concordance with quinidine and underlying cellular electrophysiological mechanism. Am J Med 1990;89:235–238.

35. Stratmann HG, Kennedy HL. Torsade de pointes associated with drugs and toxins: recognition and management. Am Heart J 1987;113:1470–1482.

36. Hutcheon DE, Scriabine A, Morris DL. Cardiovascular action of hydroxyzine. J Pharmacol Exp Ther 1956;118:451–460.

37. Burrell ZL, Gittinger WC, Martinez A. Treatment of cardiac arrhythmias with hydroxyzine. Am J Cardiol 1958;1:624–628.

38. Birdsong S, Pate JW. Effects of hydroxyzine hydrochloride on ventricular fibrillation in deep hypothermia. Am Surg 1960;26:492–493.

39. Abaza A, Delattre G, Germain G. Hydroxyzine et troubles du rhythm cardiaque. Coeur Med. Intern 1967;6:373–387.

40. Hollister LE. Hydroxyzine hydrochloride: possible adverse cardiac interactions. Psych Comm 1975;1:61–65.

41. Magera BE, Betlach CJ, Sweatt AP, Derrick CW. Hydroxyzine intoxication in a 13-month-old child. Pediatrics 1981;67:280–283.

42. Davies AJ, Harindra V, McEwan A, Ghose RR. Cardiotoxic effect with convulsions in terfenadine overdose. Br Med J 1989;298:325.

43. Monahan BP, Ferguson CL, Killeavy ES, Lloyd BK, Troy J, Cantilena LR. Torsade de pointes occurring in association with terfenadine use. JAMA 1990;264:2788–2790.

44. MacConnell TJ, Stanners AJ. Torsade de pointes complicating treatment with terfenadine. Br Med J 1991;302:1469.

45. Transcript of proceedings. Pulmonary-Allergy Drugs Advisory Committee, June 11, 1990. Department of Health and Human Services. Public Health Service. Rockville, MD: Food and Drug Administration, 1990.

46. Matthews DR, McNutt B, Okerholm R, Flicker M, McBride G. Torsade de pointes occurring in association with terfenadine use. JAMA 1991;266: 2375–6.

47. Anonymous. Safety of terfenadine and astemizole. Med Let Drugs Ther 1992;34:9–10.

48. Anonymous. Terfenadine-cardiac side effects. Lakartindningen 1991;88: 1999–2000.

49. Bastecky J, Kvasnicka J, Vortel J, Tauchman M, Wasylivova V. Severe intoxication with antihistamines complicated by ventricular tachycardia. Vnitr Lek 1990;36:266–269.

50. Guilbaud JC, Moin M, Sader R. A case of rhythm disorder during terfenadine poisoning. Reanim Soins Intensifs Med Urgence 1991;7:233–234.

51. Hanrahan JP, Choo PW, Carlson W, Greineder D, Faich GA, Platt R. Antihistamine-associated sudden death, ventricular arrhythmias, syncope, and QT-interval prolongation: a comparison of terfenadine and other antihistamines. Post Mark Surveill 1992;6:23–24.

52. Hansten PD, Horn JR. A clinical perspective and analysis of current developments: terfenadine drug interactions. Drug Interact Newslett 1992;Aug: 586–588.

53. Kintz P, Mangin P. Toxicological findings in a death involving dextromethorphan and terfenadine. Am J Forensic Med Pathol 1992;13:351–352.

54. Biglin KE, Faraon MS, Constance TD, Lieh-Lai M. Drug-induced torsades de pointes: a possible interaction of terfenadine and erythromycin. Ann Pharmacother 1994;28:282.

55. Venturini E, Borghi E, Maurini V, Vecce R, Carnicelli A. Prolongation of the Q-T interval and hyperkinetic ventricular arrhythmia probably induced by terfenadine use in liver cirrhosis patients. Recenti Prog Med 1992;83: 21–22.

56. Eller M, Russel T, Ruberg S, Okerholm R, McNutt B. Effect of erythromycin on terfenadine metabolite pharmacokinetics. Clin Pharmacol Ther 1993; 53:161.

57. Bastecky J, Kvasnicka J, Vortel J, Tauchman M, Wasylivova V. Zavazna intoxikace antihistaminiky komplikovana komorovou tachykardii. Vnitrni lekarstvi 1990;36:266–269.

58. Zimmermann M, Duruz H, Guinand O, Broccard O, Levy P, Lacatis D, Bloch A. Torsade de pointes after treatment with terfenadine and ketoconazole. Eur Heart J 1992;13:1002–1003.

59. Marion Merrell Dow, Inc. Important Drug Warning. Cincinnati, OH; Marion Merrell Dow, Inc. August 6, 1990.

60. Anonymous. Reports of dangerous cardiac arrhythmias prompt new contraindications for the drug Hismanal. FDA Med Bull 1993;23:2.

61. Anonymous. Janssen's Hismanal (astemizole) labeling adopts drug interaction warnings: new contraindications against erythromycin, ketoconazole, and itraconazole use. FDC Rep 1992;Nov:3–4.

62. Wiley JF, Gelber ML, Henretig FM, Wiley CC, Sandhu S, Loiselle J. Cardiotoxic effects of astemizole overdose in children. J Pediatr 1992;120: 799–802.

63. Craft TM. Torsade de pointes after astemizole overdose. Br Med J 1986; 292:660.

64. Committee on Safety in Medicines: Cardiotoxicity of astemizole in overdose-dosing is critical. Curr Prob 1987;18.

65. Garteiz DA, Hook RH, Walker BJ, Okerholm RA. Pharmacokinetics and biotransformation studies of terfenadine in man. Arzneimittelforschung 1982;32:1185–1190.

66. Gonzalez FJ. Human cytochromes P450: problems and prospects. Trends Pharmacol Sci 1992;13:346–352.

67. Cholerton S, Daly AK, Idle JR. The role of individual human cytochromes P450 in drug metabolism and clinical response. Trends Pharmacol Sci 1992; 13:434–439.

68. Loose DS, Kan PB, Hirst MA, Marcus RA, Feldman D. Ketoconazole blocks adrenal steroidogenesis by inhibiting cytochrome P450-dependent enzymes. J Clin Invest 1983;71:1495–1499.

69. Yun C-H, Okerholm RA, Guengerich FP. Oxidation of the antihistaminic drug terfenadine in human liver microsomes. Drug Metab Disp 1993;21: 403–409.

70. Hunt CM, Watkins PB, Saenger P. Heterogeneity of CYP3A isoforms metabolizing erythromycin and cortisol. Clin Pharmacol Ther 1992;51:18–23.

71. Breimer DD, Schellens JHM, Soons PA. Nifedipine: variability in its kinetics and metabolism in man. Pharmac Ther 1989;44:445–454.

72. Watkins PB. Drug metabolism by cytochromes P450 in the liver and small bowel. Gastro Clin North Am 1992;21:511–526.

73. Amacher DE, Schomaker SJ, Retsema JA. Comparison of the effects of the new azalide antibiotic azithromycin and erythromycin estolate on rat liver cytochrome P450. Antimicrob Agents Chemother 1991;35:1186–1190.

74. Brown MW, Moldonado AL, Meredith CG, Speeg, KV, Jr. Effect of ketoconazole on hepatic oxidative metabolism. Clin Pharmacol Ther 1985;37: 290–297.

75. Honig PK, Woosley RL, Zamani K, Conner DP, Cantilena LR, Changes in the pharmacokinetics and electrocardiographic pharmacodynamics of terfenadine with concomitant administration of erythromycin. Clin Pharmacol Ther 1992;52:231–8.

76. Honig PK, Wortham DC, Zamani K, Conner DP, Mullin JC, Cantilena LR. Terfenadine-ketoconazole interaction. Pharmacokinetic and electrocardiographic consequences. JAMA 1993;269:1513–1518.

77. Honig PK, Zamani K, Woosley RL, et al. Erythromycin changes terfenadine pharmacokinetics and electrocardiographic pharmacodynamics. Clin Pharmacol Ther 1992;51:156.

78. Eller MG, Okerholm RA. Pharmacokinetic interaction between terfenadine and ketoconazole. Clin Pharmacol Ther 1991;49:130.

79. Honig P, Wortham D, Zamani K, Conner D, Cantilena L. Effect of erythromycin, clarithromycin, and azithromycin on the pharmacokinetics of terfenadine. Clin Pharmacol Ther 1993;53:161.

80. Honig P, Wortham D, Zamani K, Conner D, Cantilena L. The pharmacokinetics and cardiac consequences of the terfenadine-ketoconazole interaction. Clin Pharmacol Ther 1993;53:206.

81. Benton R, Honig P, Zamani K, Hewett J, Cantilena LR, Woosley RL. Grapefruit juice alters terfenadine pharmacokinetics resulting in prolongation of QTc. Clin Pharmacol Ther 1994;55:146.

82. Chen T, Gillis RA, Woosley RL. Block of delayed rectifier potassium current, Ik, by terfenadine in cat ventricular myocytes. J Am Coll Cardiol 1991; 17:140A.

83. Chen Y, Woosley RL. Electrophysiologic action of astemizole. American Heart Association Meetings. November, 1993. Atlanta, GA.

84. Pinney WP, Scanlon RT, Koller BS, Woosley RL. Terfenadine increases QT segment in isolated guinea pig hearts. Ann Allergy 1993;70:64.

85. Rampe D, Wible B, Brown AM, Dage RC. Effects of terfenadine and its metabolites on a delayed rectifier K + channel cloned from human heart. Mol Pharmacol 1993;44:1240–1245.

86. Baker B. Risk of arrhythmia with Seldane monotherapy appears very small. Int Med New Cardiol News 1992;Dec 15:6.

87. Honig P, Smith J, Wortham D, Zamani K, Cantilena L. Pharmacokinetic variability of terfenadine biotransformation in healthy volunteers. Clin Pharmacol Ther 1994;55:138.

88. Sanders RL, Dockhorn RJ, Alderman JL, McSorley PA, Wenger TL, Frosolono MF. Cardiac effects of acrivastine compared to terfenadine. J Allergy Clin Immunol 1992;89:183.

89. Eller M, Stoltz M, Okerholm R, McNutt B. Effect of hepatic disease on terfenadine and terfenadine metabolite pharmacokinetics. Clin Pharmacol Ther 1993;53:162.

90. Cleary JD, Taylor JW, Chapman SW. Itraconazole in antifungal therapy. Ann Pharmacother 1992;26:502–508.

91. Crane JK, Shih HT. Syncope and cardiac arrhythmia due to an interaction between itraconazole and terfenadine. Am J Med 1993;95:445–446.

92. Pohjola-Sintonen S, Viitasalo M, Toivonen L, Neuvonen P. Torsades de pointes after terfenadine-itraconazole interaction. Br Med J 1993;306:186.

93. Honig PK, Wortham DC, Hull R, Zamani K, Smith JE, Cantilena LR. Itraconazole affects single-dose terfenadine pharmacokinetics and cardiac repolarization pharmacodynamics. J Clin Pharm 1993;33:1201–1206.

94. Honig PK, Wortham DC, Zamani K, Mullin JC, Conner DP, Cantilena LR. The effect of fluconazole on the steady state pharmacokinetics and electrocardiographic pharmacodynamics of terfenadine in humans. Clin Pharmacol Ther 1993;53:630–636.

95. Honig PK, Wortham DC, Zamani K, Conner DP, Mullin JC, Cantilena LR. Effect of concomitant administration of cimetidine and ranitidine on the pharmacokinetics and electrocardiographic effects of terfenadine. Eur J Clin Pharmacol 1993;45:41–46.

96. Simons FER, Kesselman MS, Giddins NG, Pelech AN, Simons KJ. Astemizole-induced torsade de pointes. Lancet 1988;2:624.

97. Snook J, Boothman-Burrell D, Watkins J, Colin-Jones D. Torsade de pointes ventricular tachycardia associated with astemizole overdose. Br J Clin Pract 1988;42:257–9.

98. Barnes CL, McKenzie CA, Webster KD, Poinsett-Holmes K. Cetirizine: a new, nonsedating antihistamine. Ann Pharmacother 1993;27:464–470.

99. Sale ME, Woosley RL, Barby JT, Yeh J, Chung M. Lack of electrocardiographic effects of cetirizine in healthy humans. J Allergy Clin Immunol 1993;91:258 (abstr.)

100. Tanaka H, Habuchi Y, Nishimura M, Sato N, Watanabe Y. Blockade of Na + current by promethazine in guinea-pig ventricular myocytes. Br J Pharmacol 1992;106:900–905.

101. Santillan A, Almanza J, Valenzuela F. Effects of cyproheptadine on the electrophysiological characteristics of the sinus node, ventricular myocytes and papillary muscles of the guinea pig heart. Arch Inst Cardiol Mex 1990; 60:361–368.

102. Riccioppo NF. Effects of cyproheptadine on electrophysiological properties of isolated cardiac muscle of dogs and rabbits. Br J Pharmacol 1983;80: 335–341.

103. Sanchez-Perez S, Pastelin G, Mendez R. Disparity between the antihistaminic and antiarrhythmic activities of chlorpheniramine and its isomers. Life Sci 1978;22:1179–1188.

104. Molyvdas PA, James FW, Sperelakis N. Azelastine effects on electrical and mechanical activities of guinea pig papillary muscles. Eur J Pharmacol 1989;164:547–553.

105. Claritin® (Loratadine) Product Labelling. Schering Corporation. Kenilworth, NJ, 1994.

106. Hey JA, del Prado M, Chapman RW, Egan RW, Siegel MI, Sherwood J, Kreutner W. Loratadine produces antihistamine activity without adverse CNS or cardiovascular effects in guinea pigs: comparative studies with sedating and nonsedating H₁-antihistamines. J Allergy Clin Immunol 1994;93: 163.

107. Seldane® (Terfenadine) Product Labelling. Marion Merrell Dow, Inc. Kansas City, MO, 1994.

108. Hismanal® (Astemizole) Product Labelling. Janssen Research Foundation. Titusville, NJ, 1994.

109. Burke TG, Mutnick AH. Ventricular fibrillation and anoxic encephalopathy secondary to astemizole overdose. Ann Pharmacother 1993;27:239–240.

110. Simonsen L. What are pharmacists dispensing most often? Pharmacy Times 1991;April:57–70.

111. 1991 National Prescription Audit. Plymouth meeting. Pennsylvania: IMS America, 1991.
112. Meltzer EO. Performance effects of antihistamines. J Allergy Clin Immunol 1990;86:613–9.
113. White JM, Rumbold, GR. Behavioural effects of histamine and its antagonists: a review. Psychopharmacology 1988;95:1–14.
114. Rombaut NEI, Hindmarch I. Psychometric aspects of antihistamines: a review. Human Psychopharmacology Clin Exp 1994;9:157–169.
115. Simons FER. H_1-receptor antagonists: comparative tolerability and safety. Drug Safety 1994;10:350–380.
116. Simons FER. The therapeutic index of newer H_1-receptor antagonists. Clin Exp Allergy 1994;24:707–723.
117. Peck CC, Temple R, Collins JM. Understanding consequences of concurrent therapies. JAMA 1993;269:1550–1552.

Epilogue

Decades of research have led to a second generation of H_1-receptor antagonists which are considerably safer than their predecessors with regard to reduced central nervous system toxicity. Still, no existing H_1-receptor antagonist is perfectly free from central nervous system effects in all situations, for example, if the manufacturers' recommended dose is exceeded. Some H_1-receptor antagonists with an improved margin of safety in the central nervous system cause adverse cardiovascular system effects under certain circumstances. Also, the most effective H_1-receptor antagonists available currently do not provide complete relief of symptoms in all patients with allergic rhinoconjunctivitis or urticaria and, in ordinary doses, are even less effective in patients with asthma or atopic dermatitis. Therefore, there is an urgent need for an H_1-receptor antagonist that is even safer and more effective than the "best" H_1-receptor antagonists now in use.

During the next few years, additional second-generation H_1-receptor antagonists such as fexofenadine, setastine, mizolastine, epinastine, and emedastine may become available. These medications are being studied far more rigorously than their predecessors were with regard to safety and efficacy. Some of them will become widely used in the treatment of allergic disorders, because they will have a risk/benefit/cost ratio as good as, or better than, the best H_1-receptor antagonists in current use, but others may fall by the wayside.

As we enter the twenty-first century, additional new H_1-receptor antagonists will enter phase I or phase II development. These will include other active metabolites of existing H_1-receptor antagonists, enantiomers of existing compounds, and medications with combined H_1-/H_2-receptor antagonist activity which have the potential to be extremely useful in urticaria and anaphylaxis treatment.

Eventually, in the development of clinically useful H_1-receptor antagonists, instead of starting with the medication itself and searching for a safer, more effective metabolite, enantiomer, or other "relative," development of new medications will likely begin with the molecular target, the H_1-receptor, and H_1-receptor antagonists will be custom-designed to fit the target. Although this is not yet a reality for the histamine H_1-receptor, the structure of the receptor has been deduced, and minor changes in H_1-receptor antagonist molecule configuration have been demonstrated to relate to specific activity at the H_1-receptor site.

The recent developments in H_1-receptor and H_1-receptor antagonist research described in this book therefore have important practical implications for people suffering from allergic disorders. In years to come, our knowledge of H_1-receptor antagonists gained painstakingly during the past 50 years will take a quantum leap forward, thanks to application of molecular biology techniques. More perfect H_1-receptor antagonists will become a reality.

F. Estelle R. Simons

Index

Acrivastine, 93, 102
  adult dosage, 187
  allergic rhinitis, 235
  chronic urticaria and, 285–6
  clinical pharmacology of, 194
  pediatric dosage, 330
  pharmacokinetics of, 181, 193–194
Acute anaphylaxis, treatment with
    antihistamines, 310–311
α-Adrenergic agonists, 234
β_2-Adrenergic agonists, 251
β_2-Adrenergic receptor, 44, 45
Adverse effects of H_1-receptor
    antagonists, 233–234
  see also Cardiovascular system;
    Central nervous system
    (CNS)
Alcohol,
  interaction with first-generation
    H_1-receptor antagonists,
    372–3
Alkylamines, 94, 101

Allergic diseases,
  histamine in, 76–81
    allergic rhinitis, 78–79
    anaphylaxis, 80–81
    asthma, 76–78
    histamine receptors, 66–69
    histamine synthesis and
      metabolism, 61–66
    overview of histamine effects,
      69–76
    urticaria, 79–80
  mechanisms of, 118–119
  sedative effects of, 373–375
Allergic rhinitis, 78–79
Allegic rhinoconjunctivitis, 221–233
  H_1-receptor antagonists in
    children with, 335–338
Amitriptyline, 120
Amselamine, 5
Anaphylaxis, 80–81, 297–327
  antihistamines and, 308–319
    newer H_1-antagonists, 309

415

About the Editor

F. ESTELLE R. SIMONS is the Bruce Chown Professor and Deputy-Head of the Department of Pediatrics and Child Health, and Head of the Section of Allergy and Clinical Immunology at the University of Manitoba, Winnipeg, Canada. She is past President of the Canadian Society of Allergy and Clinical Immunology, past Chair of the Allergy Section of the Canadian Pediatric Society, past Chair of the Royal College of Physicians and Surgeons of Canada Specialty Committee in Clinical Immunology and Allergy, and past Chair of the Royal College Examining Board in Clinical Immunology and Allergy. She is a Fellow of the American Academy of Allergy, Asthma, and Immunology (AAAAI), and has chaired several important AAAAI committees, including the Committee on Drugs, and the Asthma and Rhinitis Interest Section. She is also a Fellow of the American College of Allergy, Asthma, and Immunology, and the American Thoracic Society. Dr. Simons is the author or coauthor of over 200 peer-reviewed papers in allergy and clinical immunology, and serves on the editorial boards of numerous journals. A diplomate of the American Board of Pediatrics and the American Board of Allergy and Clinical Immunology, she received the B.Sc. (1965) and M.D. (Honours) degree (1969) from the University of Manitoba, Winnipeg, Canada.